Natural History of Varicella-Zoster Virus

Editor

Richard W. Hyman, Ph.D.
Professor of Microbiology
The Pennsylvania State University
College of Medicine
Hershey, Pennsylvania

CRC Press
Taylor & Francis Group
Boca Raton London New York

CRC Press is an imprint of the
Taylor & Francis Group, an **informa** business

First published 1987 by CRC Press
Taylor & Francis Group
6000 Broken Sound Parkway NW, Suite 300
Boca Raton, FL 33487-2742

Reissued 2018 by CRC Press

Library of Congress Cataloging-in-Publication Data

Natural history of varicella-zoster virus.

Includes bibliographies and index.
1. Chicken-pox. 2. Varicella-zoster virus.
I. Hyman, Richard W., 1941-
RC125.N38 1987 616.9'14 86-24406
ISBN 0-8493-6584-8

A Library of Congress record exists under LC control number: 86024406

Publisher's Note
The publisher has gone to great lengths to ensure the quality of this reprint but points out that some imperfections in the original copies may be apparent.

Disclaimer
The publisher has made every effort to trace copyright holders and welcomes correspondence from those they have been unable to contact.

ISBN 13: 978-1-315-89582-6 (hbk)
ISBN 13: 978-1-351-07492-6 (ebk)

Visit the Taylor & Francis Web site at http://www.taylorandfrancis.com and the
CRC Press Web site at http://www.crcpress.com

THE EDITOR

Richard W. Hyman, Ph.D., is Professor of Microbiology at The Pennsylvania State University College of Medicine in Hershey.

Dr. Hyman was graduated in 1962 from the University of California, Berkeley, with a B.S. degree in chemistry and obtained his Ph.D. degree in chemistry in 1970 from the California Institute of Technology.

Dr. Hyman is a member of the American Society of Biological Chemists, the American Society for Microbiology, the American Society for Virology, the Biophysical Society, and the Protein Society.

Dr. Hyman is the principal author of numerous publications specializing in the molecular pathology of human herpesvirus infections. Dr. Hyman is a devoted tennis enthusiast.

CONTRIBUTORS

Ann Arvin, M.D.
Associate Professor
Department of Pediatrics
Stanford University Medical School
Stanford, California

Charles Grose, M.D.
Associate Professor and Head
Division of Infectious Diseases
Department of Pediatrics, and
Associate Professor
Department of Microbiology
University of Iowa
Iowa City, Iowa

Richard W. Hyman, Ph.D.
Professor
Department of Microbiology
The Pennsylvania State University
College of Medicine
Hershey, Pennsylvania

Michiaki Takahashi, M.D., D.Med.Sci.
Professor
Department of Virology
Research Institute for Microbial
Diseases
Osaka University
Osaka
Japan

Richard J. Whitley, M.D.
Professor of Pediatrics and
Professor of Microbiology
Departments of Pediatrics and
Microbiology
University of Alabama at Birmingham
Birmingham, Alabama

This book is dedicated to Norman Davidson on the occasion of his seventieth birthday

TABLE OF CONTENTS

Chapter 1

VARICELLA ZOSTER VIRUS: PATHOGENESIS OF THE HUMAN DISEASES, THE VIRUS AND VIRAL REPLICATION, AND THE MAJOR VIRAL GLYCOPROTEINS AND PROTEINS

Charles Grose

TABLE OF CONTENTS

I. PATHOGENESIS OF THE HUMAN DISEASES

A. Introductory Comments

Chickenpox is a common childhood exanthem which most boys and girls contract while they are enrolled in kindergarten or the early years of grade school (Figure 1). Although the disease has been mentioned since ancient times, in the Middle Ages there was much confusion between the different pox afflictions, which included the scourages of large pox (syphilis) and smallpox (variola). Somewhere in between was an exanthem called *varicella* in French because it appeared to be a diminutive or benign form of the most malignant variola. In English-speaking countries, varicella was usually called *chickenpox*, possibly to better describe the size of the vesicular lesions as being similar to that of a chick pea (from the French, *pois chiche*).[1] The early descriptions of zoster also showed confusion with other vesicular diseases, in particular, recurrent herpes simplex infections. The term *zoster* was selected to describe the dermatomal exanthem because the usual location of the disease corresponded to the site of the belt (zoster) worn by a Greek warrior to secure his armor. The term *herpes* also has a Greek derivation and indicates something which creeps. Herpes zoster, therefore, identified an exanthem which crept around the midsection of an afflicted individual (Figure 2). In a similar vein, *shingles* is the anglicization of the middle French word *chingle*, which is a belt.

An astute Hungarian pediatrician by the name of Bokai first clarified the association of chickenpox and shingles.[2-3] He carefully observed nine situations in which younger family members or close associates apparently contracted chickenpox from another individual with shingles (Figure 3). He also documented the incubation period of chickenpox in the secondary cases and noted that the time interval was usually about 2 weeks. Two situations involved exposure on a children's ward. In one instance, three children developed chickenpox between 15 and 18 days after contact with a case of zoster. In the second instance, four cases of chickenpox occurred in patients 14 to 17 days after admission to the ward with a case of zoster. From these compelling clinical observations, Bokai correctly concluded in 1909 that the two diseases of chickenpox and zoster were etiologically related.

Substantiation of Bokai's hypothesis did not come until 1953. With the advent of tissue culture technology, Weller was able to isolate virus from the vesicles of both chickenpox patients and zoster patients.[4] When the isolates from the two groups of patients were compared, he could not define any differences in their cytopathology or growth characteristics. Weller, therefore, concluded that the same viral agent caused chickenpox and zoster. It was also established that these viruses most closely resembled members of the herpesviruses and not the pox viruses. In summary, therefore, contrary to its appellation, chickenpox is associated with neither the chicken nor the pox. Nevertheless, it is a fascinating disease because of the ability of the etiologic agent, varicella-zoster virus (VZV), to remain latent within the nervous system before inexplicably and capriciously reactivating to cause shingles many decades later.

B. Incubation Period of Chickenpox

As mentioned above, Bokai was very perceptive when he noted the time interval between the index case of zoster and the secondary cases of chickenpox. An analysis of the mean number of days in his 15 observations indicates a 14.5-day incubation period.[3] The validity of this result is confirmed by two larger analyses of the incubation period of chickenpox. One study from England included 184 cases and another from the U.S. contained 67 observations.[5] In both instances, the authors determined that the usual range was 10 to 20 days, with an average incubation period of 14 to 15 days.

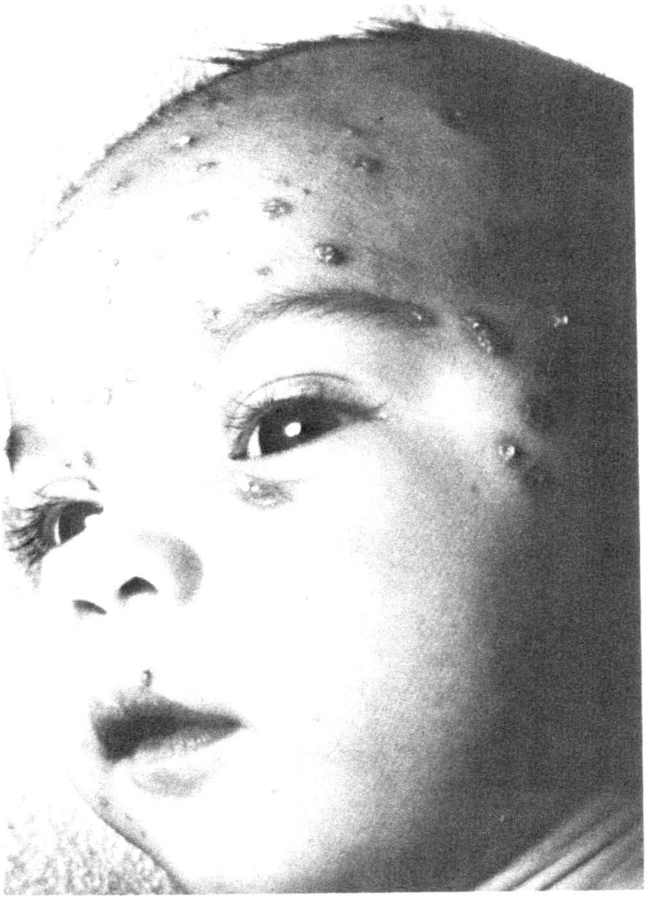

FIGURE 1. Chickenpox in a young child. The vesicular lesions are just beginning to erupt about the face and will soon spread across the trunk and over the extremities.

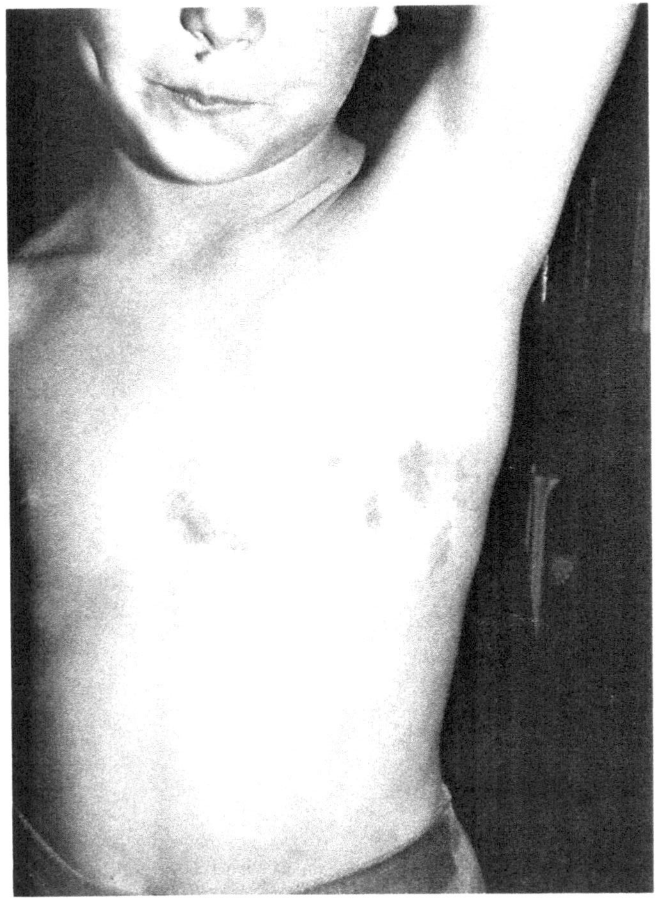

FIGURE 2. Zoster (shingles) in a boy. The vesicular rash began on the back and spread across the left chest in the distribution of the fifth thoracic dermatome. The rash stopped at mid chest.

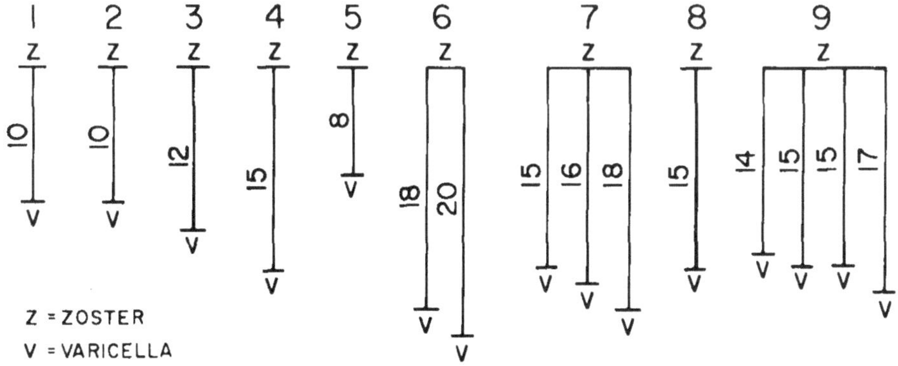

FIGURE 3. Episodes of chickenpox following zoster as recorded by Bokai in Budapest. The nine index cases of zoster were followed by chickenpox in several close contacts. The incubation period (in days) is indicated by the number on the vertical lines. The average interval varied from 14 to 15 days.

Individuals who acquire chickenpox by unusual routes, e.g., intrauterine transmission, may have an incubation period closer to 10 days (*vide infra*).

C. Infectious Period (Period of Contagion)

The period of contagion indicates that time interval when the disease can be transmitted from the index patient to a secondary nonimmune individual. In the case of chickenpox, contagion is frequently equated with the presence of vesicles on the skin. This is an erroneous assumption. The virus is most easily transmitted immediately before and during the first few days of the rash. Since contagion has been documented as early as 4 days *before* the appearance of rash, the route of transmission is considered to be respiratory.[6,7] Infectivity may be transferred during the first 4 to 5 days after onset of the rash; again the virus is presumably transmitted by a respiratory route even at this time. Although children are customarily isolated during the second week of illness, when the lesions are crusting and the fever has abated, there is no evidence for contagion at this time.

An airborne mode of transmission of VZV is also documented by two outbreaks of chickenpox within wards of children's hospitals.[8,9] In both reports, the index case was an immunocompromised child with progressive varicella infection, including viral pneumonitis. Both children were housed in rooms which had a positive air flow with respect to the corridor, i.e., air was carried from the hospital room of the index case into the hallway, whence it entered adjacent hospital rooms housing VZV-susceptible patients. Under both circumstances, VZV nonimmune children in the same ward subsequently contracted chickenpox from the index case (Figure 4). Since both index cases had varicella pneumonitis, it is presumed that they exhaled a large number of VZV-laden water droplets which were carried in the prevailing air currents. There was no contact between any of the secondary cases and the index cases, nor did they share any medical equipment.

D. Pathogenesis of Chickenpox

As a contagious disease, chickenpox is more easily understood from the perspective of its pathogenesis (Figure 5). The model for chickenpox is based on the schema for the pathogenesis of mousepox, which was initially described by the Australian virologist Fenner.[10] Fenner considered most exanthematous diseases to represent the second stage of a dual viremic model of infection. As he predicted, chickenpox is most likely an example of this same pathogenetic schema.[11] The infection is acquired by small

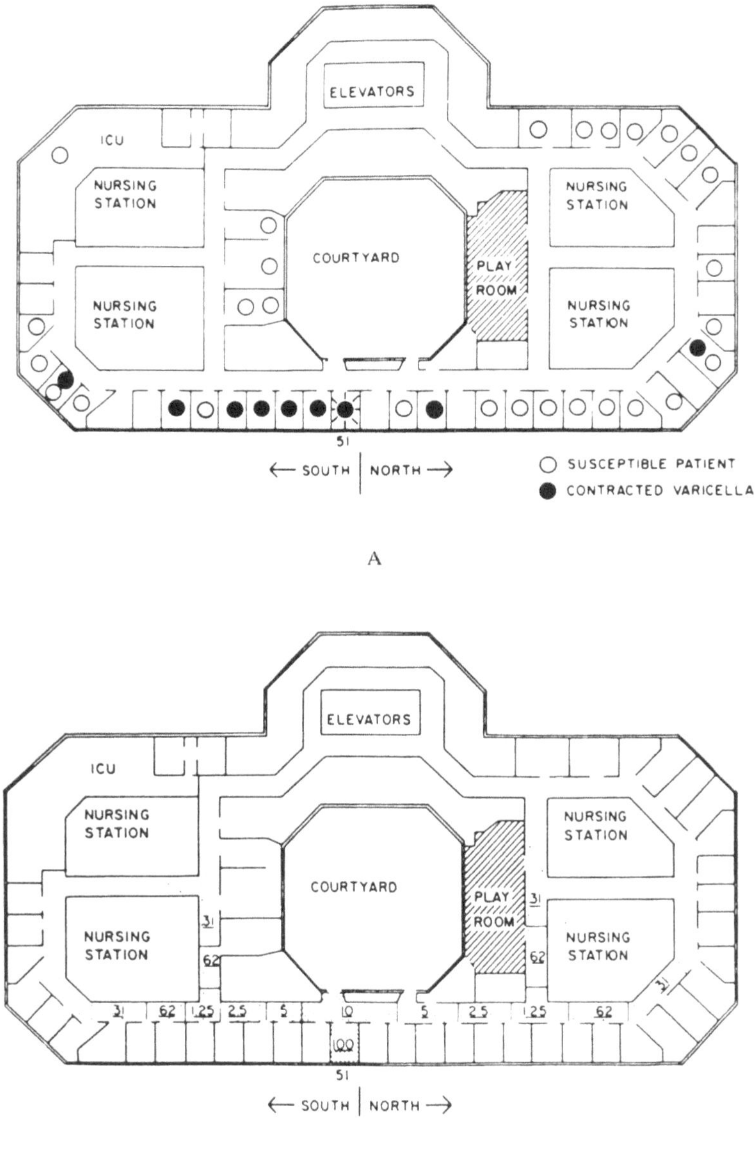

FIGURE 4. Diagram of an outbreak of airborne nosocomial chickenpox in a children's ward. The figure illustrates the floorplan of a pediatric ward where an outbreak of nosocomial varicella occurred. Panel A, location of 36 susceptible patients present between noon and midnight on the day of exposure. The index patient occupied room 51. Panel B, concentration gradients of a tracer gas which was released into the corridor from the room of the index case. (Reproduced from Gustafson, T. L., Lavely, G. B., Brawer, E. R., Hutcheson, R. H., Wright, P. F., and Schaffner, W., *Pediatrics*, 70, 550, 1982. With permission.)

droplets containing live virus, which fall upon the conjunctivae and upper respiratory tract of the index case. The virus invades, perhaps via the lymphatic drainage, and enters the regional cervical lymph tissue, where local replication occurs. Another possible site of primary replication is the salivary tissue. After a 4 to 6 day period, a sufficient quantity of infectious virus is released to cause a primary viremia. Occasion-

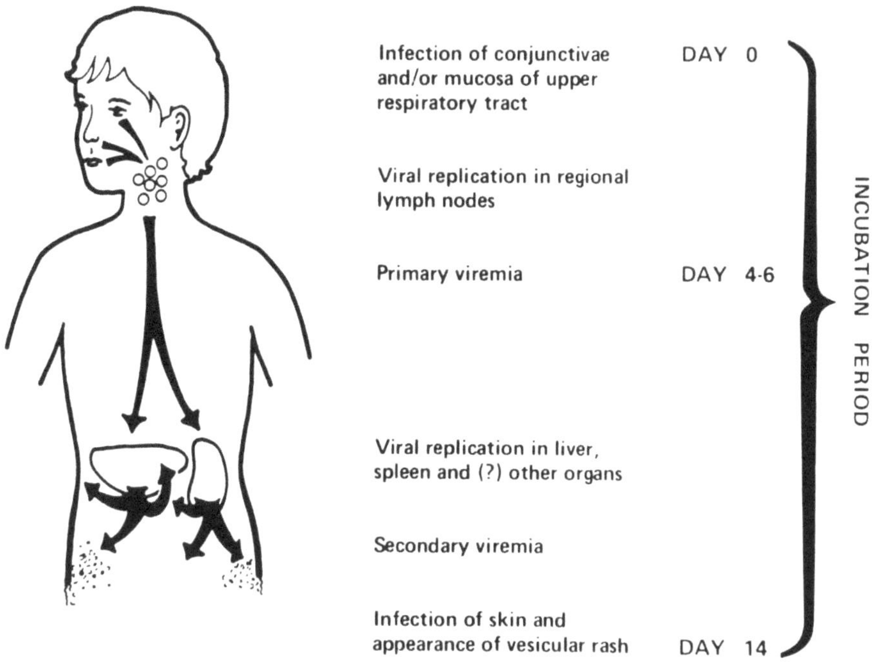

FIGURE 5. Pathogenesis of chickenpox. Chickenpox appears to follow the Fennerian schema of pathogenesis for exanthematous diseases with dual viremic phases. Prior to the first viremia, varicella-zoster virus replicates at a local site in the head and neck. During the first viremia, the virus is disseminated throughout the body where a second cycle of replication occurs. The exanthem follows the secondary (but major) viremia by a few days and signals the end of the 2-week incubation period. (Reproduced from Grose, C., *Pediatrics,* 68, 735, 1981. With permission.)

ally the child will exhibit a brief scarlatiniform rash during this viremia; on rare occasions a vesicle or two may be noted, especially about the head or neck.

During the primary viremia, the virus is disseminated throughout the body. The secondary sites of viral replication are not known for certain, but probably include epithelial derivatives, such as the columnar epithelium of the respiratory tree and the gastrointestinal tract; the glandular epithelium of the liver, pancreas, and adrenals; and the epithelial cells of the renal tubules, pulmonary alveoli, and Hassall's bodies. The endothelial lining of the blood vessels may also be a secondary site of virus multiplication. Approximately 1 week after the primary viremia, the secondary viremia takes place. It should be emphasized that the word *secondary* is used only in a temporal sense, because this latter viremia is the major viremia. At this time the virus exits the capillaries and travels to the epidermis, forming the characteristic unilocular vesicle. These vesicles were well described by Tyzzer in 1906, after he excised and examined lesions at 1, 2, 3, 4, 5, 6, and 7 days of age.[12] The pathological alterations were characterized by enlarged cells with eosinophilic inclusions in their nuclei. This swelling has also been called a ballooning degeneration. An electronmicrograph of a vesicle is illustrated in Figure 6.

During the final days of the incubation period, the infected child frequently develops a low-grade fever, presumably as a result of the second viremia and the induction of interferon formation. The child may also manifest symptoms of nausea and vomiting, which correlate with mild signs of hepatic dysfunction. The end of the incubation period is heralded by the appearance of the characteristic vesicular lesions, usually appearing first along the hairline (see Figure 1). The vesicles appear in successive crops which progressively cover the face, upper torso, and lower torso, and finally the ex-

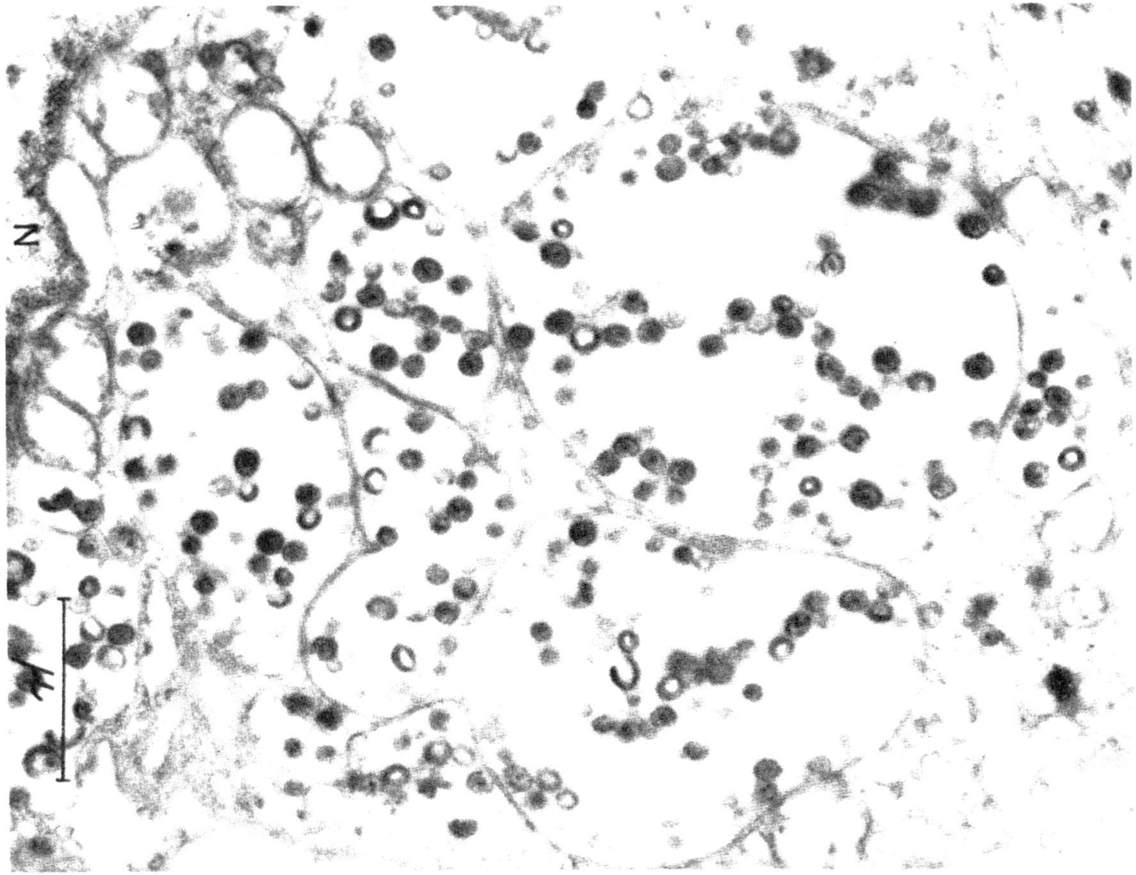

FIGURE 6. The chickenpox vesicle upon examination by electron microscopy. The vesicle contains numerous viral particles in a cell-free state. Most of the viral particles have electron-dense cores.

tremities over a 1 week period. Each lesion begins as a macule which evolves into a maculopapule and finally into a vesicle a few millimeters in diameter. The fluid is initially clear, but by day 4 to 5, the fluid is cloudy and the vesicle becomes a crusted lesion. The scab may persist for 1 to 3 weeks. Because the vesicular lesions appear in successive crops, it is characteristic of the exanthem of chickenpox that lesions of all stages are present at different locations on the body at the same time. It is also typical of the rash to be more confluent over parts of the skin which have previously been abraded or traumatized, e.g., by sunburn or diaper rash.

The average course of chickenpox in middle-class American children has been meticulously documented by Ross.[13] The prodrome usually consists of malaise, anorexia, rhinitis, and mild pharyngitis. The child has the highest temperature elevations during the first 3 days of the rash, but the fever rarely goes above 102°F. The average number of skin lesions in the index case within a household is 207 to 258 pocks. The secondary cases within the same household — usually other siblings — tend to have somewhat more extensive exanthems with a total pock count between 310 and 510. By the end of the first week of illness, the skin lesions begin to crust and the child's general disposition returns to normal.

The cutaneous lesions of erythema multiforme are occasionally seen together with the rash of chickenpox. The concentric circles of erythema multiforme (red, white, and blue from outside to inside) are interspersed among the vesicular lesions of chickenpox; occasionally a chickenpox vesicle forms the center of one of the circular lesions of erythema multiforme.[14]

E. Pathogenesis of the Neurologic Manifestations

Chickenpox is a systemic viral infection which frequently invades the central nervous system (CNS). The single most common CNS complication of chickenpox is acute cerebellar ataxia, although the reason for this propensity to localize in the cerebellum is not known. The signs of CNS disease may precede the appearance of the vesicular exanthem or occur shortly after the onset of the rash. Most CNS disease occurs, however, during the second week of clinical illness. These three different temporal phases of CNS disease correlate, respectively, with CNS invasion by virus during the period of local replication or after either of the two viremias.[1] CNS disease preceding appearance of the rash strongly supports the Fennerian concept of pathogenesis because it offers an explanation for CNS pathology during the incubation period before there is a detectable immune response. Presumably the virus is carried from the site of local replication in the head and neck directly into the brain via emissary veins and lymphatics. Once in the brain, the virus replicates concurrently with virus elsewhere in the body (Figure 7).

Several examples of preeruptive CNS disease have been published, but one case in particular was well documented.[15] A 5-year-old boy first developed malaise, headache, and nausea; 2 days later he was vomiting and walked with an ataxic gait. When he was subsequently examined, the physician noted signs of cerebellar pathology including slow and slurred speech, inability to perform coordinating and alternating movements, and ataxic gait. An examination of his spinal fluid demonstrated an increased number of cells: 38 lymphocytes and 14 granulocytes per milliliter. The diagnosis of chickenpox was revealed when a vesicular rash appeared on the 12th day after onset of the CNS symptoms. The most likely explanation for this 12 day interval between CNS signs and systemic rash is invasion of the CNS coincident with local replication. Other examples of preeruptive disease run the gamut of CNS signs and symptoms from aseptic meningitis to meningoencephalitis with seizures to isolated facial paralysis. In most instances these examples of preeruptive disease occur about 3 days prior to the appear-

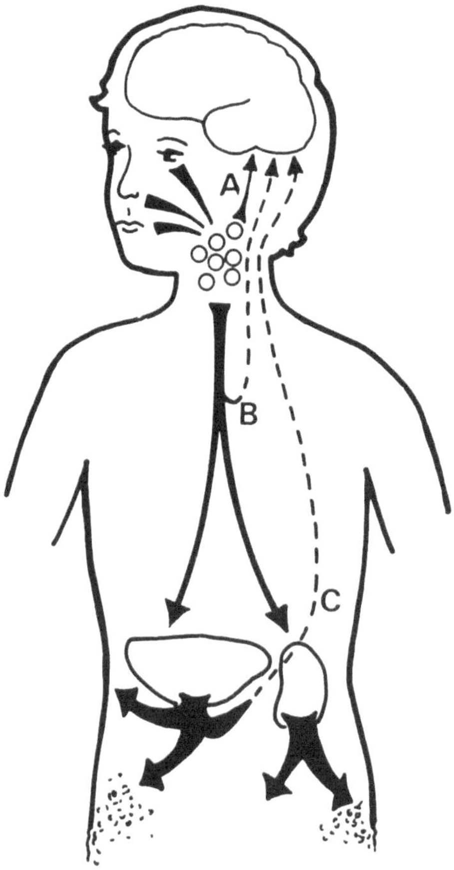

FIGURE 7. Pathogenesis of CNS manifestations of chickenpox. VZV can invade the central nervous system at three different times: (A) during local replication, (B) during the primary viremia, (C) during the secondary viremia. In the first situation, the CNS manifestations arise more than a week before the vesicular rash. (Reprinted from Grose, C., in *Human Herpesvirus Infestions,* Reference 1, chap. 3, by courtesy of Marcel Dekker, Inc. With permission.)

ance of the rash; thus, they are CNS manifestations of viral invasion during the primary viremia. In contrast to the above examples of preeruptive CNS pathology, the majority of CNS disease in association with varicella occurs during the latter half of the first week and early in the second week of the chickenpox rash.[16-18] This time of onset is temporally associated with CNS invasion after the major (but second) viremic phase. As stated above, acute cerebella ataxia is the most common neurologic complication of childhood chickenpox.

F. Reye Syndrome and Varicella

One of the intriguing associations of varicella is with Reye syndrome.[19,20] The evidence for Reye syndrome has been assessed in children presenting with the abrupt onset of vomiting after acute varicella infection.[21] The patients with a recent history of chickenpox were required to have the following criteria for entrance into the study: acute onset of vomiting, absence of jaundice, elevated levels (3×) of serum aspartate (SGOT) or alanine (SGPT) aminotransferase, and cerebrospinal fluid studies which exclude infection. The patients were considered to be Grade I if they were quiet and sleepy but responded to verbal commands, Grade II were stuporous with thick speech and difficulty counting, while Grades II through V had progressive deepening coma. Eight patients with varicella followed by Grade I Reye syndrome were enrolled. The onset of vomiting usually began on the fifth day of chickenpox; most of these patients had also taken salicylates. The liver biopsies of these patients were examined by both light and electron microscopy. The criteria for Reye syndrome were panlobular accumulation of small lipid droplets in swollen and pale hepatocytes. Ultrastructural changes included mitochondria which were larger and irregular in shape with a decreased number of matrix granules. No correlation was observed between the severity of the ultrastructural lesions and the duration of vomiting or the levels on admission of the liver enzymes or serum ammonia. Very few mildly affected patients with varicella and Grade I Reye syndrome progressed further. Of 30 patients, two who developed more severe Reye syndrome had significantly higher initial levels of serum ammonia (291 vs. 53 μg/dℓ) and longer prothrombin times (3.9 vs. 1.6 sec). There was also a trend for the latter two patients with progressive disease to have higher levels of SGPT (1779 vs. 827 IU).

G. Extracutaneous Viral Manifestations of Chickenpox

That chickenpox is an infectious disease with viremic phases has now been repeatedly documented by the successful isolation of VZV from the blood of both immunocompromised and healthy children with early chickenpox.[22-25] Because it is a systemic disease, several extracutaneous manifestations of chickenpox have been documented. In most instances, these represent replication of virus at a relatively sequestered site. The complications have included arthritis, orchitis, appendicitis, hepatitis, pneumonitis, pericarditis, and glomerulonephritis.

Arthritis often occurs about a week after the onset of chickenpox.[26,27] Presumably, therefore, the viral invasion of the joint occurred during the second viremic phase. The affected sites include the large joints of the extremities. Most commonly, the joint is warm and swollen, with limitation of movement. The joint fluid usually contains an increased number of cells with a marked predominance of lymphocytes. Virus has been isolated from the joint fluid. After a brief convalescence, the symptoms resolve with no permanent sequelae.

Acute glomerulonephritis is an uncommon association of chickenpox.[28] In addition, there is always the problem of an unappreciated coincident Group A streptococcal infection. The varicella-associated nephritis occurs after onset of the rash. Examination of the kidney biopsy specimen demonstrates a proliferative glomerulonephritis. The renal abnormalities usually spontaneously abate within a few weeks after acute chickenpox. An example of the association of chickenpox and rapidly progressive glomerulonephritis occured in a 12-year-old boy.[29] In the second week of chickenpox, he developed microscopic hematuria and mild proteinuria. He progressed to anuria, with edema about the face and feet. His clinical chemistries included a blood urea nitrogen of 150 mg/dℓ and a creatinine of 12.5 mg/dℓ, while prothrombin time and partial thromboplastin time were normal. Serum antistreptolysin O titers and complement levels were normal. Because of azotemia, the patient was treated by intermittent peritoneal dialysis. A renal biopsy demonstrated hypercellularity and epithelial proliferation of the glomeruli with formation of crescents. There was focal necrosis of glomerular tufts, tubules were dilated, and the renal interstitium showed focal infiltration with inflammatory cells. After onset of the renal complaints, he began to pass urine again within 6 weeks, and by 8 weeks his renal function had returned to normal.

Pericarditis has been described in rare instances, especially in adults with chickenpox.[30] The symptoms of pericarditis include rapid onset of severe chest pains shortly after the onset of chickenpox. An electrocardiogram may show S-T- and T-wave changes compatible with diagnosis of pericarditis. These electrocardiographic alterations may persist for several days before returning to normal.

Orchitis has been reported in both boys and adult men with chickenpox.[31] The involved testicle may be very painful and swollen. Although atrophy of the affected testicle has been described, these reports may reflect only the most severe cases and, therefore, do not represent the normal course of varicella-associated testicular disease.

Children with chickenpox may complain of diffuse abdominal discomfort or pain. Occasionally, one of these children develops signs of peritoneal inflammation.[1] At time of surgical exploration, appendicitis is apparent. This association of appendicitis with early chickenpox may represent viral replication in appendiceal tissues following the viremic phase.

One of the most serious manifestations of the extracutaneous replication of virus is varicella pneumonitis.[32,33] Lung involvement is rarely seen in otherwise healthy children; however, pulmonary disease occurs more frequently in teenagers and young adults, as well as in immunocompromised children with chickenpox.[34] Pregnant women also seem to be at special risk to develop varicella pneumonitis. The radiologic changes in the lung follow the skin rash by a few days and tend to develop in patients with an extensive exanthem. Post-mortem examination of lung tissue demonstrates viral antigens throughout the parenchyma.

H. Deaths Caused by Chickenpox

Chickenpox is a rare cause of mortality in the U.S. Therefore, it is noteworthy to review the circumstances surrounding deaths due to chickenpox. When the medical records at Children's Hospital in Philadelphia were examined retrospectively, 96 cases of children hospitalized with chickenpox were found between 1975 and 1980.[35] Of the 96 children, 81 were considered to be healthy while 15 were immunocompromised. The latter group included children on chemotherapy for leukemia (6), lymphoma (2), and brain tumors (2). In addition, there were 2 children with thalassemia and 1 each of children with severe nephrosis, asthma, and trisomy 21 syndrome. A total of 10 deaths was recorded, seven due to Reye syndrome and one each because of varicella pneumonitis, neonatal chickenpox, and secondary bacterial sepsis. The pneumonitis occurred in a 19-year-old woman who acquired respiratory symptoms 8 days after onset of chickenpox. The fatal case in an infant was the result of chickenpox which first appeared at day 7 of life. A 2-year-old boy with chickenpox developed septicemia with *Staphylococcus aureus;* he died following rupture of a mycotic aneurysm of the middle cerebral artery.

The above medical record survey clearly indicates that Reye syndrome is the most life-threatening sequela of children hospitalized with chickenpox. Altogether, 17 children with chickenpox-associated Reye syndrome were admitted between 1975 and 1980. The 17 children ranged in age from 3.5 to 10.0 years, with a mean age of 5.9 years. Besides the 7 deaths, an additional 4 patients required prolonged convalescent care because of persistent neurologic impairment.

The number of deaths following chickenpox has also been calculated for the U.S. between the years 1972 to 1978.[36,37] During that time there was an annual estimated total of 2,823,055 chickenpox cases for the entire country. During the 7 year period there were a total of 735 deaths attributed to chickenpox, of which more than half (389) were considered to be chickenpox encephalitis (Reye or non-Reye). At least 70% of the deaths occurred in children less than 10 years old, although the age group above

Table 1
STATISTICS OF CASES OF CHICKENPOX AND RELATED DEATHS

Age (years)	Estimated no. annual cases	Av. no. deaths	Deaths:cases
<1	98,980	7	1:13,140
1—4	910,686	18	1:50,593
<5	1,002,666	25	1:40,107
5—9	1,409,889	38	1:37,102
10—14	312,785	10	1:31,278
15—19	55,362	2	1:27,681
>19	42,353	29	1:1,460

Note: Data derived from Reference 36.

20 years was at much higher risk (Table 1). Based on all reported cases, approximately one of every 40,000 people (mainly children) with chickenpox per year will die. If only adults 20 or older with chickenpox are considered, about 1 of every 1460 cases will have a fatal conclusion. Although there is no information about underlying conditions in this survey, other data suggest that less than 20% of the deaths occur in patients with known malignancies.

From data presented in the Table, it is apparent that the risk of death following chickenpox is three- to fourhold higher in children under 1 year of age than in older children. To further assess this risk, the number of deaths in chickenpox of infancy was examined in greater detail for the 15 year period of 1968 to 1982.[38] A total of 122 deaths was reported as shown in Table 1; the mean was 8 and the median was 7 deaths per year. These deaths usually occurred in the months of highest number of chickenpox cases — March through May. There was no sexual or racial predilection.

A more complete analysis was carried out on 92 of the chickenpox-related deaths (Figure 8). The exact age at death was distributed throughout the first year, from day one to 11 months; the median age was 5.5 months. Of the 92 cases, 5 probably died from chickenpox acquired by intrauterine exposure. The two most frequently cited complications leading to death were pneumonia (35 cases) and encephalitis (22 cases). Pneumonitis was more frequently associated with death in the neonates and infants less than 6 months of age, while fatal chickenpox encephalitis more commonly developed in the infants 6 months or older. An explanation for this dichotomy was not apparent. Barring a marked underreporting of chickenpox and related deaths in older children, these data confirm that infants are at an increased risk of death following chickenpox.

I. Pathogenesis of Intrauterine VZV Infection

For many years there was a dispute whether VZV could cause intrauterine infection. With the publication of several case reports, a consensus has emerged about a defined embryopathy syndrome secondary to intrauterine VZV infection. The congenital defects syndrome occurs when the pregnant woman contracts chickenpox early in gestation. Chickenpox obviously can also occur during late gestation. In this instance, intrauterine transmission leads to chickenpox in the newborn infant at time of birth or shortly thereafter. This latter situation is generally called neonatal chickenpox, even though the infection is acquired congenitally, i.e., prior to delivery.

Neonatal chickenpox will be discussed first, because its various manifestations can

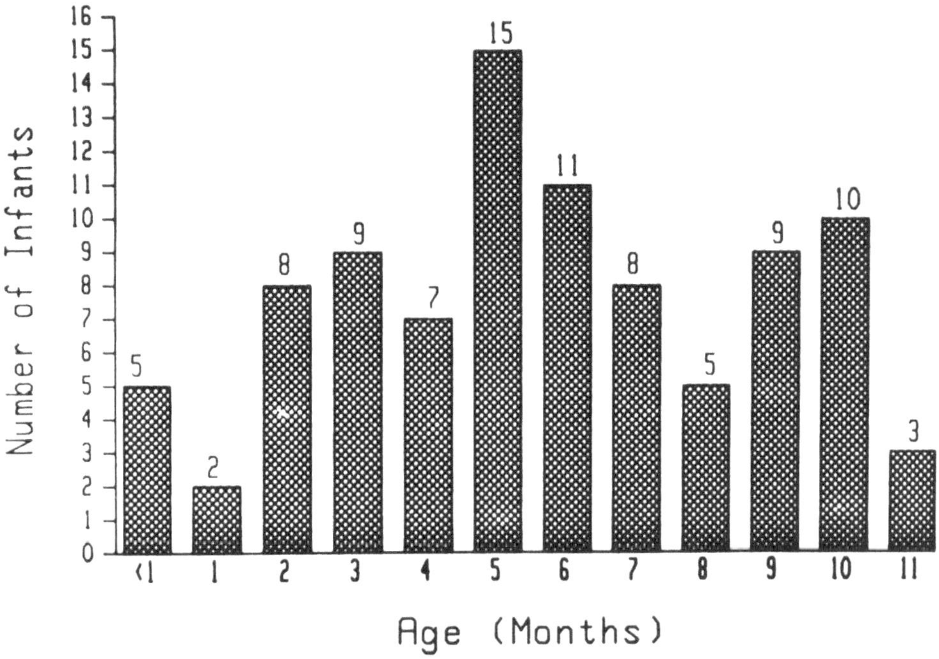

FIGURE 8. Age of death in 92 cases of chickenpox during infancy. The number of infants with chickenpox who died during each 1-month interval are indicated by the number atop each bar. (From Preblud, S. R., Bregman, D. J., and Vernon, L. L., Deaths from varicella infants, *Pediatr. Infect. Dis.*, 4, 503, 1985. © by Williams and Wilkins. With permission.)

be correlated with a dual viremic pathogenetic schema (Figure 9). When the individual case descriptions of neonatal chickenpox are screened, they can be divided into at least two groups: those which follow the primary viremia in the mother and those which follow the secondary viremia in the mother. In the former circumstance (Table 2), the peculiar situation arises where both the mother and the newborn infant have active varicella vesicles.[39,40] One case report, in particular, is most illuminating in this regard.[39] A pregnant 17-year-old woman developed chickenpox 1 day before delivery of twins. The first-delivered twin had about 15 vesicular lesions on her face; over the ensuing first week of life the cutaneous rash spread over the entire body. The second twin had a similar course of neonatal chickenpox. After her exanthem abated, however, she developed severe diarrhea and extensive bronchopneumonia, which led to her death. Thus, in this case report, both twin infants were born with vesicular lesions to a mother who herself had just developed chickenpox. If one refers to the schema of pathogenesis, it is apparent that the virus must have crossed the placental barrier during the primary viremic phase. Under these conditions, the subsequent infection in both the mother and the fetus would follow a similar temporal course with regard to invasion of internal organs, further replication with secondary viremia, and finally, skin lesions.

Rather than being infected during the primary viremia, a more common situation is for intrauterine transmission to occur during the secondary viremia (Table 3).[41-48] Under these conditions, the newborn infant develops cutaneous lesions about 10 days after the appearance of rash in the mother. The incubation period is usually somewhat shorter than 14 to 15 days because the initial replication cycle in the head and neck is circumvented by the direct intravenous route of infection.

In addition to the above two categories of neonatal infection, there is a third and

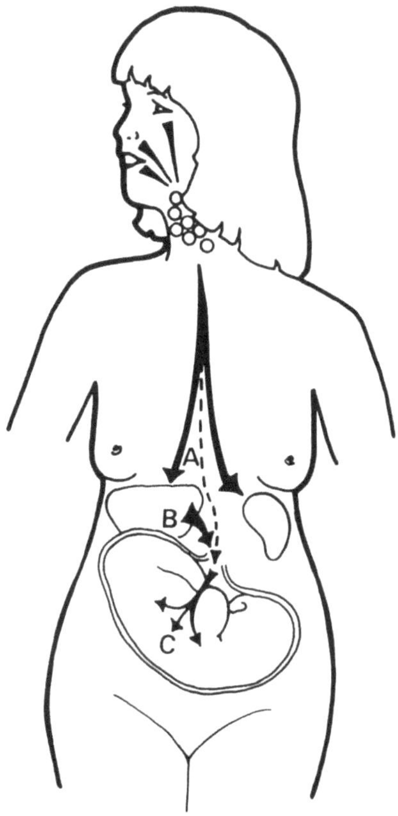

FIGURE 9. Pathogenesis of neonatal chickenpox. Based on the Fennerian schema outlined in Figure 5, the fetus can be infected by in utero transmission of varicella-zoster virus (C) during either the primary viremia (A) or the secondary viremia (B) occurring in the mother with chickenpox during the ninth month of gestation. In the former instance, the newborn infant develops chickenpox lesions at the same time as the mother. (Reprinted from Grose, C., in *Human Herpesvirus Infections,* Reference 1, chap. 3, by courtesy of Marcel Dekker, Inc. With permission.)

somewhat more obscure group of neonates who contract chickenpox during the first week post partum, but whose mothers had no manifestations of clinical chickenpox (Table 4).[49-52] The most likely explanation for these cases is to postulate a subclinical reactivation in the mother with a silent viremia. This hypothesis is supported by recent studies which demonstrate subclinical rises in anti-VZV antibodies in adults exposed to chickenpox. Presumably, these adults have had at least local replication and some may also have a viremic phase. An earlier hypothesis of a 1- to 3-day incubation period for chickenpox in the newborn is not compatible with the schema for pathogenesis.

J. Pathogenesis of the Congenital Defects Syndrome

Intrauterine VZV infection causes one of the most intriguing forms of embryopathy

Table 2
NEONATAL CHICKENPOX AFTER INTRAUTERINE INFECTION: FETAL INFECTION FOLLOWING PRIMARY VIREMIA IN MOTHER WITH CHICKENPOX

Author	Onset of maternal chickenpox (days antepartum)	Day of life rash first appeared in newborn	Ref.
Middelkamp (twins)	1	1	39
	1	1	
Harris (case 1)	3	1	40

Note: Data derived from Reference 1.

among the teratogenic viral agents. Virtually all of the structural defects secondary to congenital VZV infection are related to neurologic damage to the developing embryo with subsequent somatic consequences. These abnormalities are better understood after a brief review of the normal progression of embryogenesis. As can be seen in the timetable (Figure 10), a major portion of musculoskeletal development takes place between 6 and 12 weeks of gestation.[53] These critical events include development of the limb muscles and the formation of the limbs, as well as innervation of the extremities. At this time, the cervical plexus and the lumbosacral plexus comprise greater than 70% of the bulk weight of the spinal cord, since these two plexi innervate the developing upper and lower extremities. Early in embryogenesis, nerves from the spinal ganglia pass through the sclerotomes, which give rise to the chondroblasts and osteoblasts, in order to reach and enter the myotomes. Denervation as a consequence of zoster infection of the spinal ganglia would lead to agenesis or hypoplasia of the developing limbs and organs. In addition, destruction of one nerve cell may lead to degeneration of a connecting neuron across the synapse by a process known as transneuronal degeneration, e.g., retinal damage may lead to atrophy beyond the geniculate ganglia in the optic lobes. In this manner, almost all of the fetal malformations associated with maternal chickenpox during early pregnancy can be accounted for on the basis of neuronal degeneration secondary to viral infection during embryogenesis (Table 5).

These malformations were tabulated by Williamson,[54] who noted that the majority of defects involved the skin, eyes, and limbs. Savage et al.[55] had previously described the zig-zag skin lesions which were located mainly on the hypoplastic limb of their patient. In addition to documenting a loss of sensation, they also demonstrated by electromyographic and nerve conduction studies that their patient had functional denervation of the affected limb. Srabstein et al.[56] found severe neurologic abnormalities upon autopsy of their case: extensive scarring within the dorsal ganglion with possible reduction in the number of anterior horn cells, chorioretinitis with destruction of the optic nerve, and necrotizing encephalomyelitis. Similar cases have also been described by others.[57,58]

If maternal chickenpox occurs in the second trimester of pregnancy, an infected fetus may exhibit mainly ocular pathology. One child born to a mother who had chickenpox during her fourth month of pregnancy had microphthalmus of one eye and chorioretinal scarring of the other eye.[59] The same child also had numerous hypopigmented areas of scar formation on both legs. Another case report describes a newborn

Table 3
NEONATAL CHICKENPOX AFTER INTRAUTERINE INFECTION:
FETAL INFECTION FOLLOWING SECONDARY VIREMIA IN MOTHER WITH CHICKENPOX

Author	Onset of maternal chickenpox (days antepartum)	Day of life rash first appeared in newborn	Incubation period (days)	Ref.
Odessky (case 2)	10	1	10	41
Freud	8	2	10	42
Shuman	8	2	10	1
Ehrlich	2	7	9	1
Cutter	5	3	8	43
Librach	10	4	14	44
Steen	4	5	9	45
Henderson	<1	8	8	46
Oppenheimer	5	5	10	47

Note: Data derived from Reference 1.

with zig-zag cicatricial scarring and hypopigmentation of one extremity as the sole manifestation of a congenital chickenpox infection. This situation most likely represents a forme fruste disease when the intrauterine infection occurs after the critical period of embryogenesis, i.e., late second trimester or early third trimester of gestation.

A final variation on the theme of congenitally acquired VZV infection is the infant born with zoster such as that described in a 4-day-old boy with vesicular lesions involving the right flank, thigh, and penis.[60] This case is doubly thought provoking because the expectant mother, who remained well during her entire pregnancy, was visited in the third gestational month by a close friend who had active zoster. Presumably, therefore, the mother had a subclinical VZV reactivation with a viremic phase during which the infant was infected. The infection in the fetus did not disseminate beyond the lumbosacral ganglia, whence it reimerged a few days post partum.

All of these case reports of the embryopathy syndrome suggest that the malformations are secondary to the neurologic sequelae. Thus, intrauterine transmission appears to cause what might better be called "congenital zoster" rather than "congenital chickenpox". The extent of the neurologic sequelae appears to depend on the month of gestation in which in utero transmission occurs. Continued examination of a child with congenital defects syndrome has demonstrated a normal maturation of the immune system at 5 years of age.[61] This finding suggests that the increased severity and number of infections observed in other cases of this syndrome were due to aspiration secondary to neurologic malfunction rather than to immune incompetence.

K. Epidemiology of Chickenpox in Primitive and Urban Cultures

Chickenpox is an ancient disease whose origins are obscure. Nevertheless, the disease has been found even in the most primitive and isolated populations.[1] These population studies provide fascinating insights into the persistence of viral diseases throughout the millennia. One of these investigations included the tribes on the periphery of the Amazon basin in Brazil; these tribes usually number about 300 members and have little or no contact with neighboring peoples. When members of some of these tribes were

Table 4
NEONATAL CHICKENPOX AFTER INTRAUTERINE INFECTION: FETAL INFECTION FOLLOWING SUBCLINICAL VZV INFECTION IN MOTHER

Author	Day of life rash first appeared in neonate	Ref.
Hubbard	1	49
Pridham	1	50
Joel	1	51
Hyatt	5	52

Note: Data derived from Reference 1.

tested for antibodies to different viruses, they were found to have serologic evidence of past infection with chickenpox (as well as the other herpesviruses, i.e., herpes simplex virus, cytomegalovirus, and Epstein-Barr virus). The prevalence of these antibodies was sometimes higher than commonly found in more advanced cultures. The diseases did not seem to be unusually virulent in the tribal members nor did they appear to threaten the continued existence of the tribe. The above studies imply that herpesviruses, including chickenpox, were present since the origin of man in South America, and have been successfully transmitted from generation to generation.

The mode of transmission of chickenpox in very isolated populations is best illustrated by examples from the Shetland Islands.[62] On some of these remote islands, the inhabitants live on their own small enclosed farmyards; there are no groups of houses large enough even to be called a village. In one reputed instance, a school teacher developed zoster over the right buttock and right thigh. Her nephew, who lived with her, contracted chickenpox 16 days later. In another 2 weeks, other children in the same classroom developed chickenpox. In this manner, the virus was perpetuated from an index case of zoster to many other children inhabiting the isolated island. Thus, the usual mode of transmission in primitive or isolated societies may depend upon reactivation of the disease in adults.

The epidemiology of the disease is remarkably different in industrialized society, although again the mode of transmission is affected by the size of the populations. Since chickenpox has been a reportable disease in Massachusetts since 1907, the epidemiology of the disease has been studied over long periods of time.[63] Recurring outbreaks of chickenpox took place every 2 to 3 years; during epidemic years the rates were about double those in interepidemic periods. The disease exhibits a marked seasonality, with most cases occurring between the months of January to May. Generally, the highest number of cases were reported in March of each year. That the reason for seasonality is not directly related to climate has been demonstrated in several ways, e.g., in London, during the Second World War, there were few outbreaks of chickenpox when school was cancelled because of the German Blitz.[63] Also, zoster, which exhibits no seasonality itself is equally contagious in all seasons of the year.[64]

Rather than climate, the premier mechanism underlying the seasonality in the transmission rates of chickenpox is most likely related to seasonal changes in social behavior, in particular, children returning to school in autumn after the summer-long recess.[65,66] Other factors can also affect the annual periodicity in transmission rates.

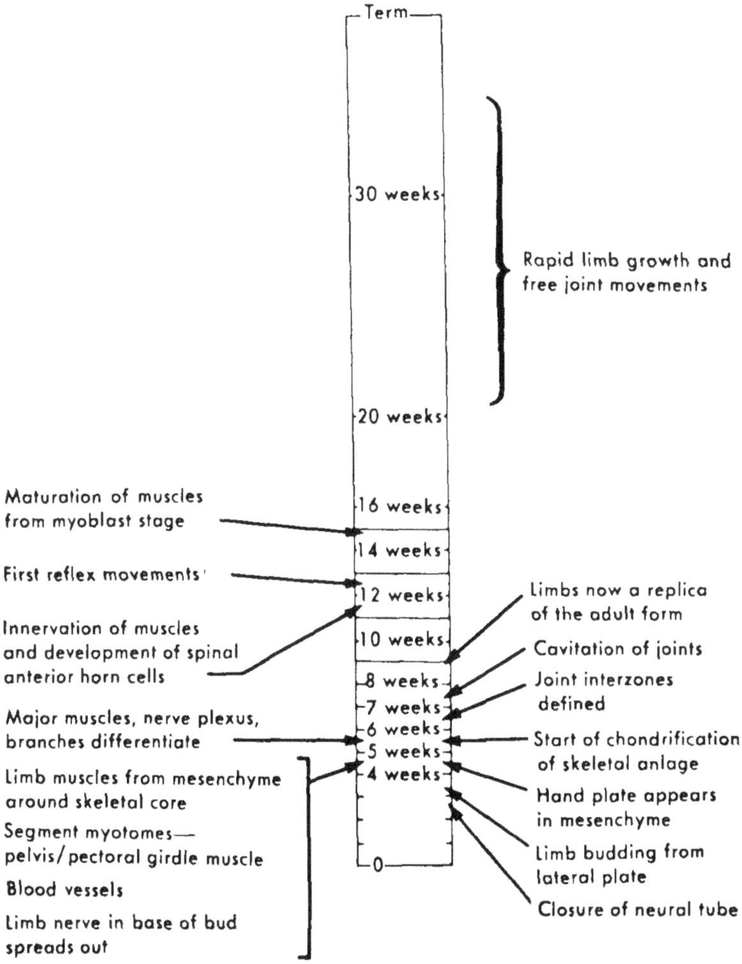

FIGURE 10. Timetable for musculoskeletal embryogenesis. The major muscle groups are formed and innervated during the first trimester of gestation.

These variables within a metropolitan area include the size of total population and the mean contact rate, i.e., the fraction of the susceptibles that an average infective child successfully exposes each day. With the beginning of each school year, there is a marked increase in the mean contact rate. Based on computer simulations, these population data suggest that chickenpox, with a "generation time" of about 2 weeks, would require 7 to 8 months in order to build up the number of infectives, and then deplete the number of susceptibles until the outbreak can no longer be sustained. Thus, an increased mean contact rate in September and October would lead to a peak outbreak of chickenpox in the spring of the following year. It should be noted that these computer-simulated disease patterns relate only to large rather congested urban areas, such as New York City. As discussed earlier, chickenpox more commonly recurs in 2- to 3-year epidemics in regions of the U.S., which have a mixed urban-surburban-rural character, e.g., many parts of Massachusetts. It is also apparent that zoster plays little role in the transmission of chickenpox in a modern urban society.

L. Pathogenesis of Zoster

Zoster or shingles represents the clinical reactivation of the VZV strain which caused

Table 5
SEQUELAE OF FETAL INFECTION WITH VARICELLA-ZOSTER VIRUS DURING EARLY GESTATION

Damage to sensory nerves
Cutaneous manifestations
Zig-zag (cicatricial) skin lesions
Hypopigmentation
Damage to optic stalk, optic cup, and lens vesicles
Microphthalmia
Cataracts
Chorioretinitis
Optic atrophy
Damage to cervical and lumbosacral cord
Hypoplasia of upper/lower extremities
Motor/sensory deficits
Absent deep tendon reflexes
Anisocoria/Horner's syndrome
Anal/vesical sphincter dysfunction
Damage to brain
Encephalitis
Microcephaly

Note: Data derived from Reference 1.

the original case of chickenpox in chickenpox. The association of chickenpox and zoster was first reported by Bokai, as described in the Introduction and in Figure 3, when he recognized that close contacts of zoster patients often developed chickenpox. Even though it became clear that these two diseases were caused by the same agent, many questions remain to be answered about the epidemiology, biology, and molecular biology of zoster. The epidemiology of zoster has been studied in great detail by Hope-Simpson at the Epidemiological Research Unit in Cirencester, England.[64] Within a large general practice of about 3500 people, he recorded all cases of zoster which were diagnosed in the 16-year period 1947 to 1962. He observed a total of 192 cases, which represented an annual attack rate of 3.4 per 1000 people (Table 6).

Hope-Simpson carefully noted the anatomical localization of the zoster rashes. Although he found no predilection for either right or left side, he did observe that the ganglia are not involved in a random distribution (Figure 11). Rather, the two most commonly involved sites are the fifth cranial nerve and the lower thoracic/upper lumbar dermatomes (T5—L2). Relatively few cases occurred in the cervical or lumbosacral dermatomes. He also made the interesting observation that one attack of zoster does not necessarily confer immunity against subsequent attacks. Indeed, 8 of the 192 cases recorded by Hope-Simpson were second episodes and one was a third attack. An unexpected finding was the recurrence of 4 of the 9 multiple episodes in the same sensory dermatome which had been involved in the first attack. He also looked at the age distribution of the zoster patients. The attack rate was low in children and then reached an annual plateau of about 2.5 cases per 1000 adults ages 20—49. However, from age 50 upward, there was a marked rise in the attack rate to 10 cases per 1000 adults 80 years of age or older. In other words, after the age of 80, there will be one case of zoster per year for every 100 octogenarians.

Hope-Simpson did not observe any seasonality to the distribution of cases, i.e., zoster occurred in every month of the year. He also looked to see whether contacts of zoster patients or contacts of chickenpox patients shortly developed zoster, that is to

Table 6
OCCURRENCE OF ZOSTER ACCORDING TO AGE

Age (years)	No. of cases	Rate/ 1000/ year
0—9	6	0.74
10—19	10	1.39
20—29	17	2.58
30—39	18	2.29
40—49	23	2.92
50—59	37	5.09
60—69	38	6.79
70—79	27	6.42
80—89	16	10.10

Note: Data derived from Reference 64.

say, if close association with active VZV infection is responsible for zoster. In neither situation did he find epidemiologic evidence of any direct causal role; therefore, he considers transmission to be a one-way event from zoster to chickenpox and not vice versa. He did offer a mild disclaimer, however, because of the numerous anecdotal situations cited in the medical literature in which a close contact of a zoster patient has contracted clinical zoster within a 2 to 3 week period.

Having decided that direct transmission is not the route by which zoster occurs, Hope-Simpson presented an hypothesis of reactivation in the presence of waning anti-VZV immunity. He postulated that VZV originally enters the sensory nerves via the vesicular skin lesions during chickenpox. The virus is transported to the sensory nerve ganglion where it resides in a latent state for several decades. Subclinical reactivation may occur periodically, but the reemerging virus is neutralized by the host immune system. With increasing years and waning immunity, the human host eventually is unable to adequately restrict the replication of a reactivated virus. The virus then continues to move down the sensory nerve to the skin where it causes the typical dermatomal exanthem called zoster (shingles).

Zoster occurs more often in the immunodeficient patient than in the healthy adult. Recognition of this increased incidence was apparent in the population of renal transplant patients.[67] In an early study of 73 patients, six (8%) developed zoster within 3 years after transplantation. The onset of zoster could not be correlated with splenectomy, thymectomy, or regimen of immunosuppressive chemotherapy, which included azathioprine (Imuran®) and prednisone. In contrast to the preceding results, the onset of zoster was strongly correlated with local irradiation (3 doses of 150 rads each) over the implanted kidney in four of the six transplant patients. The lesions appeared 6 to 23 days after the radiotherapy. In a fifth patient, zoster appeared 5 days after diagnostic hip and pelvic roentgenograms. In all five cases, the involved dermatome was in proximity to the irradiated area of the body. The sixth patient with zoster had no irradiation therapy.

Zoster also occurs more commonly in adults with lymphoproliferative malignancies, in particular, Hodgkin's lymphoma. In a 20-year retrospective study at the Clinical Center of the National Institutes of Health, a total of 107 cases of zoster were recorded in 104 patients.[68] The single most common underlying disease was Hodgkin's lym-

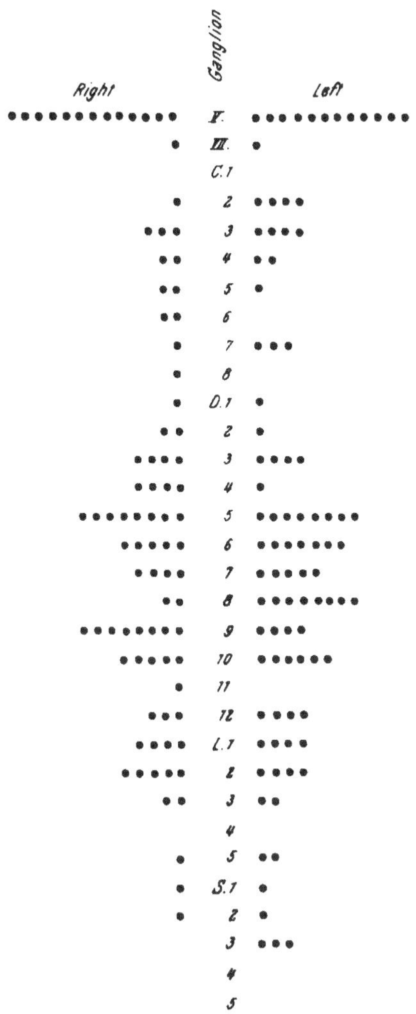

FIGURE 11. Dermatomal distribution of zoster (shingles) as observed by Hope-Simpson.[64] Most cases of zoster in adults occur in either the fifth cranial nerve or the lower thoracic-upper lumbar regions. Both right and left sides are affected equally. (From Hope-Simpson, R. E., *Proc. R. Soc. Med.*, 58, 9, 1965. With permission.)

phoma in 27 patients. Less common diagnoses included acute and chronic lymphoblastic leukemia, non-Hodgkin's lymphoma and acute myelogenous leukemia, as well as a variety of other chronic debilitating illnesses. There was a strong association between onset of zoster and radiation therapy to the affected site within the prior 3 months. When the effect of the chemotherapy regimen was examined, zoster was not associated with any particular treatment protocol. However, disseminated zoster appeared to occur more commonly in patients on high-dose corticosteroid therapy. Patients with systemic lupus erythematosis also develop zoster more frequently than the general population, especially if they receive immunosuppressive agents, such as prednisone, cyclophosphamide, or azathioprine.[69]

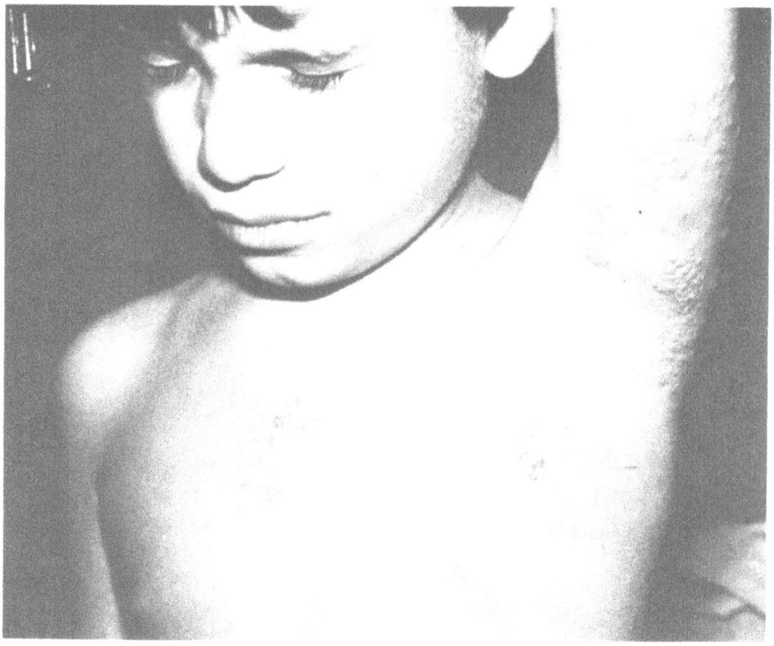

FIGURE 12. Zoster in a young boy. In contrast to zoster in adults, zoster in children who have had chickenpox during the first year of life frequently occurs in the general distribution of the cervical or lumbosacral dermatomes. In this case, the left proximal arm is involved.

M. Zoster in Infants and Children

The dermatomal distribution of zoster in children, especially very young children, differs from that outlined for the adult patient.[1] In contrast to the preferential localization to dermatomes T5—L2, zoster in early childhood usually develops in the cervical or lumbosacral dermatomes.[70-75] In other words, the lesions are often found on either the upper or lower extremity (Figure 12). This distribution is the same as that which occurs in the fetus which is infected in utero, e.g., as manifestated by the zigzag scarring along the extremities which are maldeveloped. The reason for this divergence in localization between infantile zoster and adult zoster probably relates to the development of the spinal cord at time of the initial chickenpox infection. As mentioned earlier, during adulthood about 70% of the spinal cord is contained within the thoracic and upper lumbar regions. In contrast, the major portions of spinal cord of the rapidly developing fetus and growing newborn include the cervical and lumbosacral plexi which innervate the developing extremities. Thus, zoster in the young child appears to be a milder version of the embryopathy syndrome. These observations also suggest that the chance of VZV seeding a particular sensory ganglion is proportional to its mass relative to the remainder of the spinal cord.

II. THE VIRUS AND VIRAL REPLICATION

A. Structural Properties of the Virus

Varicella-zoster virus is one of the human herpesviruses, the others being herpes simplex virus (HSV) types 1 and 2, cytomegalovirus (CMV), and Epstein-Barr virus (EBV). These viruses share many morphological characteristics, including an electron-dense core, an icosahedral nucleocapsid, a tegument surrounding the capsid, and an envelope of variable dimensions. The size of the VZV nucleocapsid is approximately 95 nm in diameter, whereas the whole enveloped virion measures anywhere from 150 to 200 nm in diameter.[76-79] There is considerable pleomorphism in viral particles grown in cultured cells, dependent mainly on differences in the amount of tegument and envelope components of the virus. In some cell substrates, the cytoplasmic viral particle assumes an onion-like appearance because of the varying number of envelope lamellae. The average diameter for the intracytoplasmic particle is about 150 nm, while those in the extracellular space range about 175 nm or greater. Unusual smaller electron-dense particles, called "dense bodies" and lacking nucleic acid, are frequent constituents of infected cultures also. The role of these dense bodies remains obscure. Properties of the genomic VZV DNA are discussed in Chapter 3 of this book.

B. Isolation and Propagation in Cell Culture

VZV replication occurs only in a limited number of cell substrates, most of which are of human origin. These include primary cell cultures established from human embryo tissues, e.g., fetal kidney and fetal lung cells, and human amnion cell lines, as well as human newborn foreskin cells. In his original reports,[4,80] Weller described the serial propagation of VZV in roller tube cultures of human embryonic skin-muscle tissue and foreskin tissue. In their monograph, Taylor-Robinson and Caunt list more than 20 human and primate cell cultures which support the growth of VZV to some extent.[81] Nevertheless, they conclude that VZV replicates efficiently in only a few human embryonic tissues. A more recent report describes the isolation and serial propagation of VZV in cell cultures derived from human malignant melanoma tumors.[82] Interestingly, the melanoma cell is the neoplastic derivative of the melanocyte, which arises embryologically from the neural crest ectoderm.

VZV can be isolated in human embryonic cells or human melanoma cells (HMC) within 3 to 5 days after inoculation of vesicular fluid samples from patients with chickenpox or zoster. Since vesicular fluid contains large quantities of infectious virus, samples of only a few microliters are usually sufficient for virus isolation. The cytopathic effect (CPE) in HMC cultures is characterized by small multinucleate cells initially and later by large syncytia (Figure 13). Virus isolation is equally successful at incubation temperatures ranging from 31 to 37°C. An isolate from a culture incubated at 32°C has been propagated in HMC cultures and used as stock virus for further experiments described below; this virus was designated the "VZV-32" strain.

C. Cell-Free Varicella Zoster Virus

A major obstacle to laboratory investigation of VZV has been the lack of infectious virions released into the medium overlying infected cells. Because of this close cell association, serial propagation is usually accomplished by passage of trypsin-dispersed infected cells. The most successful method of obtaining VZV relatively free of cellular contamination ("cell-free virus") has been sonic disruption of VZV-infected cells. These techniques were pioneered by Caunt and Robinson,[83] who were able to obtain titers between 4 $\log_{10}$ and 5 $\log_{10}/m\ell$ by sonically disrupting VZV-infected human thyroid cells, which had been removed from the glass surface by 0.02% EDTA. Similar or higher yields of cell-free virus were also achieved by sonication of VZV-infected human embryo lung fibroblast cells which had first been mechanically dislodged from the culture flask.[84,85]

The yields of cell-free virus from the medium and cellular components of VZV-infected human melanoma cells have been investigated in detail.[79] Cell-free virus was prepared from the infected cell substrate by three different methods:

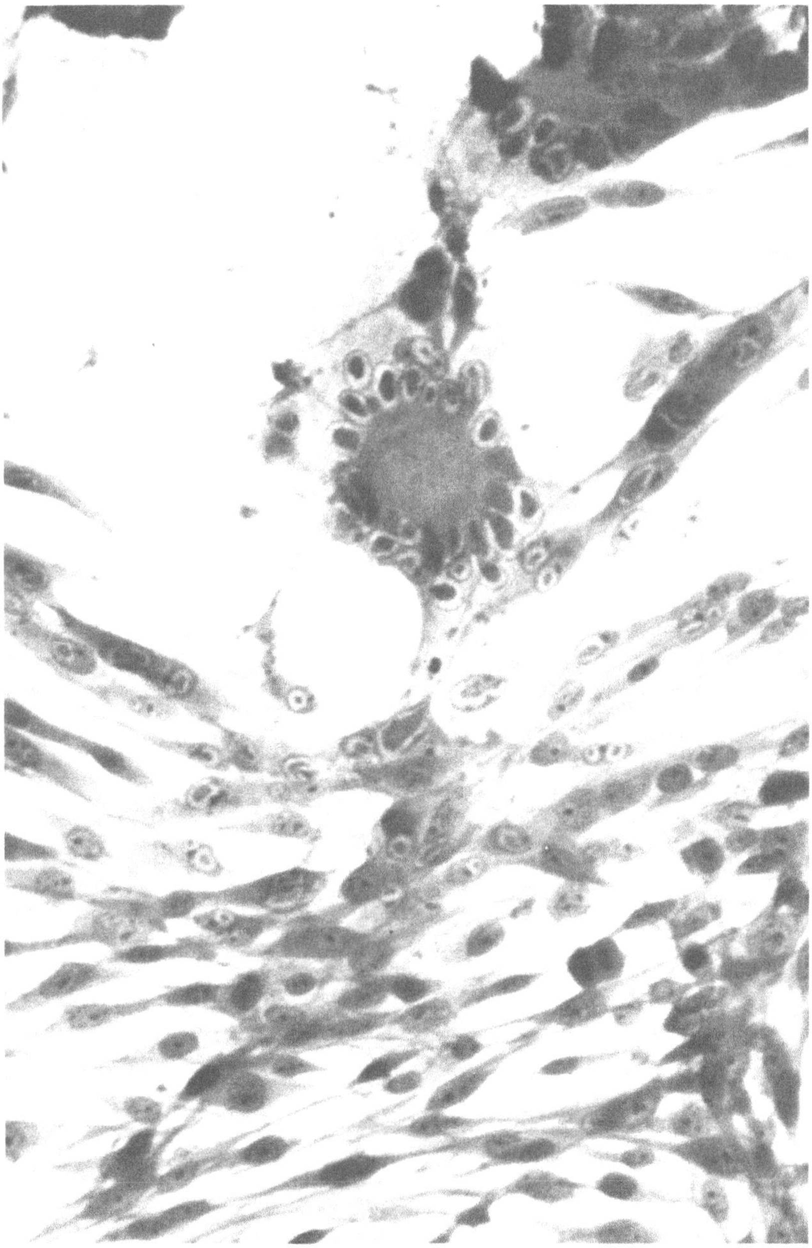

FIGURE 13. Cytopathology of VZV infection in human melanoma cell cultures. In the left half of the field are numerous uninfected cells, while in the right half of the figure there is a large virus-induced syncytium with about 20 nuclei, each with a prominent dark inclusion body. A smaller polykaryocyte is seen in the right lower corner.

1. The infected monolayer was dislodged with a rubber policeman into 10 m*l* Eagle minimum essential medium. The cell suspension was centrifuged at 1500 r/min for 15 min to pellet the cells, which were resuspended in 1 m*l* of sorbitol buffer (10% sorbitol in 0.05 *M* Tris, pH 7.4, and 0.001 *M* $MgCl_2$) and sonically disrupted (20 Hz) for 30 sec. The sample was centrifuged again to remove cellular debris from the supernatant.
2. The infected monolayer was dislodged from the surface of the flask by the use of solid glass beads 3 mm in diameter. About 10 g of sterile beads in 4 m*l* minimum essential medium were placed into a 150 cm^2 flask and gently rolled over the monolayer for 2 min. The cell suspension was removed and sonicated as described in the first method.
3. The infected cells were subjected to Dounce homogenization before sonication. After mechanical removal the infected cells were resuspended in 2 m*l* of reticulocyte standard buffer (RSB: 0.01 *M* NaCl, 0.0015 *M* $MgCl_2$, 0.01 *M* Tris, pH 7.4) and placed in an ice bath for 5 min. The swollen cells were ruptured in a glass Dounce homogenizer. The cells were observed by microscopy before and after the Dounce procedure to assure >95% cell breakage. After homogenization, 60% sucrose (w/w) was added immediately to yield a final concentration of 0.25 *M*. Nuclei were removed by centrifugation for 5 min at 2000 r/min, resuspended in 2 m*l* sorbitol buffer, and sonically disrupted.

Similar yields of cell-free virus were achieved by each of the first two methods of cellular disruption (Figure 14). Within 60 hr after inoculation, when VZV-infected human melanoma cells exhibit advanced cytopathology, the overlying culture medium contains about 100 plaque-forming units (PFU)/m*l*. The cell pellet (2 m*l*) from a sonically disrupted 150 cm^2 monolayer yields 100,000 PFU/m*l*. When a VZV-infected monolayer of similar size is first pelleted and then subjected to Dounce homogenization, most of the infectivity is found in the nuclear fraction. Upon electron microscopic visualization, it was apparent that the nuclei frequently included cytoplasmic remnants containing vacuoles filled with virus particles.

Concentration of cell-free virus has been carried out by three methods, which include ultracentrifugation, dialysis against hydrophilic compounds (sucrose granules or polyethylene granules), and liquid polymer phase separation. The first two procedures caused considerable loss of biologic activity. However, precipitation of infectious virus in VZV culture medium by the addition of 8% polyethylene glycol in the presence of high salinity was accompanied by as much as a 50-fold increase in titer (Table 7).

D. Cryopreservation of the Enveloped Virion

As mentioned in the preceding section, VZV is an extremely labile virus. When VZV is released from infected cells by methods such as rapid freezing-and-thawing or sonic disruption, the infectivity of the preparation is quickly lost if improperly handled. Techniques for cryopreservation of VZV virions without loss of either structural integrity or infectivity have been described.[85] For these studies, several lyophilization media were investigated (Table 8). In a four-part experiment, the suspension media — SPGA, SPA, Tris-sorbitol, and PBS were compared. The mean titer of infectious virus lyophilized in PBS fell from 4.7 $\log_{10}$ to 2.0 $\log_{10}$ PFU/m*l*. In contrast, the titers in virus samples suspended in Tris-sorbitol, SPGA and SPA media remained above 4 $\log_{10}$ PFU/m*l* after freeze-drying and reconstitution.

The components of the buffers were further analyzed to determine which most influenced recovery of infectivity. Virus samples lyophilized in phosphate-buffered medium containing from 0.1 to 1.0 *M* sucrose demonstrated similar preservation of titer above

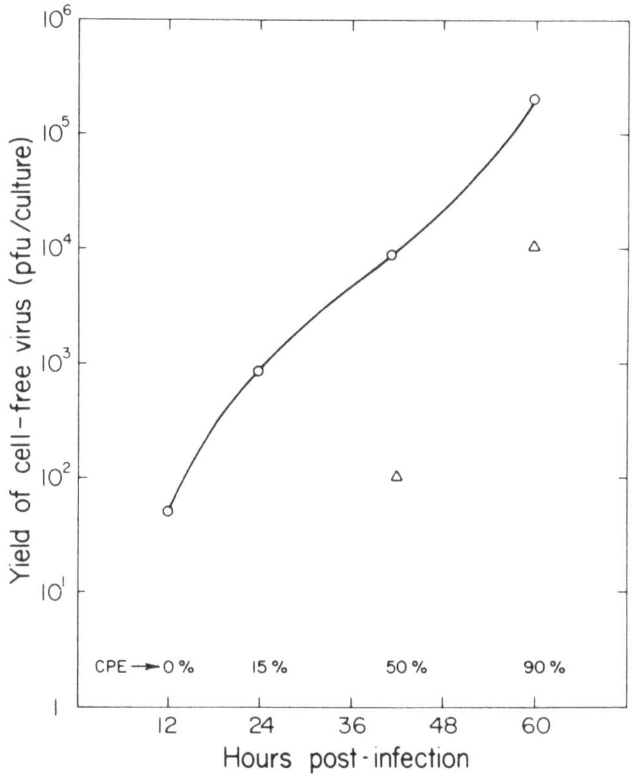

FIGURE 14. Yields of cell-free virus from infected monolayers and culture medium. Replicate cultures were inoculated with VZV-infected cells at a ratio of 1 inoculum cell to 4 uninfected cells, and the cultures were incubated at 32°C. Before the infected monolayers were harvested, the culture medium (25 m*l*) was removed and assayed for infectious virus. The cells were dislodged with a rubber policeman and sonically disrupted. Virus titer in cell pellet (o—o); virus titer in culture medium (Δ); c.p.e. = cytopathic effect. (From Grose, C., Perrotta, D. M., Brunell, P. A., and Smith, G. C., *J. Gen. Virol.*, 43, 15, 1979. With permission.)

4 $\log_{10}$ PFU/m*l*. When the residual moisture content of virus lyophilized in media with and without sucrose was determined, almost twice as much moisture was retained in sucrose-containing lyophilization media. Visualization by electron microscopy demonstrated large numbers of enveloped virions in virus preparations freeze-dried in sucrose-containing media, whereas virus subjected to lyophilization in PBS (without sucrose) consisted mainly of naked nucleocapsids. Thus, the loss in infectivity in the latter preparation was correlated with physical disenvelopment of the virion, while the retained infectivity of virus freeze-dried in sucrose-containing solutions was due to retention of the envelope on the virion.

E. Virus Purification

Purification of infectious VZV particles by density gradient sedimentation is not a very successful technique because of the marked lability of the enveloped virion. Purification by equilibrium sedimentation in cesium chloride yields two virus bands, an upper band containing 85% enveloped particles at a density of 1.274 g/cm^3 and a lower band containing 95% unenveloped particles at a density of 1.294 g/cm^3.[87] The

Table 7
CONCENTRATION OF CELL-FREE VARICELLA-ZOSTER VIRUS

Method[a]	Initial vol (ml)	Initial titer (PFU/ml)	Final vol (ml)	Final titer (PFU/ml)	Conc.
1	10	0.6×10^5	1.4	1.5×10^4	None
2	10	0.6×10^5	1.4	1.0×10^4	None
3	90	2.5×10^2	10	1.5×10^3	6-fold
4	150	2.5×10^2	2	1.25×10^4	50-fold

Note: Data derived from Reference 79.

[a] 1, Dialysis against sucrose granules; 2, dialysis against polyethylene glycol flakes; 3, separation with 10% dextran-10% polyethylene glycol; and 4, precipitation with 8% polyethylene glycol.

virus has also been purified by sedimentation in potassium tartrate medium.[79] In a 10 to 50% linear potassium tartrate gradient, infectivity and radioactivity peaks were confined to a fraction with a density of 1.180 g/cm^3. Likewise, infectious virions can also be recovered after two sedimentations in a potassium tartrate-glycerol viscosity-density gradient.[79] Although there is a considerable drop in titer during the first centrifugation, nearly all of the infectivity present in the first gradient is recovered in the second gradient (Figure 15).

Finally, virus has been subjected to centrifugation in 15 to 45% linear metrizamide gradients.[88] When a virus-enriched sample is sedimented for 18 hr at 70,000 g, three bands are visible in the gradient. Of the three (designated upper, middle, and lower, respectively), the middle band at a buoyant density of 1.156 to 7 g/cm^3 is enriched for enveloped virions (Figure 16). Electron microscopic visualization and enumeration have demonstrated a total of 10.04 $\log_{10}$ enveloped particles and 8.26 $\log_{10}$ unenveloped particles from middle bands representing the yield from a 150 cm^2 VZV-infected monolayer (Table 9). Although high numbers of intact enveloped particles are present in the middle band of a metrizamide gradient, little or no infectivity is detected in the middle fraction (Figure 17). The reason for this drastic loss of biologic activity is unknown.

F. Polypeptide Constitution of the Virion

Several laboratories have published data about the polypeptide constitution of the VZ virion. Shemer et al. purified VZV by successive sedimentations in glycerol and potassium tartrate-glycerol gradients.[89] When they characterized the molecular composition of the virion, they found a total of 33 proteins ranging in molecular weight from 16,000 to 244,000 after analysis of [^{35}S]methionine-labeled VZV preparations in either 11% or 20% polyacrylamide gels. Additionally, 5 of the 33 polypeptides incorporated [^{14}C]glucosamine. Shiraki et al. purified VZ virions by rate zonal centrifugation in sucrose followed by sedimentation to equilibrium in cesium chloride.[90,91] They enumerated 32 [^{35}S] methionine-labeled structural proteins which ranged in size from 21,500 to 280,000 daltons, among which were six glycosylated species. In the purification procedure in metrizamide outlined by Grose et al., 30 polypeptides were detected in the electrophoretic profile of the twice-sedimented virion-enriched middle band.[88] The smallest constituent of the polypeptide profile was 30,000 daltons and the largest was 200,000 daltons (Figure 18). The most prominent of the higher molecular weight polypeptides labeled with [^{35}S]methionine had an estimated molecular weight

Table 8
LYOPHILIZATION MEDIA

Medium	Constituents and Concentration	pH
SPGA	0.2 *M* sucrose	6.6
	0.01 *M* potassium phosphate	
	0.005 *M* sodium glutamate	
	1% (v/v) bovine serum albumin	
SPA	0.2 *M* sucrose	6.6
	0.01 *M* sodium phosphate	
	1% (v/v) bovine serum albumin	
SP	0.2 *M* sucrose	6.6
	0.01 *M* sodium phosphate	
Tris-sorbitol	0.5 *M* sorbitol	7.4
	0.01 *M* Tris (hydroxymethyl-aminomethane)	
	0.001 *M* magnesium chloride	
PBS	0.01 *M* sodium phosphate	7.2
	0.15 *M* sodium chloride	

Note: Data derived from Reference 86.

of 155,000 daltons and was designated as the major capsid protein. A major structural protein of similar molecular weight has been found in nearly all herpesviruses studied to date.[92] At least 3 prominent glycosylated proteins were also detected in the purified VZV virion preparation. These structural glycoproteins are discussed in detail in later sections of this chapter.

G. Viral Replication

To study viral replication,[82] growth kinetics experiments have been conducted at the two incubation temperatures of 32° and 36°C. Since VZV remains nearly completely cell associated, most experiments are carried out with an inoculum of infected cells rather than cell-free virus. In one set of experiments, several HMC monolayers were inoculated with infected cells at a ratio of 1 infected cell to 8 uninfected cells in a fresh monolayer incubated at 36°C. At 12 hourly intervals after infection, cultures were harvested and assayed for infectivity. The amount of infectious cell-free virus was quantitated after sonic disruption of the monolayer. Cell-free virus was first detected at 12 hr after infection and reached a maximal titer at 36 hr postinfection, when CPE involved 50% of the monolayer. When similar experiments were carried out at an incubation temperature of 32°C, the highest yield of cell-free virus (10^5 PFU/m*ℓ*) was found in cell cultures with far advanced CPE about 60 hr postinfection. In addition, the titer of the sonically treated viral preparations at 32°C was about $^1/_2$ log higher than that from a comparable infection at 36°C. Another interesting temperature-related phenomenon was observed when cell-free virus derived from VZV-infected HMC cultures grown at 36° was simultaneously assayed at 36° and 32°C. The titer of the same virus inoculum was higher by 0.5 to 1.0 log when the plaque assay was incubated at the lower temperature (Figure 19). The same observation had also been made with VZV grown in human embryonic lung fibroblast cultures and assayed at both 31° and 37°C.[93]

The events which occur during the replicative cycle include attachment to the host cell and penetration of the cell membrane. The cellular receptor for the herpesvirus is unknown, but it appears that the envelope of the virion first fuses to the cell membrane; then the virus enters the cell. Uncoating occurs in the cytoplasm and the capsid

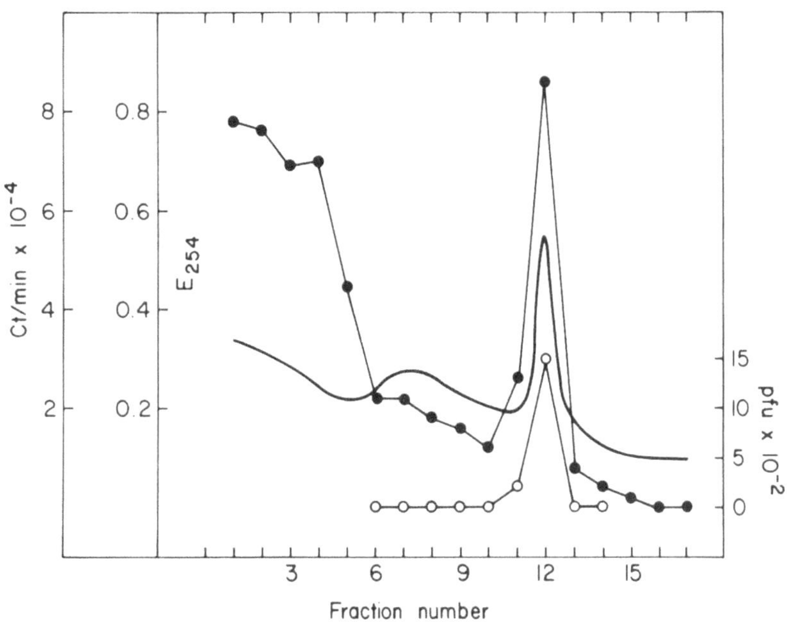

FIGURE 15. Sedimentation of virus in potassium tartrate-glycerol gradients. Radiolabeled cell-free virus (titer = 10^4 PFU) was layered onto a 16 m*l* potassium tartrate-glycerol gradient and centrifuged for 16 hr at 108,000 × g. When the gradient was fractionated, a coincident peak of infectivity (o—o), radioactivity (•—•), and adsorbence (—) was found in fraction 12. (From Grose, C., Perrotta, D. M., Brunell, P. A., and Smith, G. C., *J. Gen. Virol.*, 43, 15, 1979. With permission.)

travels to the nucleus. Transcription of viral DNA takes place in the nucleus, after which the RNA transcripts travel to the cytoplasm where they are translated while attached to cytoskeletal elements. The nascent gene products return to the nucleus where they are assembled into capsids while attached to the nuclear matrix. Electron microscopic visualization of actively replicating VZV demonstrates the presence of nucleocapsids within the nuclear matrices.[94] Newly synthesized DNA is inserted into the capsid and the progeny nucleocapsids exit the nucleus, acquiring their initial envelope at the nuclear membrane. Further envelopment of the nucleocapsid appears to take place within the cytoplasm; in particular, envelopment may occur as the nucleocapsids enter cytoplasmic vacuoles presumably derived from the *trans* region of the Golgi.[95] In turn, the virion-laden cytoplasmic vacuoles travel to the outer membrane, where the enveloped viral particles attach to the exterior of the cell. Because of the low titer of infectivity in the culture medium, it is uncertain whether virions produced in cultured cells normally detach from the infected cell plasma membrane.

H. Regulation of VZV Replication

Because of the very low titer of cell-free virus obtained from infected cells, studies of the VZV replicative cycle are difficult to perform. Nevertheless, several attempts have been made by fluorescence microscopy to estimate the duration of one cycle of replication. In an early study, human embryo lung fibroblast cultures were inoculated with infected cells and the cultures were observed periodically for cell-to-cell spread.[96] Between 8 to 16 hr passed before infectivity was noted to spread beyond the site of the initial infected-cell inoculum to a neighboring focus. In a similar study monitored by fluorescence immunostaining,[97] viral antigens were first detected in newly infected cells by 2 hr postinfection, and reactivity appeared in the nuclei within 4 hr postinfection.

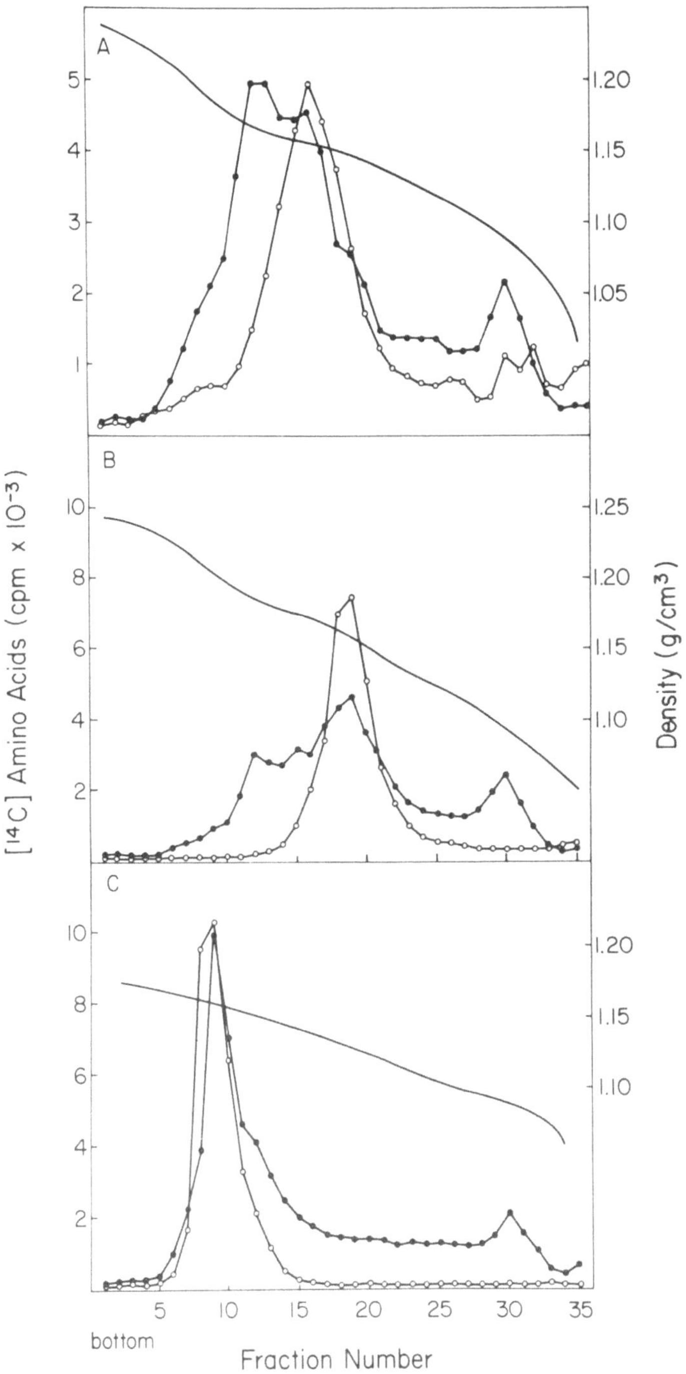

FIGURE 16. Sedimentation profiles of cell-free VZV in three different metrizamide solutions. Cell-free virus labeled with [C_{14}] amino acids was sedimented in 15 to 45% linear metrizamide gradients prepared in three different buffers. The solutions included (A) 0.001 *M* sodium phosphate (pH 7.0), 0.2% BSA; (B) 0.05 Tris (pH 7.4), 0.15 *M* NaCl, 0.001 *M* EDTA, 0.01% gelatin; and (C) 0.001 *M* phosphate (pH 7.0), 0.01% gelatin. The first sedimentation profile (-●-) consisted of upper, middle, and lower bands. The virion-enriched middle band was subjected to a second sedimentation (-○-) in metrizamide dissolved in the same buffer. Density as calculated from the refractive index of alternate fractions is represented by the continuous line.

Table 9
SEDIMENTATION OF VIRAL PARTICLES PRESENT IN VZV-INFECTED MONOLAYERS AND MEDIA[a]

Location in metrizamide density gradient	Enumeration of viral particles ($\times \log_{10}$)		
	Enveloped	Unenveloped	Total
Upper band	8.72	8.88	9.11
Middle band	10.04	8.26	10.11
Lower band	9.82	10.49	10.57

Note: Data derived from Reference 88.

[a] Cell-free virus from 150 cm² VZV-Infected monolayer consisted of 10.92 and 10.51 $\log_{10}$ enveloped and unenveloped particles, respectively, after precipitation with 8% polyethylene glycol in 0.5 *M* NaCl. The pellet was distributed among six metrizamide density gradients. After sedimentation, the three major bands from each gradient were removed and pooled prior to enumeration by electron microscopy.

Subsequently, the pattern of fluorescence spread to the cytoplasm by 14 hr and to neighboring cells by 18 hr postinfection. Thus, a single VZV replicative cycle lasts somewhere between 8 and 18 hr, depending on the cell substrate and the temperature of incubation of the infected cells.

The regulatory mechanisms affecting VZV replication have not been well delineated but probably will closely resemble those already described for HSV replication.[98] Based on an operational definition, there are a minimum of three phases for transcription of viral DNA and these three phases roughly correspond to the synthesis of three groups of HSV polypeptides, which are designated immediate early (alpha), early (beta), and late (gamma) proteins. Their synthesis is said to be coordinately regulated and sequentially ordered in a cascade fashion, i.e., groups of viral polypeptides appear in the infected cell in a predictable pattern and the synthesis of later appearing proteins is dependent upon the prior synthesis of earlier proteins. The alpha or immediate early ("IE") proteins are important regulatory gene products whose synthesis is not dependent on either viral or host protein synthesis after infection. Formation of beta viral polypeptides is dependent upon the formation of functional IE proteins but does not require prior viral DNA synthesis. These beta proteins include viral specified enzymes involved in DNA replication, e.g., DNA polymerase and thymidine kinase. Finally, the gamma or late gene products appear in the infected cell along with the onset of viral DNA synthesis. The structural proteins and glycoproteins of the virion belong to this group. As mentioned earlier, it is very likely that VZV replication will closely mimic the triphasic growth pattern of HSV in cell culture, as described above. Because the VZV late gene products have been most extensively characterized to date, this review will also emphasize structural VZV glycoproteins and proteins.

III. MAJOR VIRAL GLYCOPROTEINS AND PROTEINS

A. Viral Glycoproteins

Glycoproteins are widely distributed amongst the enveloped viruses, including the herpesviruses. The structures of the viral glycoproteins encompass those seen elsewhere

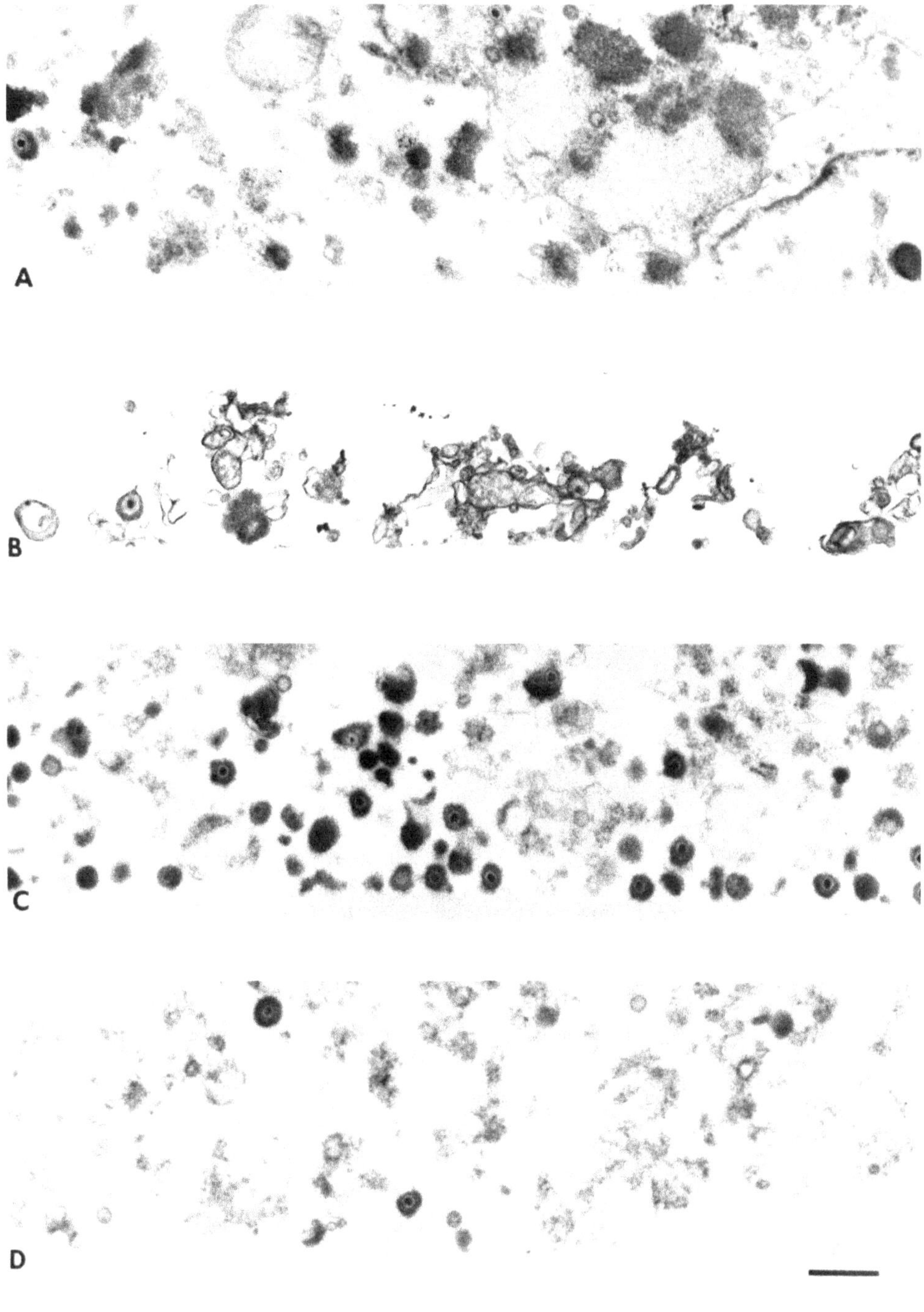

FIGURE 17. Electron microscopic examination of virus before and after sedimentation in a metrizamide gradient. The four panels represent specimens from a virus-enriched pellet after polyethylene glycol precipitation of medium and sonicate (A), and from the upper (B), middle (C), and lower (D) bands of a metrizamide density gradient after centrifugation of the virus-enriched pellet. Specimens were sedimented onto Millipore filters®, after which they were postfixed in chrome-osmium, dehydrated in graded ethanols, and embedded in Epon®. Filters with sedimented virus were cut perpendicularly to the plane of the membrane, and pale gold sections were mounted on 300 mesh copper grids. The Millipore filter® is visible near the bottom of each panel. Magnification of the electron micrographs was determined using a carbon grating replica. Scale: 0.5 μm. (From Grosse, C., Friedrichs, W. E., and Smith, G. C., *Biken J.*, 26, 1, 1983. With permission.)

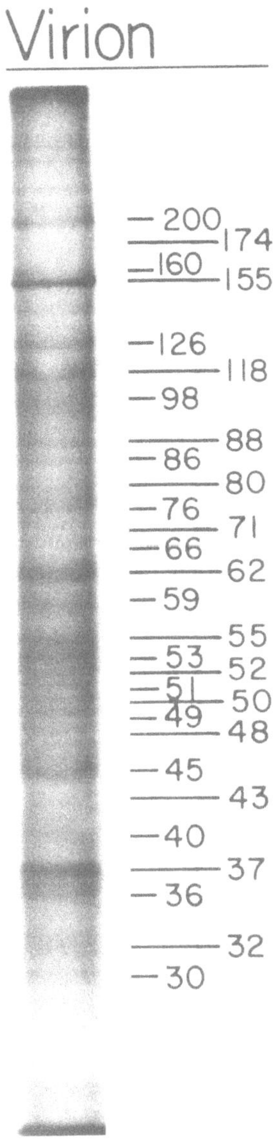

FIGURE 18. Structural polypeptides of the varicella-zoster virion. Cell-free virus labeled with [^{35}S]methionine was recovered from an infected cell monolayer and concentrated by polyethylene glycol precipitation. After two sedimentations in 15 to 45% linear metrizamide gradients (see Figure 16), the virion-enriched middle bands were collected and solubilized. A sample containing approximately 40,000 cpm was subjected to electrophoresis in a 10% acrylamide slab gel. The molecular weights of the structural polypeptides between 30,000 and 200,000 are enumerated in the right margin.

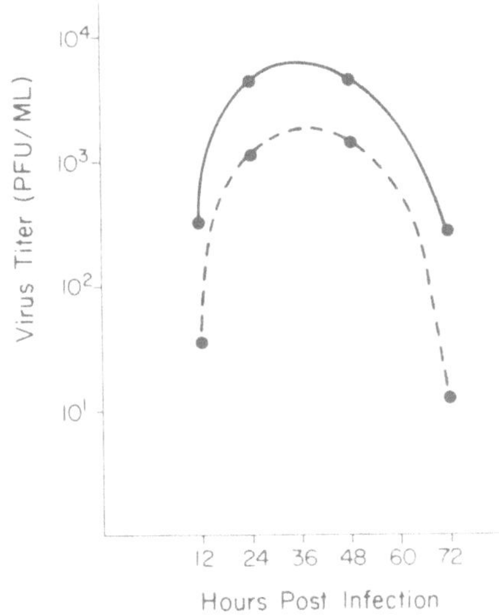

FIGURE 19. Titration of cell-free virus at two incubation temperatures. Replicate human melanoma monolayers were inoculated with VZV-infected cells and harvested at 12-hourly intervals after infection. At each time point, cell-free virus was prepared and simultaneously titrated in two plaque assays, one of which was incubated at 32°C (—) while the other was incubated at 36°C (---). The titer was higher in the plaque assay incubated at 32°C.

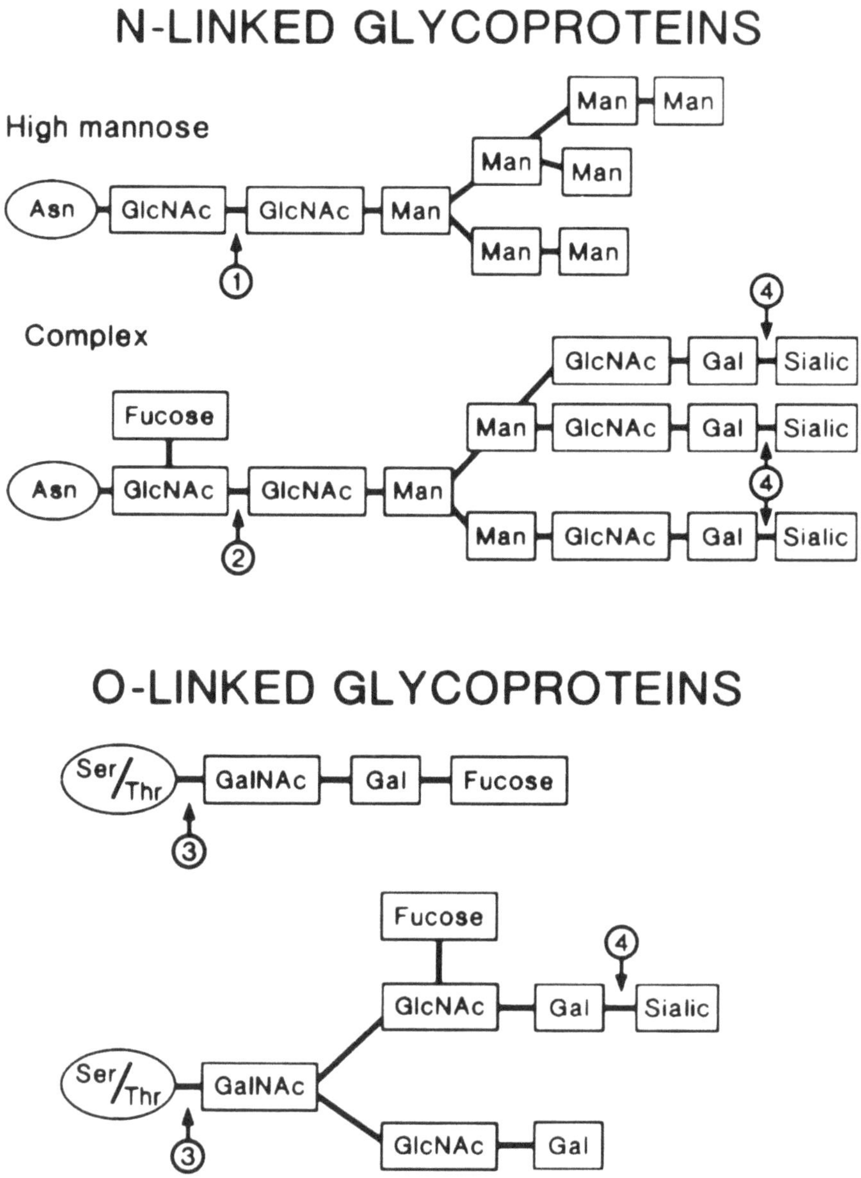

FIGURE 20. Hypothetical examples of asparagine-linked (N-linked) and serine/threonine-linked (O-linked) glycoproteins. The N-linked glycoproteins include both high-mannose and complex types. The arrows indicate the site of cleavage of four glycosidases: (1) endo-beta-*N*-acetylglucosaminidase H (endoglycosidase H, Endo H); (2) endo-beta-*N*-acetylglucosaminidase F (endoglycosidase F, Endo F); (3) endo-alpha-*N*-acetylgalactosamidase; and (4) neuraminidase.

in nature and include the two broad categories of N-linked and O-linked glycoproteins (Figure 20). The N-linked glycoproteins have the oligosaccharide N-glycosidically linked from the *N*-acetylglucosamine (GlcNAc) to the amide nitrogen of asparagine (Asn) residues in the polypeptide backbone. The N-linked glycoproteins are further subdivided into high-mannose (oligomannosidic) glycans and complex glycans. High-mannose oligosaccharides consist of mannose (Man) and *N*-acetylglucosamine residues only, while complex type glycoproteins contain additional sugars such as galactose (Gal), fucose (Fuc), and sialic acid (NeuNAc). The O-linked glycoproteins have

Table 10
NOMENCLATURE FOR VZV GLYCOPROTEIN GENES AND THEIR GLYCOPROTEIN PRODUCTS

Gene	New nomenclature	Old names of glycoproteins	Ref.
gpI	gpI(98), gpI(62)	gp98, gp62	103—108
	gpI(94), gpI(83), gpI(55), gpI(45)	gp2	109—112
	gpI(92), gpI(83), gpI(55), gpI(45)	gC: gp92, gp83, gp55, gp45	113, 114
	gpI(90), gpI(80), gpI(60)	90K, 80K, 60K	115, 116
	gpI(92), gpI(59), gpI(47)	gp92, gp59, gp47	117
gpII	gpII(140), gpII(66)	gp140, gp66	118—120
	gpII(116), gpII(106), gpII(64)	gp3	111, 112
	gpII(115), gpII(62), gpII(57)	gB: gp115, gp62, gp57	113
	gpII(130), gpII(125), gpII(62)	gp1, gp3	121, 122
	gpII(120), gpII(118), gpII(65)	120K, 118K, 65K	116
	gpII(125), gpII(63)	gp125, gp63	117, 123
gpIII	gpIII(118)	gp118	103, 104, 105, 107, 124, 125
	gpIII(115)	gp1	
	gpIII(105)	gA: gp105	113
	gpIII(118)	118K	115, 116
gpIV	gpIV(55), gpIV(45)	gp55, gp45	126
gpV	Hypothetical gene		

Note: Table based on proceedings described in Reference 102.

the sugar O-glycosidically linked from the *N*-acetylgalactosamine (GalNAc) to the oxygen of a serine (Ser) or threonine (Thr) residue in the polypeptide backbone. The O-linked glycoproteins have few common features; they vary considerably in size, from a simple core consisting of only *N*-acetylgalactosamine in branched oligosaccharides containing several other sugars such as galactose, *N*-acetylglucosamine, fucose, and sialic acid.

The biosynthesis of all N-linked glycoproteins involves the formation of intermediate lipid-linked oligosaccharides.[99,100] The lipid carrier is dolichol phosphate (Dol-P), a phosphorylated polyprenol containing carbon chains of varying length. The sequence of reactions by which the common precursor oligosaccharide is assembled is called the "dolichol phosphate cycle". These events take place in the endoplasmic reticulum, where the lipid-linked oligosaccharide is also transferred en bloc to the asparagine residue that is part of the sequence Asn-X-Ser/Thr in a nascent polypeptide. Processing of the glycoprotein begins in the rough endoplasmic reticulum (RER) and continues as the glycoprotein is transported from the RER through the stacks of the Golgi apparatus from the *cis* face to the medial (central) compartment and finally to the *trans* region, whence the mature glycoprotein exits to its final destination. In VZV-infected cells, trafficking may be to either the plasma membrane of the cell or, alternatively, the envelope of the new virion. In contrast to the abundant information about biosynthesis of N-linked glycoproteins, relatively little is known about the formation of O-linked glycoproteins.[101] The synthesis of O-linked glycoproteins does not involve lipid-linked intermediates; rather the first step is the direct transfer of *N*-acetylgalactosamine to a hydroxyamino acid (serine or threonine) in the polypeptide backbone. This initial step probably occurs in the RER or *cis* compartment of the Golgi.

VZV is an enveloped virus which directs the synthesis of several glycosylated products. At the Tenth International Herpesvirus Workshop in Ann Arbor, Michigan (August 1985), a consensus nomenclature was agreed upon by several investigators in the field. Under the new nomenclature, which is described in Table 10, each VZV glyco-

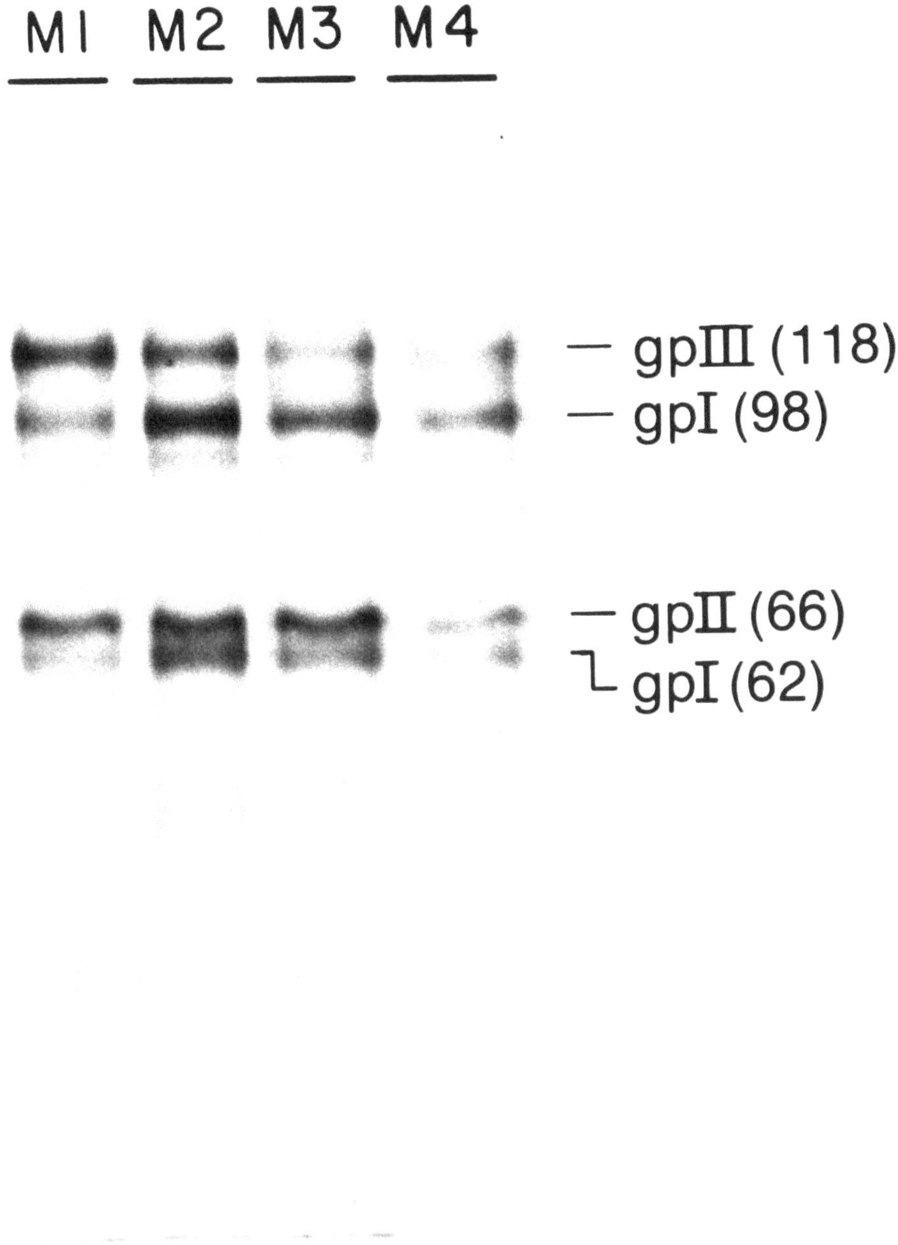

FIGURE 21. The major glycoproteins of varicella-zoster virus as defined by radioimmune precipitation. Detergent-solubilized extracts of VZV-infected cells intrinsically labeled with tritiated fucose were mixed with high-titer VZV antiserum prepared in four hyperimmunized mice (M1 to M4). The radioimmune precipitates were subjected to electrophoresis under reducing conditions in polyacrylamide slab gels. The major fucosylated viral products are designated according to the new Roman numeral nomenclature for the VZV glycoproteins described in Table 10.

protein family will be designated by a Roman numeral.[102] The Table also contains a listing of publications which refer to each of the VZV glycoprotein families. In our initial studies, we designated the individual VZV glycoproteins on the basis of their estimated molecular weight ($\times 10^{-3}$) when separated in polyacrylamide gels under reducing conditions (Figure 21).

B. VZV Glycoprotein gpI

VZV gpI was formerly known as the gp98/gp62 complex in reports from this laboratory.[103-108] In our studies of this glycoprotein, we mainly relied upon one monoclonal antibody clone designated 3B3.[107] This antibody was especially useful because it bound to VZV gpI during both immunoprecipitation and immunoblotting procedures. As part of our characterization of VZV gpI, we examined VZV-infected cell cultures as well as VZV-infected tissues with anti-gpI monoclonal antibody 3B3 as the probe.[106] We initially investigated the relative proportions of the VZV-specific glycoproteins in infected membranes. For this study, virion-free membranes from tritiated glucosamine-labeled infected cells were isolated in discontinuous sucrose gradients. The glycoprotein constituents were identified after electrophoresis of the radiolabeled membrane-enriched fraction in a polyacrylamide gel. Two predominant glycoprotein peaks representing VZV gpI and VZV gpII were identified in the isolated membrane fraction. By densitometric tracing, the two peaks were found to make up approximately 80% of the radiolabeled antigen in the cell-membrane fraction, with gpI being the greater of the two (Figure 22).

We next determined whether the epitope recognized by monoclone 3B3 was highly conserved, i.e., whether it was present on several VZV strains.[106] For these experiments, we inoculated cultures with our laboratory strain VZV-32, the vaccine strain VZV-Oka, and five clinical VZV isolates, and we examined the monolayers by indirect immunofluorescence and indirect immunoperoxidase techniques with antibody 3B3. In a detailed survey of more than 100 monolayer preparations infected with 1 of the 7 VZV strains, VZV gpI was found to be consistently and abundantly expressed by each VZV strain examined. The monoclonal antibody specifically stained infectious foci and did not bind to uninfected cells. As illustrated by photomicrographs of VZV-infected cultures (Figure 23), a consistent progression of intracellular expression of gpI became apparent. In newly infected cells, there was bright membrane staining, presumably representing the attachment of the infected cell inoculum. Rapidly thereafter, reactivity was concentrated in the compact, often triangular juxtanuclear regions, which include the RER and the Golgi apparatus. Later, gpI was present more diffusely throughout the cytoplasm of cells in the center of areas of advanced cytopathology. Large, multinucleated, giant cells, found only in areas with marked cellular degeneration, consistently stained more heavily in the central zone and the fused membranes than in the cytoplasm. Reactivity was always lacking within the nuclei of the syncytia.

We also examined the localization of VZV glycoproteins within human tissues.[106] For these studies, skin biopsy samples were collected from patients with chickenpox and zoster. The specimens of new vesicular lesions were snap frozen and cut with a cryostat at −20°C into sections 5 μm in thickness. When acetone-fixed thick sections of macules, papules, vesicles, and pustules were examined by indirect immunofluorescence and immunoperoxidase assays, we observed that the intraepidermal vesicles were strongly immunostained by both methods (Figure 24). By the immunofluorescence technique, abundant amounts of gpI were detected in the cells and the intercellular spaces of the basal layer (also known as the germinal layer) and the Malpighian layer of the skin. When the same samples were examined by immunoperoxidase methods, no gpI antigen was detected in either the overlying superficial keratin layer or the underlying dermis. Only in denuded vesicles did the area of immunostaining extend to the skin surface.

In addition to biopsy samples, cells obtained from samples of vesicular fluid (Figure 25) were examined. By both immunofluorescence and immunoperoxidase techniques, VZV glycoprotein antigen was detected in all vesicular lesions which dated from day 1 to day 7 of rash; gpI antigen was also detected in older lesions which had pustulated.

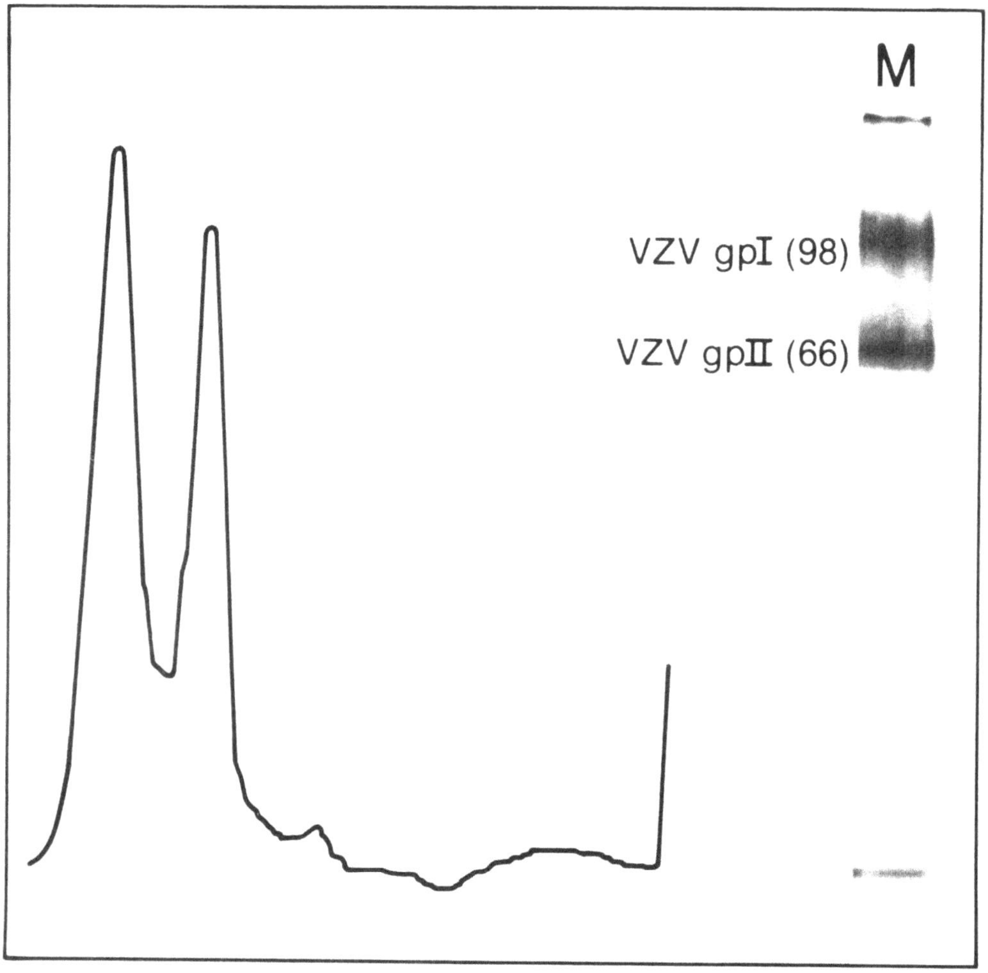

FIGURE 22. Membrane glycoproteins of VZV-infected cells. Tritiated glucosamine-labeled, VZV-infected cells were hypotonically disrupted and their cell membranes were isolated in discontinuous sucrose density gradients. The glycoprotein constituents were identified after fractionation of the radiolabeled membrane (M) in an acrylamide gel. The densitometrogram on the left depicts the relative amounts of radioactivity within the two peaks observed in the fluorogram. The viral glycoproteins are designated by the new nomenclature.

The patterns of immunostaining in the vesicular fluid samples closely paralleled those seen in the biopsy samples, viz., VZV gpI antigen was most abundant in the infected cell membranes, although it was also present throughout the cytoplasm. Even though vesicular smears from all stages of infection containing immunoreactive cells, differences were noted. For example, samples from fresh vesicles often containing sheets of well-demarcated immunoreactive epithelial cells, whereas samples from older lesions consisted of clumps of stained cells surrounded by a speckled fluorescence ("starry sky") of extracellular glycoprotein antigen. Immunofluorescence was also observed along an occasional hair follicle. Thus, these localization studies clearly demonstrated that VZV gpI was synthesized in abundance in infected cell cultures as well as in vesicular lesions excised from patients with chickenpox and zoster.

VZV gpI contains both N-linked and O-linked glycans on a 73,000-dalton polypeptide backbone.[108,112,117] The carbohydrate composition of this glycoprotein has been determined by the use of inhibitors of glycosylation, e.g., tunicamycin and monensin,

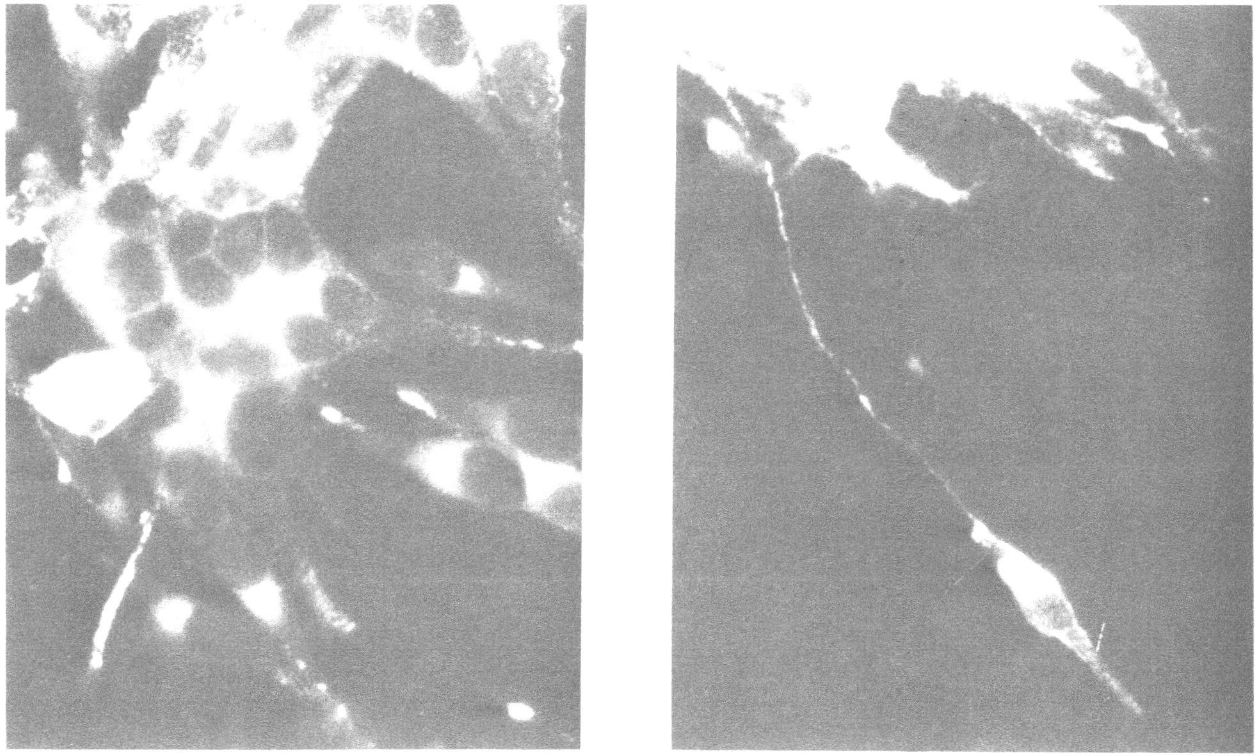

FIGURE 23. Intracellular expression of VZV glycoprotein gpI(98). VZV-infected monolayers were immunostained with monoclonal antibody 3B3, which binds to VZV glycoprotein gpI(98), followed by addition of the second reactant goat anti-mouse antibody conjugated with fluorescein. Left: glycoprotein gpI accumulates in the central area of a nascent polykaryocyte, whose nuclei are easily demarcated by the absence of immunostaining. Within recently infected cells near the periphery of the focus, cytoplasmic staining is more intense in the compact, V-shaped juxtanuclear zones, while membranes are lined with stippled fluorescence. Right: glycoprotein gpI is present in the cell processes that extend outword from an established viral focus to a newly infected cell within an human melanoma cell monolayer.

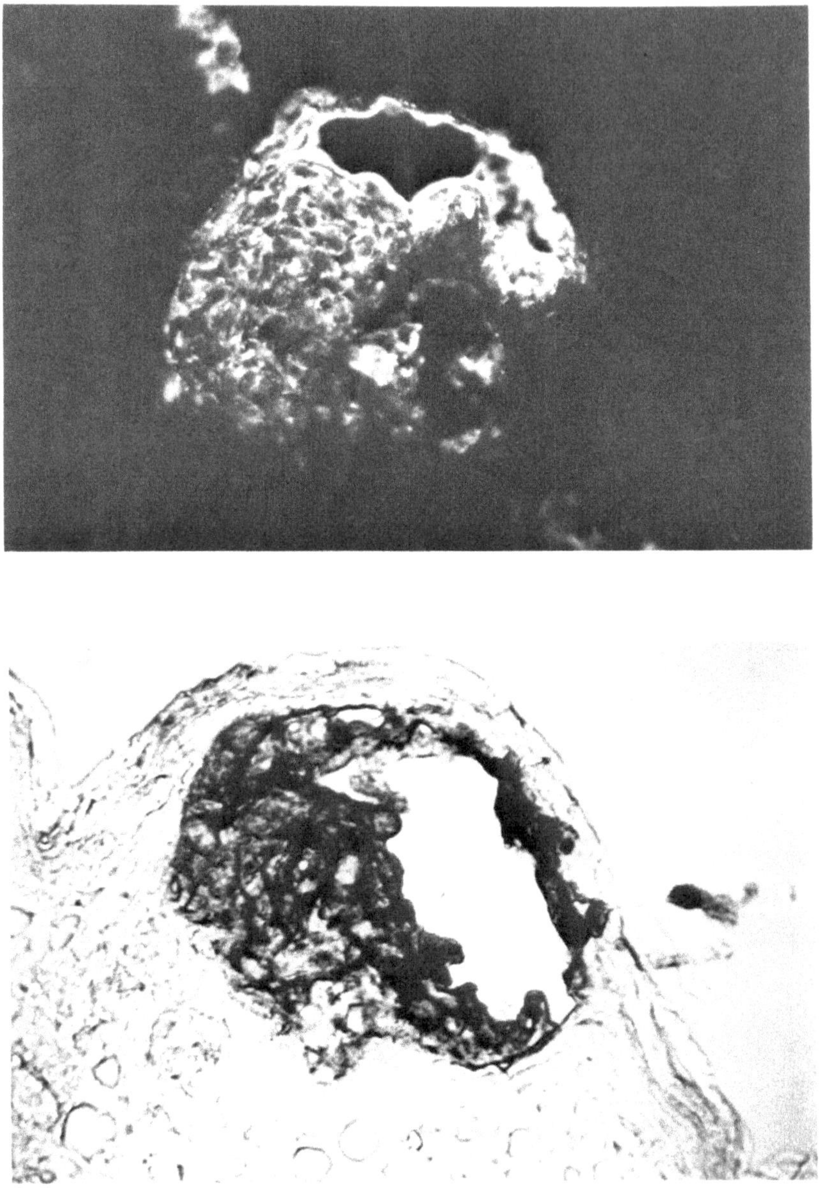

FIGURE 24. Distribution of VZV glycoprotein gpI(98) in a vesicular lesion of a patient with zoster. Thin sections of a vesicle biopsy specimen were examined by indirect immunofluorescence (upper panel) and indirect immunoperoxidase (lower panel) with monoclonal antibody 3B3 as the primary reactant. The photomicrograph shows that the viral glycoprotein is present throughout the basal and Malpighian layers of the epidermis, while the outer granular and keratin layers overlying the emergent vesicle are devoid of this VZV-specified glycoprotein. Immunoperoxidase staining reveals no viral glycoprotein in the dermis beneath the dermoepidermal junction, which is well demarcated in the lower panel.

and by several exo- and endoglycosidases, including neuraminidase, endoglycosidase H, and endoglycosidase F (see Figure 20). Tunicamycin is an antibiotic which inhibits the formation of lipid-linked *N*-acetylglucosamine compounds but does not affect the biosynthesis of O-linked glycans. In the presence of tunicamycin, the formation of all high-mannose and complex-type oligosaccharides is inhibited. When tunicamycin was added into the culture medium at 5 and 10 hr postinfection, the antibiotic completely blocked the formation of the mature product gp98 (Figure 26). In its place, a prominent protein was newly recognized at M_r 73,000. This nonglycosylated protein was presumably the polypeptide backbone of gpI. A glycoprotein with an M_r 88,000, which was synthesized in the presence of tunicamycin, most likely represents the addition of O-linked sugars to the nascent glycoprotein in the presence of tunicamycin.

Endo-beta-*N*-acetyl-glucosaminidase H (Endo H) cleaves predominantly between the two proximal *N*-acetylglucosamine residues of high-mannose oligosaccharides (see Figure 20). It does not digest either complex type or O-linked glycans, but hybrid type moieties show intermediate sensitivity. When gpI labeled with either [^{3}H]fucose, [^{14}C]glucosamine, or ^{14}C-amino acids was digested with Endo H, the relative migration of gpI remained unchanged. Thus, gpI(98) contained no high-mannose residues. However, an intermediate form gpI(81) was sensitive to Endo H digestion with a fall in molecular weight to 73,000.

When the mature glycoprotein was digested with neuraminidase, the molecular weight fell from 98,000 to approximately 90,000. These results indicate that sialic acid residues are present on several of the oligosaccharide chains. Interestingly, when the virus is grown in the presence of monensin, the synthesis of the mature gpI is also blocked, and a 90,000-dalton product accumulates in the treated cultures. Monensin is a sodium ionophore which attaches to the trans region of the Golgi apparatus, and thereby interrupts the final processing steps of N-linked glycoproteins and, perhaps, O-linked glycoproteins. Thus, one of the effects of monensin within VZV-infected cells appears to be an inhibition of addition of sialic acid residues. This result clearly indicates that monensin exerts its effect prior to the site in the trans Golgi subfraction containing sialyltransferase activity.

The monensin-treated VZV-infected cultures have also been analyzed by both light and electron microscopy. As previously described, gpI is usually distributed throughout the cytoplasm and the outer cell membrane; in the presence of monensin treatment, however, there is an accumulation of immunoreactive glycoprotein in the juxtanuclear areas with little detectable antigen in either the cytoplasm or the outer cell membrane. When the same monensin-treated cultures are examined by electron microscopy, marked differences are noted when compared to untreated infected cells. In the latter cells, nucleocapsids are present in the nucleus and enveloped particles are clearly seen in the cytoplasm, often in clumps within intracytoplasmic vacuoles; enveloped viruses are also present along the plasma membrane. In contrast, in monensin-treated cultures, nucleocapsids enter the cytoplasm but never become enveloped, nor do they traverse the cytoplasm to the outer cell membrane. Infectivity is virtually absent in monensin-treated cultures.

The biosynthesis of gpI is summarized in the Figure 27. The mature sialated product is gpI(98) while the high-mannose intermediate form is gp81. The O-linked glycoprotein minus its complex-type glycans is gp88, and the polypeptide precursor is p73. A similar biosynthetic schema was proposed by Namazue et al.,[112] who identified the precursor polypeptide as 75K, the O-glycosidically linked product as 83K, and the final mature O-linked and N-linked glycoprotein as 94K. Some of the biologic functions of gpI can be determined by analysis of the monoclonal antibodies. In these studies, anti-gp98 antibodies exhibit neutralization of infectivity but only in the presence of an exogenous source of complement.[113,117] The same antibodies also mediate antibody-dependent cellular cytotoxicity (ADCC).[120]

C. VZV Glycoprotein gpII

VZV gpII is a disulfide-linked glycoprotein complex. In its native state the glycopro-

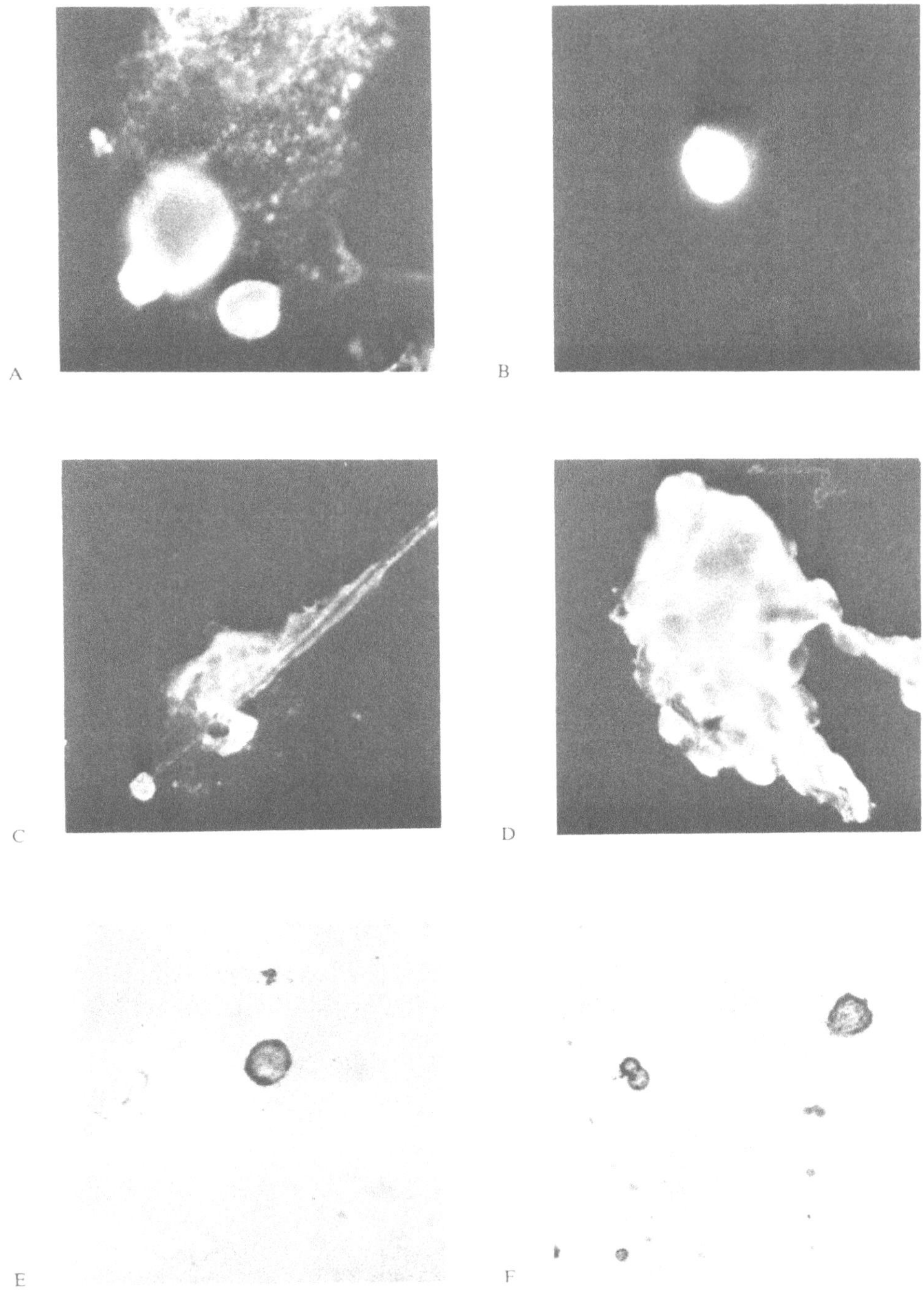

FIGURE 25. Identification of VZV glycoprotein gpI(98) in vesicular smears obtained from unrelated children with chickenpox or zoster. Cells from the bases of vesicular lesions were transferred to glass slides, fixed in acetone, and then reacted with monoclonal antibody 3B3 in either an indirect fluorescence (panels A through D) or indirect immunoperoxidase assay (panels E through H). Panel A: cluster of fluorescing cells surrounded by extracellular gpI ("starry sky" appearance); panel B: single immunoreactive cell; panel C: immunostaining cells along a hair follicle; panel D: intact cellular sheet from base of vesicle; panels E through G: discrete immunoreactive cells; panel H: sheet of reactive cells.

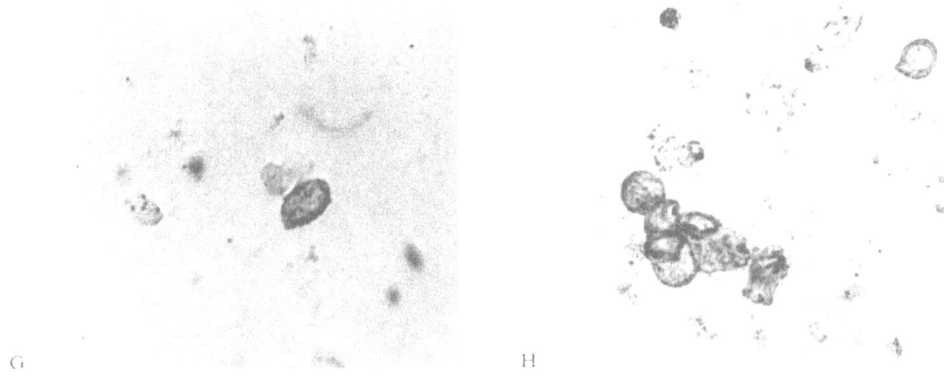

FIGURE 25 (continued)

tein migrates at approximately 140,000 daltons, while under reducing conditions it migrates mainly as one or more components of approximately 66,000 daltons (Figure 28).[119] This glycoprotein has been identified by specific monoclonal antibodies which bind glycosylated products within a detergent-solubilized [^{3}H]fucose-labeled VZV-infected cell extract. When the extract solubilized with nonidet P40 and deoxycholate is compared with the insoluble pellet, the glycoprotein composition of gp140 is more prominent in the insoluble pellet. In other words, this glycoprotein antigen is relatively insoluble under usual conditions of extraction and emulsification. The glycoprotein composition of the purified virion also includes the disulfide linked complex. When enveloped virions are purified by sedimentation in metrizamide gradients and the virion-enriched fraction is solubilized in SDS sample buffer lacking 2-mercaptoethanol, a 140,000 glycoprotein is easily identified. When the same virion-enriched sample is analyzed in the presence of a reducing agent, the molecular weight of VZV gpII falls to approximately 66,000.

Human tissues have been examined to determine whether the antigen is synthesized under conditions of naturally acquired VZV infection.[117,119] Epithelial cells collected from the base of a typical chickenpox vesicular lesion immunostain in a cytoplasmic pattern, while the nuclei are consistently devoid of any immunofluorescence. Abundant amounts of gpII also are detectable in skin biopsy samples obtained from a zoster vesicle (Figure 29). The glycosylated antigen is present in the cytoplasm of the involved cells but is most dense in the intercellular spaces of the epidermal Malpighian layer. The uppermost keratinized layer of the epithelium overlying a vesicle does not contain the viral antigen, presumably because VZV cannot replicate in cells lacking nuclei.

Preliminary structural analyses of gpII(140) have been performed. Digestion of this glycoprotein with endoglycosidase H causes a slight shift in relative mobility, corresponding to approximately 3,000 daltons. Analyses with endoglycosidase F (endo-beta-*N*-acetylglucosaminidase F) are difficult to perform because they are carried out under reducing conditions. However, an intermediate form of gp140 has been identified at about 124,000 daltons. Since this product contains fucose residues, it cannot be the high-mannose precursor form; rather it must be an intermediate form of the mature glycoprotein. These latter two products were also identified by Edson et al.[117] and designated gp135 and gp125, respectively; they have called the major cleavage product gp63. In studies by Namazue et al.,[112] the constituents of this glycoprotein family were designated gp116, gp106, and gp64. Forghani et al.[116] enumerated similar glycoproteins groups at 118 to 120K and 64 to 65K. When gp140 is labeled with [^{14}C]glucosamine rather than [^{3}H]fucose, at least two different lower molecular weight cleavage prod-

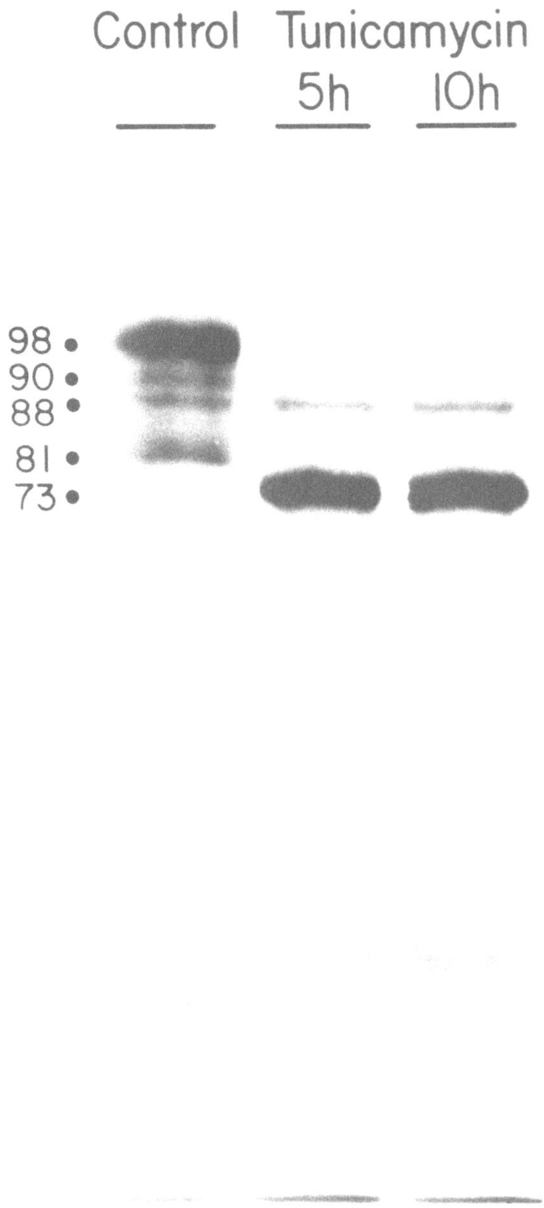

FIGURE 26. Effect of tunicamycin on VZV gpI(98) processing. VZV-infected monolayers were incubated with tunicamycin (2.5 μg/mℓ of culture medium) at 5 and 10 hr postinfection. Tunicamycin-treated and untreated cell monolayers were dislodged with a rubber policeman and washed three times with 0.15 *M* NaCl—0.01 *M* Tris (pH 7.4). The cells were suspended in 1 mℓ of the same buffer and sonically disrupted for 2 min. Electrophoretically separated proteins were transferred to nitrocellulose paper and then subjected to immunoblot analysis with antibody 3B3. The products were fractionated in a 10% acrylamide slab gel.

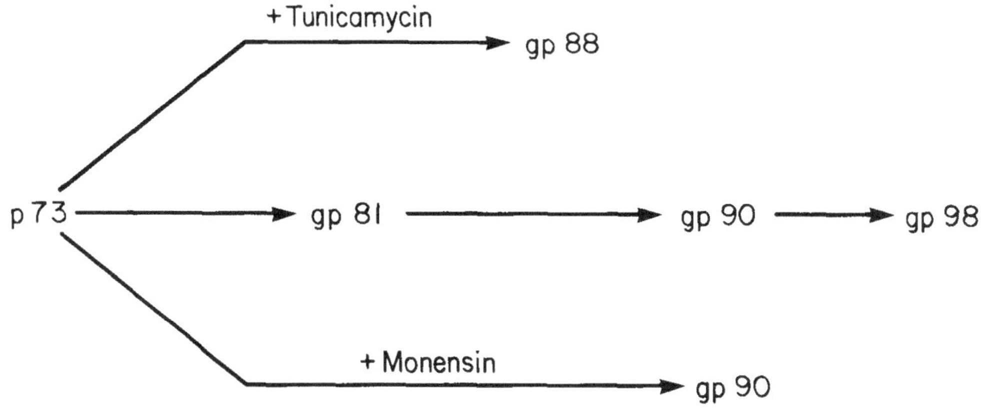

FIGURE 27. Schematic representation of the processing of VZV glycoprotein gpI(98). The unglycosylated analog of gp98 is p73. The addition of high-mannose oligosaccharides to the peptide backbone results in the generation of gp81, the high-mannose intermediate form. After processing of the N-linked sugars and the addition of O-linked residues, gp90 is formed. The final addition of sialic acid moieties to gp90 results in the mature product, gp98. In the presence of tunicamycin, a gp88 form is synthesized, which represents a product lacking N-linked oligosaccharide chains. The ionophore monensin causes a block in the addition of sialic acid to the oligosaccharides. (From Montalvo, E. A., Parmley, R. T., and Grose, C., *J. Virol.*, 53, 761, 1985. With permission.)

ucts can sometimes be found just above and below 60,000 daltons; these may represent heterodimers of native VZV gpII.[113]

In contrast to anti-gpI monoclonal antibodies, some anti-gpII monoclonal antibodies exhibit neutralization activity in the absence of an exogenous source of complement.[117,122] Cellular cytoxity was not mediated, however, by either of two anti-gpII antibodies tested.[120] A distinctive characteristic of certain anti-gpII monoclonal antibodies is their cross-reactivity with shared epitopes on the HSV-1 glycoprotein gB.[123] Previously, VZV glycoprotein gpII had been isolated from an extract of VZV-infected cells by immunoaffinity chromatography to immobilized HSV-1 polyclonal antiserum.[90]

D. VZV Glycoprotein gpIII

VZV gpIII is another VZV structural glycoprotein which contains mainly N-linked glycans. The final product has a molecular weight of approximately 118,000, as defined by studies from this laboratory.[103-107] Other laboratories have designated this same glycoprotein as 115K (gpI),[111] 118K glycoprotein,[116] gp105 glycoprotein,[113] and gp118 glycoprotein.[117]

The structural analyses of gp118 were carried out by use of both endoglycosidase H and endoglycosidase F; while Endo H hydrolyzes between the two proximal *N*-acetylglucosamine residues of high-mannose oligosaccharides, both high-mannose and complex type glycans are susceptible to enzymatic cleavage by Endo F (Figure 30). When VZV gpIII was treated with Endo H, there was a small increase in relative migration of the glycoprotein; this shift in mobility corresponded to a loss of approximately 3,000 daltons. Based on the assumption that the molecular weight of a high-mannose chain is 1400 to 1800, gpIII(118) contained 1 or 2 high-mannose residues. When the same glycoprotein was digested with Endo F, the M_r of gp118 was reduced to 79,000. Thus, the 79K protein seen after Endo F treatment most likely represented the unglycosylated precursor to gp118. To confirm the above results, we also assayed VZV-infected cells grown in the presence of tunicamycin. Under these conditions, we detected a faint 79K

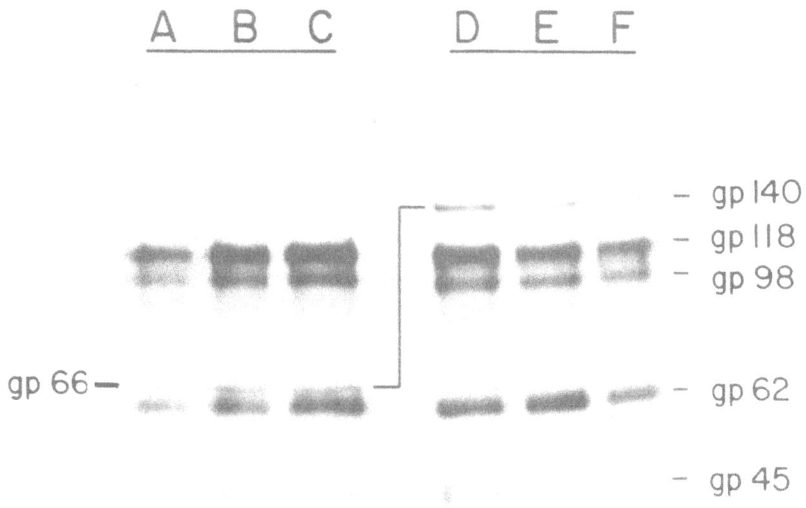

FIGURE 28. Identification of the disulfide-linked VZV glycoprotein gpII(140). Samples of anti-VZV antisera produced in rabbits or guinea pigs were reacted with aliquots (300,000) cpm) of a [^{3}H]fucose-labeled VZV antigen preparation. The antigens and the immunoprecipitates were solubilized under reducing (lanes A through C) or nonreducing (lanes D through F) conditions and subjected to electrophoresis in a 10% acrylamide slab gel. The original VZV antigen preparation (60,000 cpm) was added to lanes A and F; the rabbit serum precipitates were added to lanes B and D (25,000 cpm/lane); and caviid serum precipitates were added to lanes C and E (25,000 cpm/lane). The reduced (left side) and nonreduced (right side) forms of VZV glycoprotein gpII are indicated by the solid line between lanes C and D. (From Grose, C., Friedrichs, W. E., Weigle, K. A., and McGuire, W. L., *Virology*, 132, 138, 1984. With permission.)

band by immunoprecipitation with anti-gpIII antibodies. Since tunicamycin inhibits formation of all N-linked glycoproteins, the 79K protein most likely represents the unglycosylated analog of gp118.

The effects of neuraminidase digestion and monensin treatment were also determined for gpIII(118). In the presence of monensin in the medium overlying an infected culture, the 118,000 protein was not synthesized; rather, it was replaced by another product of M_r 98,000. The latter protein was similar in molecular weight to the neuraminidase-treated gp118. These results suggested that the 98,000 protein represented an intermediate form of gpIII lacking sialic acid residues. Similar results were discussed when gpI was examined after either monensin or neuraminidase treatment.

The kinetics of gp118 biosynthesis were examined by pulse-chase labeling experiments (Figure 31). VZV-infected cell cultures were pulse labeled with [^{35}S]methionine for 30 min and chased with unlabeled medium for either 0 min, 0.5 hr, 1 hr, 2 hr, 4 hr, or 12 hr. No prominent bands were seen in immunoprecipitates of cell lysates pulse labeled for 30 min and harvested without a chase. After a 30 min chase, a prominent band at M_r 94,000 was seen. After a 60 min chase, the 94K band was more prominent.

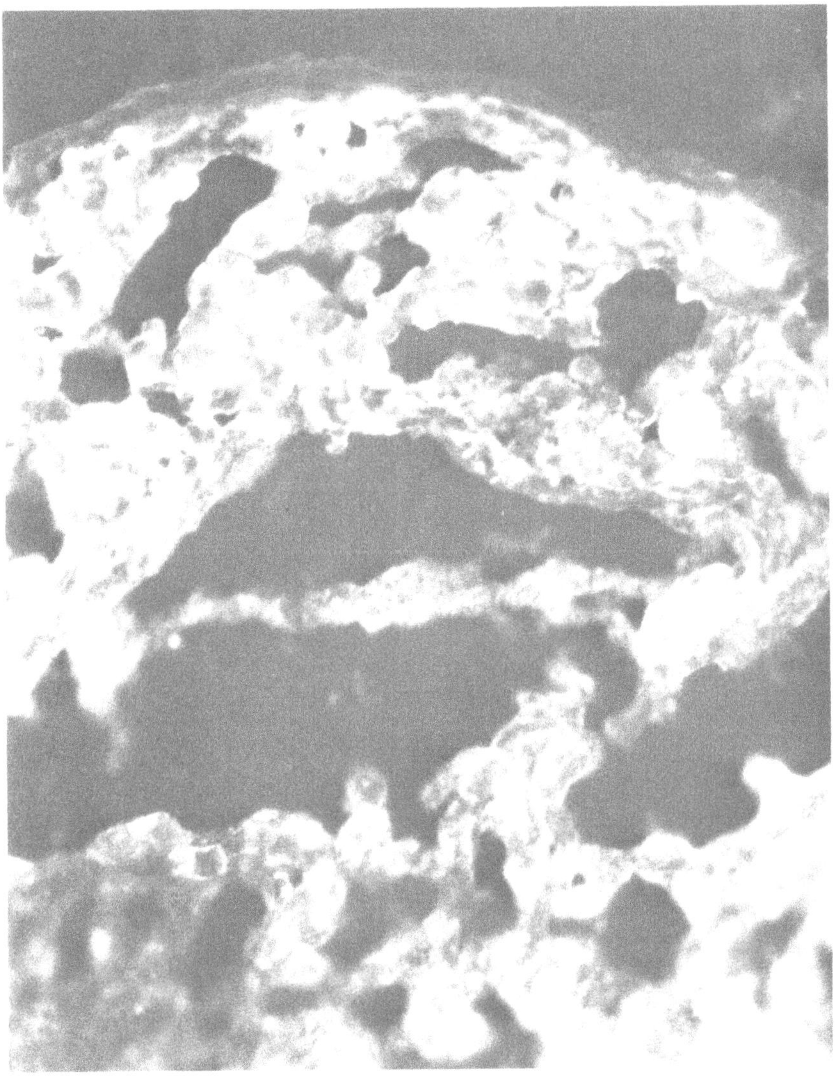

FIGURE 29. Localization of VZV-specific gpII(140) in infected tissues from a patient with zoster. Tissue samples collected from a patient with zoster were acetone-fixed on glass slides and reacted with an anti-gpII monoclonal antibody by an indirect immunofluorescence assay. The figure illustrates a thin section of a skin biopsy of a zoster vesicular lesion, with dense accumulations of gp140 around cell membranes and in intercellular spaces of the germinal and Malpighian layers of the epidermis. The uppermost keratin layer overlying the emergent vesicle is nearly devoid of reactivity and, consequently, difficult to visualize. Original magnification, × 200.

A 2 hr chase with unlabeled medium led to the appearance of a new band gp118. Concomitant with the appearance of the 118,000 band was a decrease in the intensity of the 94K band. In the 4 hr chase, the radiolabel was present mainly in the 118K band, while in samples obtained from VZV-infected cultures chased for 12 hr, only the 118K band was present. These pulse-chase labeling experiments again demonstrate that the 94,000 glycoprotein is an important precursor to the major 118,000 product. Based on the above observations, a schema for the biosynthesis of gpIII(118) is presented in Figure 32.

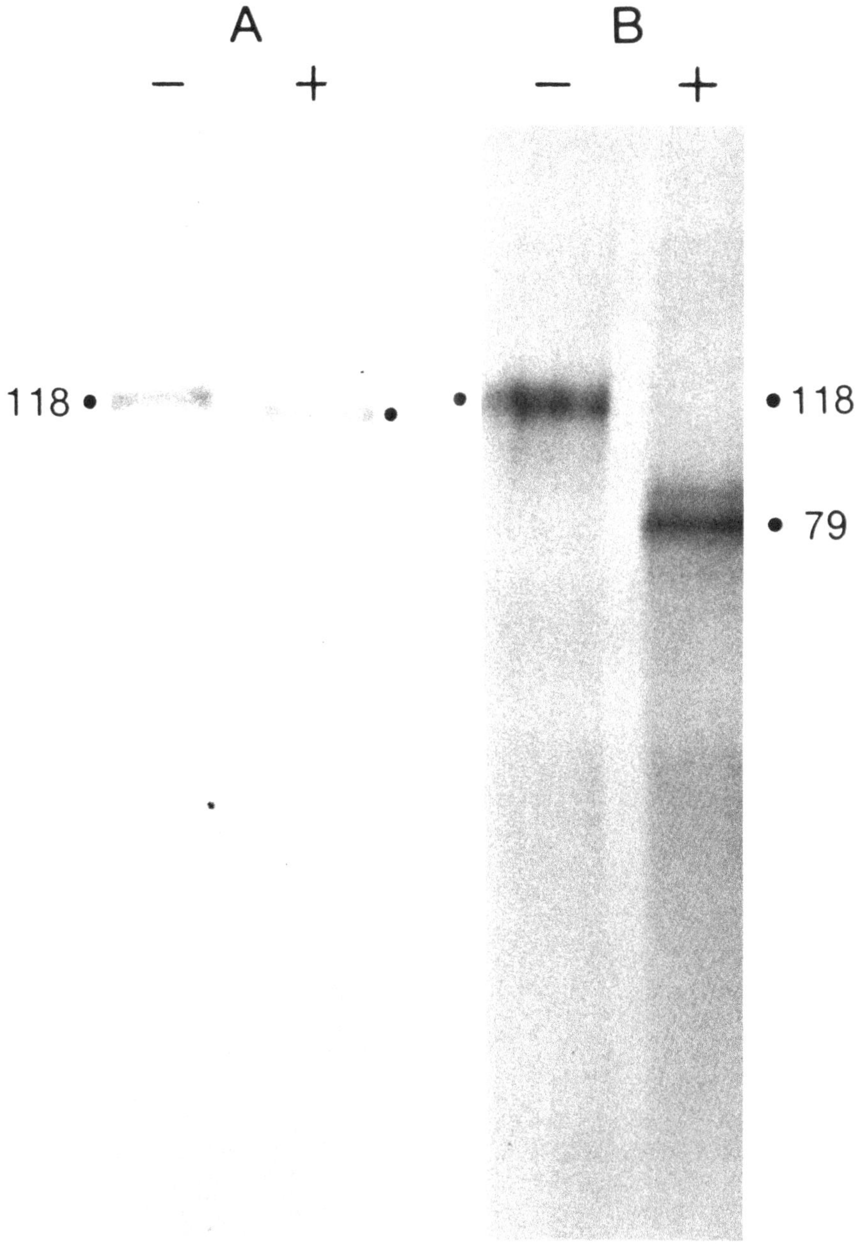

FIGURE 30. Enzymatic digestion of VZV gpIII(118) with endoglycosidase H and endoglycosidase F. (A) VZV-infected cell lysates labeled with [^{3}H]fucose were immunoprecipitated with an anti-gpIII monoclonal antibody. The precipitated polypeptides were eluted, suspended in 50 m*M* sodium citrate buffer (pH 5.5) and incubated in the absence (−) or presence (+) of Endo H (60 $\mu\ell$ of a 0.1 unit/mℓ solution). (B) VZV-infected cell lysates labeled with [^{14}C]amino acids were immunoprecipitated with an anti-gpIII monoclonal antibody. After washing, the pellets were suspended in 100 m*M* sodium phosphate buffer (pH 6.1) and incubated in the absence (−) or presence (+) of 2.5 units Endo F.

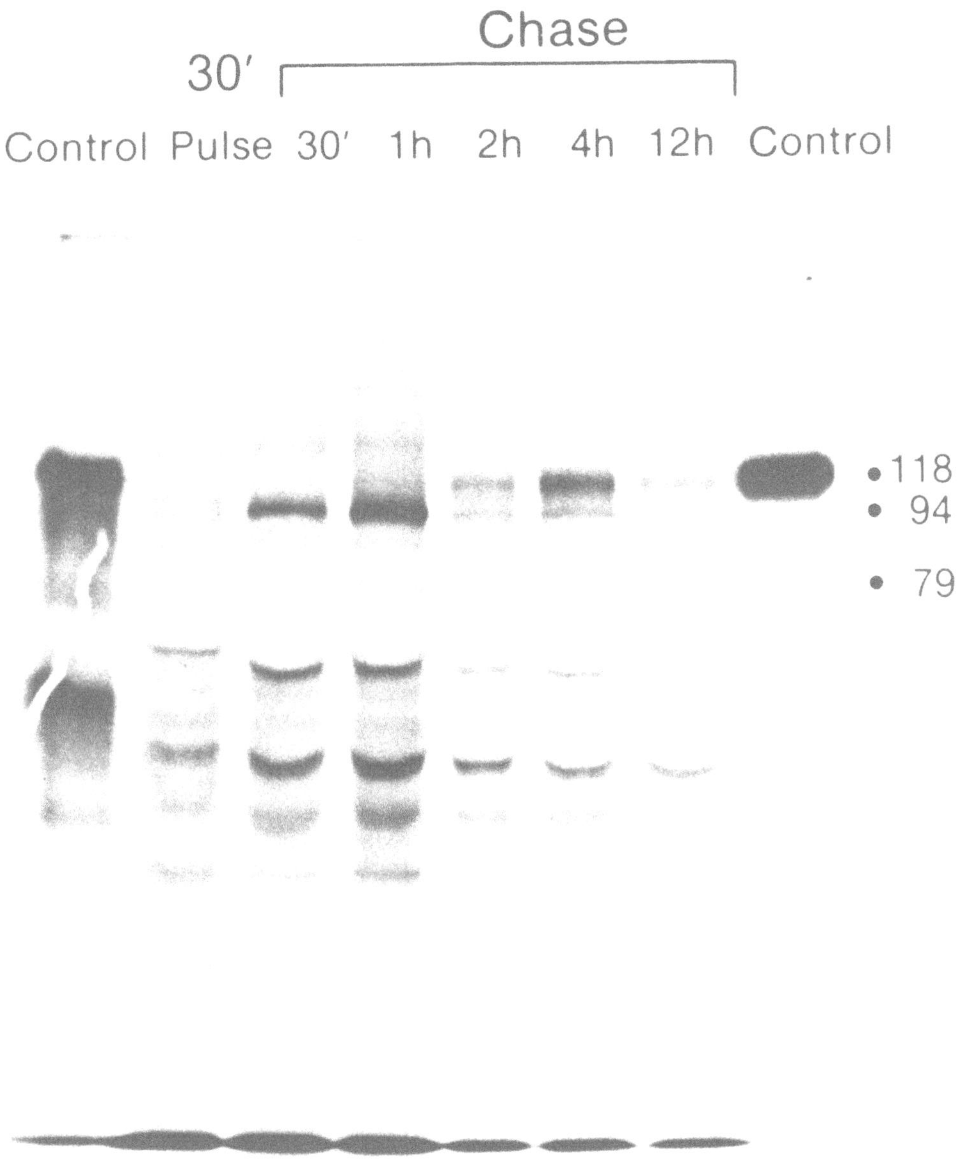

FIGURE 31. Identification of the VZV gpIII intermediate forms immunoprecipitated during pulse chase labeling experiments. VZV-infected cultures were pulse labeled with [^{35}S]methionine for 30 min, after which the radioactive medium was replaced with regular medium for increasing time intervals. VZV lysates from pulse chase labeling experiments were immunoprecipitated with an anti-gpIII monoclonal antibody. After elution, samples were subjected to electrophoresis in a 10% acrylamide slab gel. Time intervals for pulse and chases are indicated above each lane. The control in the leftmost lane is a [^{3}H]fucose-labeled VZV cell lysate and the control in the rightmost lane is an immunoprecipitated [^{3}H]fucose-labeled gp118.

In order to determine the biologic function of gp118, the panels of anti-gpIII monoclonal antibodies were assayed for neutralization activity.[125] Each antibody was tested against two strains: VZV-32 and VZV-Oka. By a plaque reduction technique, the titers of four of the antibodies were striking because of their magnitude (up to 1:40,000). These neutralization titers are considerably higher than those achieved in rabbit or guiena pig hyperimmune VZV antiserum or human convalescent postzoster serum. Since the titers were not altered by the addition of an exogenous source of complement,

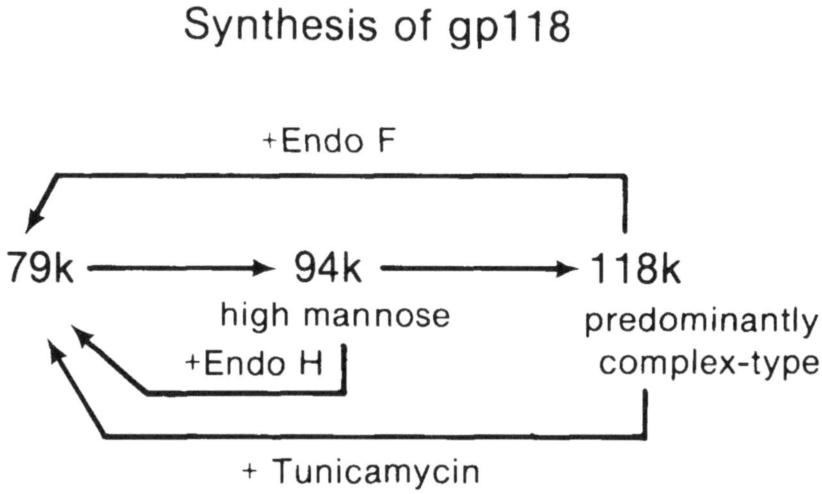

FIGURE 32. Schematic representation of the biosynthesis of VZV gpIII(118). The unglycosylated analog of gp118 is p79. A nine mannose oligosaccharide chain is transferred en bloc from a dolichylphosphate carrier to the peptide backbone, resulting in the generation of a high-mannose (Endo H sensitive) intermediate form, gp94. This step probably occurs as a cotranslational event in the rough endoplasmic reticulum.[100] Most of the oligomannosyl chains are subsequently processed in the Golgi apparatus to the sialated complex-type oligosaccharides found in the mature product gp118. These posttranslational modifications do not include formation of interchain disulfide bonds. (From Montalvo, E. A. and Grose, C., *Virology*, 149, 230, 1986. With permission.)

they were not considered to be complement-dependent. The neutralization epitope was highly conserved, because the same monoclonal antibody reduced plaque formation by both laboratory and vaccine VZV strains as well as wild-type VZV isolates.

The panel of neutralizing monoclonal antibodies were also compared by competitive antibody binding experiments (Figure 33). VZV-infected monolayers were grown on 96-well tissue culture plates. When cytopathology was seen in infected cells, aliquots of iodinated antibody and serially diluted, unlabeled antibody were added to individual wells. After an incubation period and wash, the monolayers were dislodged and the attached radioactivity counted. As can be seen in the figure, any of the panel of unlabeled neutralizing antibody could effectively compete with the binding of labeled monoclonal antibody by as much as 80% or greater. When a nonneutralizing anti-gpIII monoclonal antibody (clone 100) was tested in these assays, it failed to inhibit the binding of neutralizing antibody. The competitive antibody results suggest that the neutralizing monoclonal antibodies bind closely related or identical immunodominant epitopes, whereas the nonneutralizing antibody recognized a spatially distinct antigenic site.

These conclusions were substantiated by a comparison with immunoblotting techniques. Because antibody is not added until the glycoprotein is transferred from the polyacrylamide gel onto nitrocellulose paper, the epitopes commonly recognized are those consisting of primary amino acid sequences. When the anti-gp118 monoclonal antibodies were used as probes to detect VZV gpIII after transfer, only clone 100 reacted. When these data were compared with the neutralization data, it was apparent that the single monoclonal antibody which detected gp118 on nitrocellulose paper was the antibody which failed to inhibit VZV plaque formation. These results further sug-

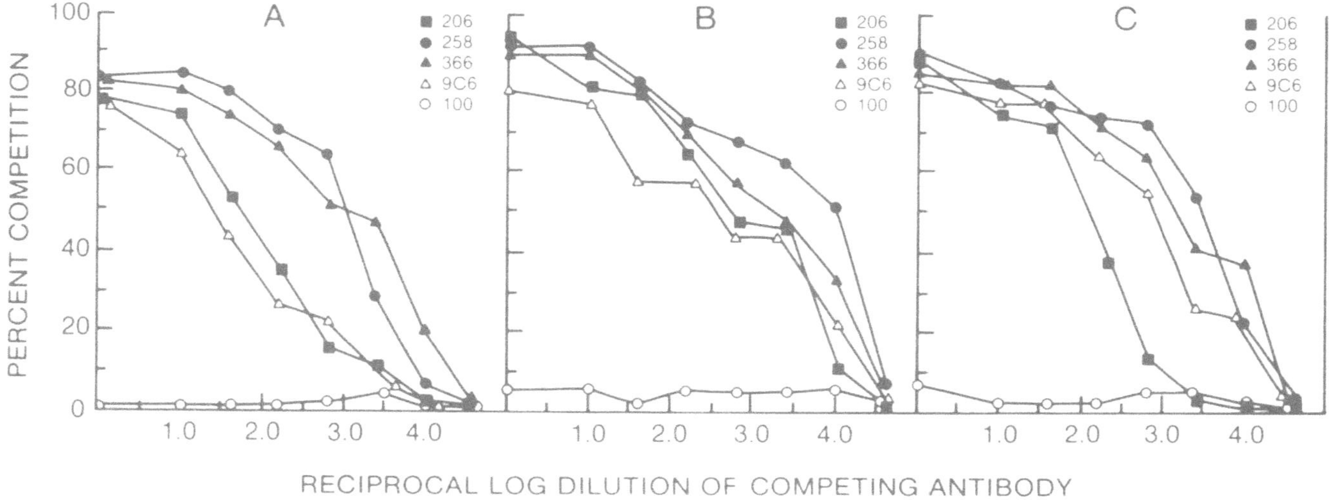

FIGURE 33. Competitive binding experiments with a panel of anti-gpIII(118) monoclonal antibodies. Iodine-125-labeled antibody and several dilutions of unlabeled competing antibody were mixed and added to VZV-infected cell monolayers. After a 1-hr incubation, the monolayers were washed and trypsin-dispersed. Percent competition was determined by comparing counts per minute (cpm) bound in the presence of competing antibody versus cpm bound in absence of competing antibody. The competitive binding profiles for each of three iodinated anti-gpIII monoclonal antibodies with neutralization activity were very similar. Another anti-gpIII monoclonal antibody (clone 100) with no neutralization activity did not interfere with binding of the other three antibodies. Each panel is designated by that anti-gpIII antibody which is radio-iodinated. (A) Clone 258, (B) Clone 366, and (C) Clone 9C6. (From Montalvo, E. A. and Grose, C., *Virology*, 149, 230, 1986. With permission.)

gest that the four monoclonal antibodies with neutralizing capacity recognize a conformational determinant dependent on the tertiary structure of the glycoprotein.

E. VZV Glycoproteins gpIV and gpV

The fourth VZV glycoprotein exists in at least two glycosylated forms with molecular weight of 45,000 and 55,000. Whereas the major three VZV glycoproteins were identified on the basis of their reactivities to polyclonal and monoclonal antisera, gpIV was recognized after preparation of antiserum against a synthetic peptide derived from a DNA sequence in the Us component of the VZV genome.[102,126] This gene, encoding a 39,000 primary translation product, had the following characteristics of a membrane protein: a hydrophobic region near the amino terminus for insertion into the endoplasmic reticulum, and a highly hydrophobic region near the carboxy terminus, followed by several basic amino acids. The antipeptide antiserum precipitated two VZV-specific glycosylated products with molecular weights of 45,000 and 55,000. The biologic functions of gpIV have not yet been defined.

On the basis of the DNA sequence of the U_L component of the VZV genome, yet another gene encoding a membrane protein has been located. This hypothetical gene and its product have been designated VZV gpV.[102]

F. Chromatographic Analyses of the VZV Glycoproteins

The VZV glycoproteins can be purified by affinity chromatography using either immobilized lectins or monoclonal antibody. The procedures have been described in detail for purification of VZV gpIII.[124] Two plant lectins Concanavalin A (Con A) and *Lens culinaris* (lentil lectin) have been extensively investigated.

The lectins were coupled to Sepharose 4B® previously activated with cyanogen bromide; the quantities of Con A and lentil lectin protein bound to the beads were 6.3 and 7.4 mg/mℓ of settled gel, respectively. Samples of [^{14}C]glucosamine-labeled VZV-infected cell extracts were chromatographed on both Con A-Sepharose® and lentil-Sepharose 4B® columns and desorbed with 0.5 *M* alpha-methylmannoside. When the recoveries were calculated for each column, 53% of the input counts per minute bound to Con A-Sepharose® and approximately one third of this amount was competitively eluted. With a lentil-Sepharose® column, 57% of the initial sample bound to the lectin and nearly half was removed by elution with 0.5 *M* alpha-methylmannoside. The elution profiles were not markedly altered by either increasing the ionic strength of the eluting buffer (up to 1 *M* NaCl) or adding a polarity-reducing agent such as 50% ethylene glycol. Minor adjustments of the concentration of immobilized lectin in the settled gel also did not substantially alter the elution profiles of either affinity adsorbent. When the eluates from both columns were subjected to electrophoretic analysis, some differences were observed between the two columns, viz., all glycoproteins bound to lentil lectin and all were eluted with 0.5 *M* alpha-methylmannoside. Although all glycoproteins bound to Con A, not all of the glycosylated products were easily eluted from this plant lectin.

VZV glycoproteins can also be purified to near homogeneity by immunoaffinity chromatography. The immunoadsorbent is prepared from ascites monoclonal antibody which has been concentrated by precipitation with saturated ammonium sulfate. The protein concentration is adjusted to 10 mg/mℓ before coupling of the monoclonal antibody to CNBr-activated Sepharose®. To avoid repetitive chromatography on immunoabsorbent columns, a detergent-solubilized VZV-infected cell extract can be applied initially to a lectin column; other preliminary purification procedures can also include heparin-Sepharose® and Affi-Gel® blue columns. As a final step, the glycoprotein-enhanced fractions are added to an immunoadsorbent column to remove the

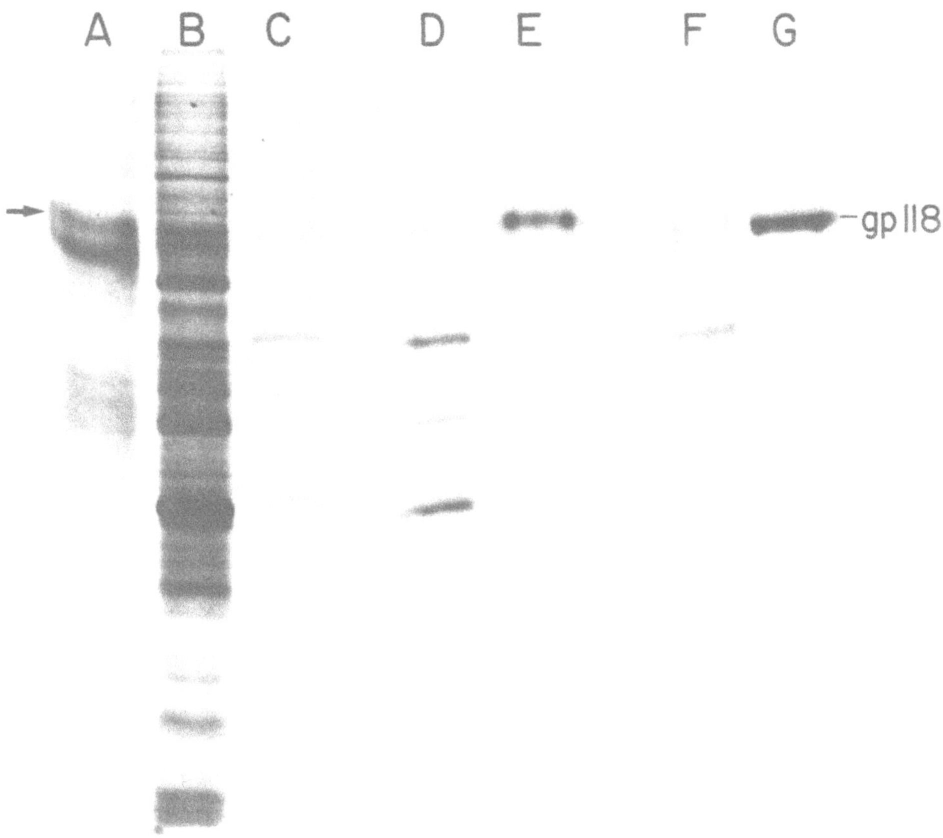

FIGURE 34. Fractionation and purification of gpIII(118) by serial affinity chromatography on immobilized Cibacron blue, lentil lectin, and immunoadsorbents. A sample of Nonidet-solubilized [^{35}S]methionine-labeled VZV-infected cell extract (7.7×10^6 cpm) was adsorbed onto an Affi-Gel® blue column for 20 min. Unbound material which represented 22% of the original sample was applied onto the lentil lectin-Sepharose® column, and further fractionation was carried out on an immunoaffinity column. Lane A, Nonidet-solubilized [^{14}C]glucosamine-labeled VZV-infected cell extract (20,000 cpm), the location of gp118 is designated by an arrow; lane B, Nonidet-solubilized [^{35}S]methionine-labeled VZV-infected cell extract, which was the initial antigen preparation of serial chromatograph (40,000 cpm); lane C, proteins unretarded by Cibacron blue-Sepharose® (6,800 cpm); lane D, proteins unretarded by immobilized *Lens culinaris* (8,600 cpm); lane E, proteins eluted from lentil lectin-Sepharose® with alpha-methylmannoside (7,000 cpm); lane F, effluent from the immunoaffinity column (10,000 cpm); lane G, proteins eluted from immobilized monoclonal antibody with 3 *M*KSCN (7,500 cpm). The location of gp118 is designated in the margin to the right of lane G. (From Friedrichs, W. E. and Grose, C., *J. Virol.*, 49, 992, 1984. With permission.)

specific glycoprotein of interest (Figure 34). By such two- and three-step chromatographic purification techniques, in which immunoadsorbent chromatography is the final step, VZV glycoproteins can be purified by as much as 1000-fold.

G. Structural VZV Protein Complex p32/p36

In addition to the glycoproteins, the virus particle is composed of several nonglycosylated products, one of which is the p32/p36 polypeptide complex. Like several other VZV proteins, the p32/p36 complex was first distinguished by its reactivity with a monoclonal antibody. As previously described, the pattern of immunoreactivity of the antiglycoprotein monoclonal antibodies includes the outer infected cell membrane, cytoplasm, and juxtanuclear areas, but no immunofluorescence is observed within the nuclei (Figure 35). In contrast, this monoclonal antibody reacts mainly with nuclear

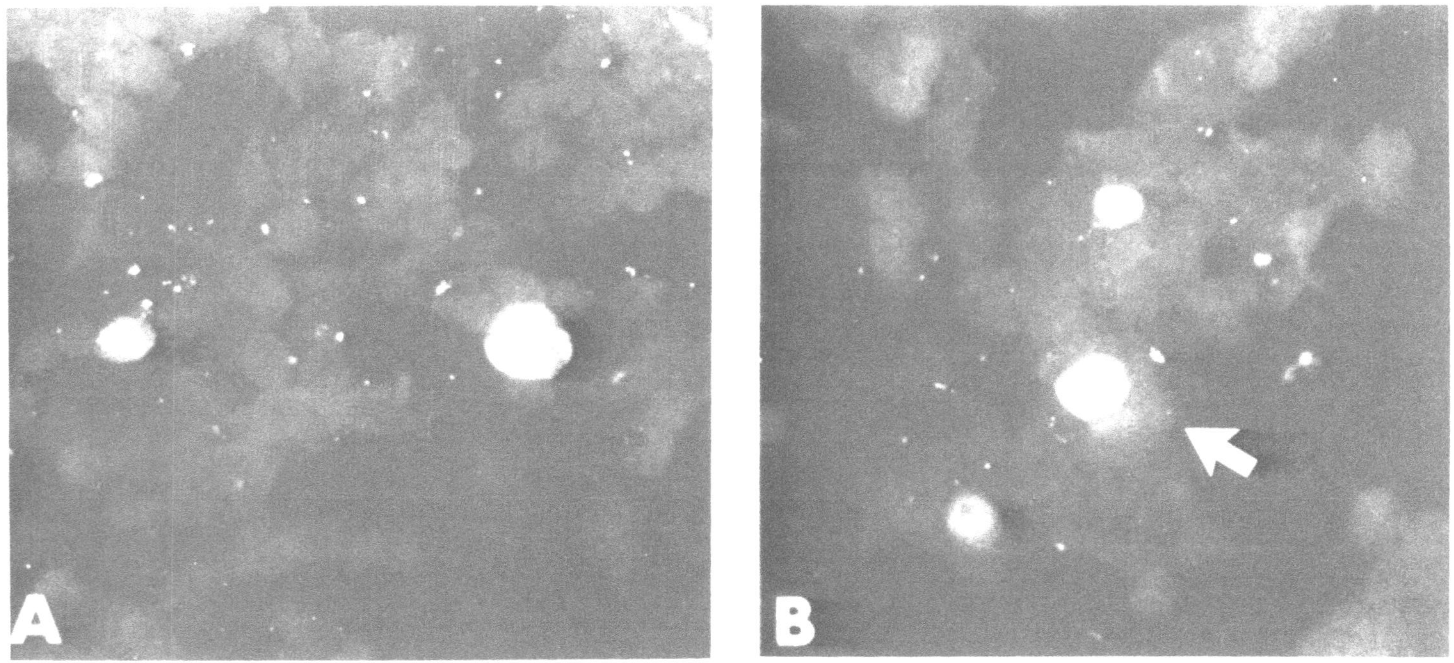

FIGURE 35. Immunostaining patterns with monoclonal antibody to VZV p32/p36 protein complex. VZV-infected cell monolayers were fixed in acetone at 4, 8, 12, and 24 hr postinfection (panels A, B, C, and D, respectively). Each monolayer was reacted first with the monoclonal antibody and then with fluorescein-conjugated goat anti-mouse antibody. The four panels demonstrate the intranuclear localization pattern of VZV-specific immunostaining. The arrows in panels B through D indicate immunoreactive nuclei within maturing and enlarging syncytial formations at increasing intervals after infection.

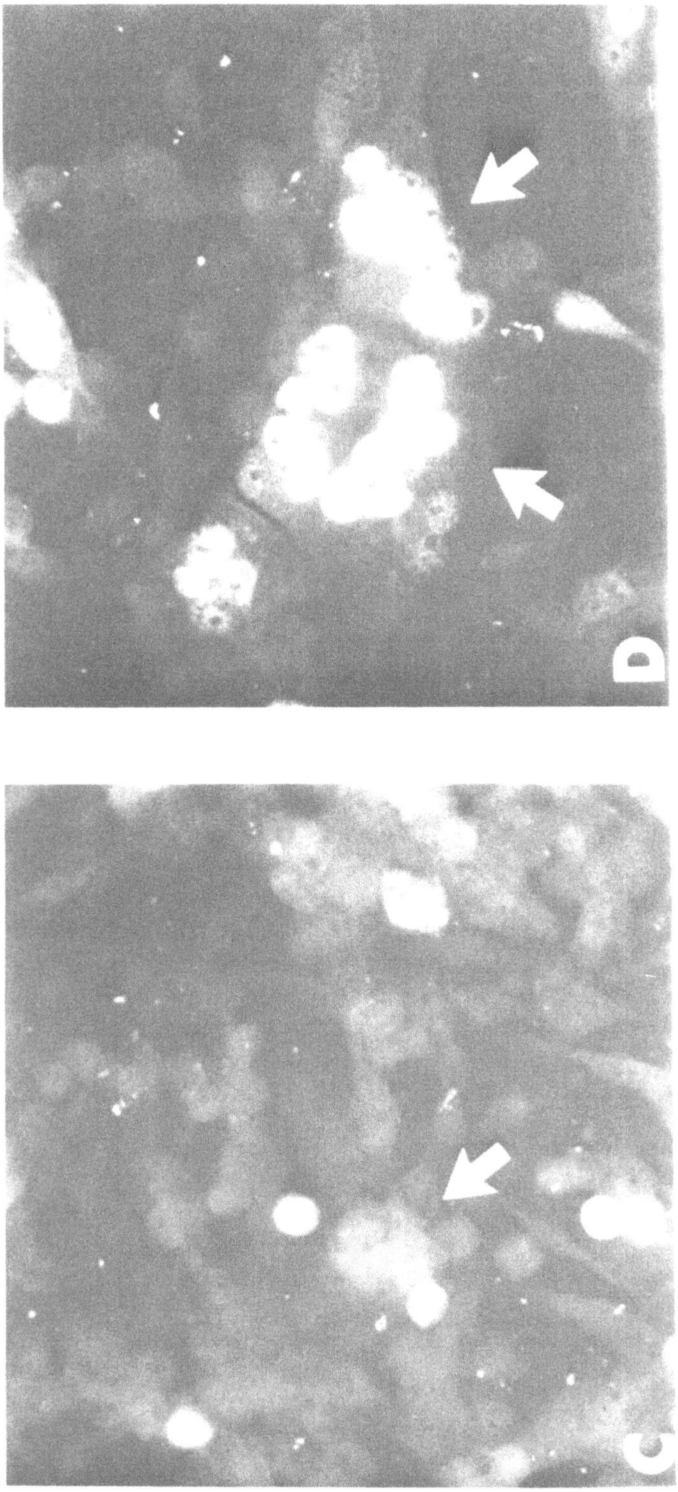

FIGURE 35 (continued)

antigens. At 4 hr postinfection, the antibody binds solely with single nuclei. By 8 and 12 hr p.i., there is evidence of transfer of infectivity to nuclei of adjacent cells. By 24 hr p.i., large syncytia are easily detected by the presence of brightly staining aggregates of immunoreactive nuclei. At no time during the infectious cycle is there noticeable reactivity in the cytoplasm of infected cells.

The VZV-polypeptide specificity of the monoclonal antibody was determined by both immunoprecipitation and immunoblotting analyses. The antibody does not precipitate any product from detergent-solubilized [^{3}H]fucose and [^{14}C]amino acid-labeled VZV antigen preparations. However, when VZV sonicates are electrophoretically separated and transferred to nitrocellulose paper (Western blotting), the monoclonal antibody binds to a 32,000 molecular weight polypeptide (p32) as well as a slightly larger protein of approximately 36,000 daltons (p36). VZV nucleocapsids also were purified by gradient sedimentation procedures to determine whether they contained the p32/p36 complex. After electrophoresis and transfer of the VZV nucleocapsid preparation, the nitrocellulose replica was probed with monoclonal antibody. Immunoblot analysis revealed that the capsids contained ample quantities of both p32 and p36 (Figure 36). In addition, less prominent proteins were occasionally seen within the complex; these were designated p34 and p38.

Since the immunostaining studies show a nuclear localization, the relationship of the p32/p36 complex to the nuclear matrix was examined. In order to isolate the nuclear matrix, the nuclei from infected cells were fractionated by the four-step procedure originally described by Berezney and Coffey.[127] The samples from the fourth and final step, which contains the nuclear matrix, also remain positive when probed with the anti-gp32/gp36 monoclonal antibody. When the same nuclear matrix fraction is examined by electron microscopy, occasional viral capsids are intertwined along the fibrogranular network (Figure 37).

Another unusual feature of the p32/p36 complex is phosphorylation of some of its constituents. In order to determine which components were phosphorylated, a nucleocapsid preparation labeled with phosphorus-32 was analyzed by both immunoblotting and autoradiography. These procedures demonstrated that p36 and p38 were phosphoproteins but p32 was not. Thus, the p32/p36 complex represents a family of VZV-specific polypeptides which is found mainly within the nucleus of the infected cells. The complex contains several components of gradually increasing molecular weight. The most prominent species have been designated p32 and p36, but other related polypeptides have molecular weights of 34,000 and 38,000. The polypeptide complex is also present in the nuclear matrix of the infected cell. Based on the above findings, constituents of the p32/p36 complex are candidate proteins for DNA insertion and encapsidation processes.

H. Other Virally Encoded Proteins

In addition to the VZV p32/p36 complex, another prominent structural protein in the VZV capsid has an estimated molecular weight of approximately 150,000. This polypeptide has been observed in preparations of purified virions and nucleocapsids by several laboratories,[88-90]; in addition, a similar prominent higher-molecular weight capsid polypeptide is a hallmark of many other herpesviruses.[92] More than 20 other proteins observed in the electrophoretic profiles of the varicella zoster virion remain completely uncharacterized.

Besides the gamma or late viral products, VZV infection also directs the synthesis of virally encoded immediate early and early proteins. Of the early VZV products, the enzyme thymidine kinase has been the most completely studied.[128] The molecular weight of the VZV-specified thymidine kinase under nondenaturing conditions is ap-

FIGURE 36. Identification of the VZV p32/p36 complex in viral nucleocapsids. The nucleocapsid pellet from a 5%/40% discontinuous glycerol gradient was layered onto a 10%/40% linear sucrose gradient for isolation of empty and full VZV capsids. Purified nucleocapsids (5 μl), empty capsids (25 μl), and full capsids (25 μl) were fractionated in a polyacrylamide slab gel, transferred to nitrocellulose paper, and subjected to immunoblot analysis. Lanes A, B, and C contain nucleocapsid, empty capsid, and full capsid, respectively, reacted with the monoclonal antigody to the VZV p32/p36 protein complex. Immunostaining patterns of empty and full capsids probed with polyclonal human anti-VZV antiserum are shown in lanes D and E, respectively.

proximately 72,000 daltons. Under reducing and denaturing conditions, VZV thymidine kinase may migrate as a band of approximately 35,000 molecular weight, thus suggesting that the 72,000 product is a dimer.[129] The substrate specificity of the VZV enzyme includes thymidine, deoxyuridine, and deoxycytidine, as well as several analogs of thymidine and deoxycytidine, e.g., iododeoxyuridine and cytosine arabinoside. The optimal pH for the VZV thymidine kinase is close to 8.0. Antiserum to this enzyme stains mainly the nuclei of VZV-infected cells, but it does not adhere to the nuclei of cells infected with thymidine kinase-deletion mutants.[130] Another enzyme relating to DNA synthesis is VZV-specified DNA polymerase, which may migrate as a 145,000-mol wt protein.[88] Among the least-studied of the VZV-encoded gene products are the immediate early proteins. Extensive analyses of the humoral immune response to individual VZV proteins during primary infection (chickenpox) and reactivation (zoster)

FIGURE 37. Electron microscopic visualization of the nuclear matrices from VZV-infected cells. After a four-step extraction procedure, the chromatin of the nucleus is completely extracted along with most of the RNA and phospholipids. The residual nucleoskeletal structure of the matrix consists mainly of acidic proteins in undulating strands, along with an occasional cluster of empty and full VZV capsids (lower right corner). Magnification × 59,500.

demonstrate the transient presence of antibodies to a number of viral polypeptides, including a 174,000-mol wt species which may represent one of the immediate early VZV gene products.[118,131]

I. Protein Kinase Activity in VZV-Infected Cells

Protein kinases catalyze the transfer of the phosphate of ATP to acceptor amino acids in proteins.[132] This reaction is an essential regulatory step for many biologic phenomena, e.g., receptor interactions and secretory processes. Protein kinases have been divided into broad categories on the basis of common biochemical properties; these include cyclic nucleotide-dependent protein kinases and cyclic nucleotide-independent protein kinases. Ever since the discovery in 1971 that a cyclic nucleotide-independent protein kinase was present in Rauscher murine leukemia virus,[133] there has been great interest in determining whether similar phosphorylation activity occurs in other viruses. The results of several investigations on the herpesviruses are summarized in Table 11. They show that at least five different herpesviruses induce the formation of protein kinases in the infected cell culture. Some of the kinases appear to be virus specified, while others closely resemble previously characterized cellular products, e.g., casein kinase II activity in cells infected with pseudorabies virus.[141] It has recently been discovered that one of the murine monoclonal antibodies reacts with a protein kinase found in VZV-infected cell cultures.[142] By immunoblot analysis, the monoclonal antibody identified a protein of approximately 50,000 daltons which was present in an extract of VZV-infected cells; no reactive protein was found in either a purified virion preparation or a control uninfected cell monolayer. In some aspects, the VZV protein kinase resembles the nonvirion pseudorabies virus product described in Table 11.[140] When the activity of the VZV protein was tested in an in vitro kinase assay with [γ-^{32}P]ATP as the phosphate donor, we observed that one of the VZV glycoproteins (VZV gpI) was selectively phosphorylated while gpII and gpIII remained unmodified.[142] Thus, phosphorylation must be added to the list of posttranslational modifications of the VZV glycoproteins mentioned earlier in this chapter.

ACKNOWLEDGMENT

Research from my laboratory described in this chapter was supported by the U.S. Public Health Service grants AI14604 and AI22795; as well as by grants from The Thrasher Research Fund, Salt Lake City, Utah; The Elsa U. Pardee Foundation, Midland, Michigan; and the Morrison Trust, San Antonio, Texas.

Table 11
PROTEIN KINASES SPECIFIED BY DIFFERENT HERPESVIRUSES

Herpes virus system	Presence in virus	pH Optimum	Cyclic AMP dependence	(γ ^{32}P) GTP donor	Mol wt	Phosphorylation of: Itself (auto-)	Casein	Phosvitin	Protamine	Histones	Viral proteins	Ref.
EAV	Yes	8.5	No	Yes	N.T.	—	—	—	1+	1+	2+	134
HSV	Yes	8.0	No	N.T.	N.T.	—	—	—	1+	—	0	135
HSV	Yes	8.0	No	Yes	N.T.	—	0	0	0	0	2+	136
HSV	N.T.	8.0	No	N.T.	N.T.	1+	1+	2+	1+	2+	N.T.	137
Tupaia	Yes	7.0—9.0	No	No	N.T.	—	0	—	0	0	2+	138
CMV	No	6.0—6.5	No	N.T.	68K	1+	1+	1+	0	0	0	139
PrV	No	7.5—9.0	No	No	68K	—	0	0	2+	1+	0	140
PrV	Yes	N.T.	No	Yes	N.T.	—	1+	—	2+	1+	2+	141

Note: 0 = No effect, 1+ = Mild effect, 2 = Strong effect; N.T. = Not tested. EAV = Equine abortion virus; HSV = herpes simplex virus; CMV = cytomegalovirus; PrV = pseudorabies virus.

REFERENCES

1. Grose, C., Varicella-zoster virus infections: chickenpox (varicella) and shingles (zoster), in *Human Herpesvirus Infections*, Glaser, R. and Gotlieb-Stematsky, T., Eds., Marcel Dekker, New York, 1982, chap. 3.
2. Bokai, J., Das Auftreten der Schafblattern unter besonderen Umstanden, *Ungar. Arch. Med.*, 1, 159, 1892.
3. Bokay, J., von, Über den ätiologischen Zusammenhang der Varizellen mit gewissen Fallen von Herpes zoster, *Wien. Klin. Wochenschr.*, 22, 1323, 1909.
4. Weller, T. H., Serial propagation *in vitro* of agents producing inclusion bodies derived from varicella and herpes zoster, *Proc. Soc. Exp. Biol. Med.*, 83, 340, 1953.
5. Hope-Simpson, R. E., Infectiousness of communicable diseases in the household (measles, chickenpox, and mumps), *Lancet*, 2, 549, 1952.
6. Gordon, J. E. and Meader, F. M., The period of infectivity and serum prevention of chickenpox, *JAMA*, 93, 2013, 1929.
7. Evans, P., An epidemic of chickenpox, *Lancet*, 2, 339, 1940.
8. Leclair, J. M., Zaia, J. A., and Levin, M. J., Airborne transmission of chickenpox in a hospital, *N. Engl. J. Med.*, 302, 450, 1980.
9. Gustafson, T. L., Lavely, G. B., Brawner, E. R., Hutcheson, R. H., Wright, P. F., and Schaffner, W., An outbreak of airborne nosocomial varicella, *Pediatrics*, 70, 550, 1982.
10. Fenner, F., The pathogenesis of the acute exanthems, *Lancet*, 2, 915, 1948.
11. Grose, C., Variation on a theme by Fenner: the pathogenesis of chickenpox, *Pediatrics*, 68, 735, 1981.
12. Tyzzer, E. E., The histology of the skin lesions in varicella, *Phillipine J. Sci.*, 1, 349, 1906.
13. Ross, A. H., Modification of chickenpox in family contacts by administration of gamma globulin, *N. Engl. J. Med.*, 267, 369, 1962.
14. Hosoya, R., Miwa, M., Nishimura, T., and Yamamoto, T., Erythema multiforme syndrome associated with varicella, *N. Engl. J. Med.*, 305, 582, 1981.
15. Goldston, A. S., Millichap, J. G., and Mitler, R. H., Cerebellar ataxia with pre-eruptive varicella, *Am. J. Dis. Child.*, 106, 197, 1963.
16. Underwood, E. A., The neurological complications of varicella: a clinical and epidemiological study, *Br. J. Child. Dis.*, 32, 83, 1935.
17. Miller, H. G., Stanton, J. B., and Gibbons, J. L., Para-infectious encephalomyelitis and related syndrome, *Q. J. Med.*, 25, 428, 1956.
18. Johnson, R. and Milbourn, P. E., Central nervous system manifestations of chickenpox, *Can. Med. Assoc. J.*, 102, 831, 1970.
19. Reye, R. D., Morgan, G., and Baral, J., Encephalopathy and fatty degeneration of the viscera, A disease entity in childhood, *Lancet*, 2, 749, 1963.
20. Mortimer, E. A., and Lepow, M. L., Varicella with hypoglycemia possibly due to salicylates, *Am. J. Dis. Child.*, 103, 583, 1962.
21. Lichtenstein, P. K., Heubi, J. E., Daugherty, C. C., Farrell, M. K., Sokol, R. J., Rothbaum, R. J., Suchy, F. J., and Balistreri, W. F., Grade I Reye's Syndrome, A frequent cause of vomiting and liver dysfunction after varicella and upper-respiratory-tract infection, *N. Engl. J. Med.*, 309, 133, 1983.
22. Feldman, S., and Epp, E., Isolation of varicella-zoster virus from blood, *J. Pediatr.*, 88, 265, 1976.
23. Meyers, M. G., Viremia caused by varicella-zoster virus: association with malignant progressive varicella, *J. Infect. Dis.*, 140, 229, 1979.
24. Ozaki, T., Ichikawa, T., Matsui, Y., Nagai, T., Asano, Y., Yaminishi, K., and Takahashi, M., Viremic phase in nonimmunocompromised children with varicella, *J. Pediatr.*, 104, 85, 1984.
25. Grose, C., Isolation of varicella-zoster virus from blood, *J. Pediatr.*, 105, 504, 1984.
26. Ward, J. R. and Bishop, B., Varicella arthritis, *JAMA*, 212, 1954, 1970.
27. Priest, J. R., Urick, J. J., Groth, K. E., and Balfour, H. H., Varicella arthritis documented by isolation of virus from joint fluid, *J. Pediatr.*, 93, 990, 1978.
28. Minkowitz, S., Wenk, R., Friedman, E., Yuceoglu, A., and Berkovitch, S., Acute glomerulonephritis associated with varicella infection, *Am. J. Med.*, 44, 489, 1968.
29. Singhal, P. C., Chugh, K. S., and Muthusethupathi, M. A., Rapidly progressive glomerulonephritis associated with varicella infection, *J. Pediatr.*, 91, 680, 1977.
30. Mandelbaum, T. and Terk, B. H., Pericarditis in association with chickenpox, *JAMA*, 170, 191, 1959.
31. Ormiston, G., Orchitis as a complication of chickenpox, *Br. Med. J.*, 1, 1203, 1953.
32. Weinstein, L., Meade, R. H., Respiratory manifestations of chickenpox, *Arch. Intern. Med.*, 98, 91, 1956.

33. Triebwasser, J. H., Harris, R. E., Bryant, R. E., Rhoades, E. R., Varicella pneumonia in adults: report of seven cases and a review of literature, *Medicine*, 46, 409, 1967.
34. Weigle, K. A., and Grose, C., Varicella pneumonitis: immunodiagnosis with a monoclonal antibody, *J. Pediatr.*, 105, 265, 1984.
35. Fleisher, G., Henry, W., McSorley, M., Arbeter, A., and Plotkin, S., Life-threatening complications of varicella, *Am. J. Dis. Child.*, 135, 896, 1981.
36. Preblud, S. R., Age-specific risk of varicella complications, *Pediatrics*, 68, 14, 1981.
37. Preblud, S. R., Orenstein, W. A., and Bart, K. J., Varicella: clinical manifestations, epidemiology and health impact in children, *Pediatr. Infect. Dis.*, 3, 505, 1984.
38. Preblud, S. R., Bregman, D. J., and Vernon, L. L., Deaths from varicella infants, *Pediatr. Infect. Dis.*, 4, 503, 1985.
39. Middelkamp, J. N., Varicella in newborn twins, *J. Pediatr.*, 43, 573, 1953.
40. Harris, L. E., Spread of varicella in nurseries, *Am. J. Dis. Child.*, 105, 315, 1963.
41. Odessky, L., Newman, B., and Wein, G. B., Congenital varicella, *N.Y. State J. Med.*, 54, 2849, 1954.
42. Freud, P., Congenital varicella, *Am. J. Dis. Child.*, 96, 730, 1958.
43. Cutter, B. G., and Hansman, D., Neonatal varicella, *Med. J. Aust.*, 1, 411, 1968.
44. Librach, I. M., Congenital chickenpox, *Postgrad. Med. J.*, 35, 575, 1959.
45. Steen, J., and Pedersen, R. V., Varicella in a newborn girl, *J. Oslo City Hosp.*, 9, 36, 1959.
46. Henderson, W. B., Chickenpox in an eight-day-old infant, *J. Pediatr.*, 4, 668, 1934.
47. Oppenheimer, E. H., Congenital chickenpox with disseminated visceral lesions, *Bull. Johns Hopkins Hosp.*, 74, 240, 1944.
48. Ehrlich, R. M., Turner, J. A. P., and Clarke, M., Neonatal varicella: a case report with isolation of the virus, *J. Pediatr.*, 53, 139, 1958.
49. Hubbard, T. W., Varicella occurring in an infant twenty-four hours after birth, *Br. Med. J.*, 1, 822, 1878.
50. Pridham, F. C., Chickenpox during intrauterine life, *Br. Med. J.*, 1, 1054, 1913.
51. Joel, L., Varicella in Sydney, *Med. J. Aust.*, 2, 728, 1966.
52. Hyatt, H. H., Neonatal varicella: report of a case in five-day-old infant and review of the literature, *J. Natl. Med. Assoc.*, 59, 32, 1967.
53. Fuller, D. J. and Duthie, R. B., The timed appearance of some congenital malformations and orthopedic abnormalities, in *Instructional Course Lectures of the American Academy of Orthopedics Surgeons*, C. V. Mosby, St. Louis, 1974, chap. 3.
54. Williamson, A. P., The varicella-zoster virus in the etiology of severe congenital defects, *Clin. Pediatr.*, 14, 553, 1975.
55. Savage, M. O., Moosa, A., and Gordon, R. R., Maternal varicella infection as a cause of fetal malformations, *Lancet*, 1, 352, 1973.
56. Srabstein, J. C., Morris, N., Larke, R. P. B., DeSa, D. J., Castelino, B. B., and Sum, E., Is there a congenital varicella syndrome?, *J. Pediatr.*, 84, 239, 1974.
57. Laforet, E. G. and Lynch, C. L., Multiple congenital defects following maternal varicella, *N. Engl. J. Med.*, 236, 534, 1947.
58. McKendry, J. B. J. and Bailey, J. D., Congenital varicella associated with multiple defects, *Can. Med. Assoc. J.*, 108, 66, 1973.
59. Frey, H. M., Bialkin, G., and Gershon, A., Congenital varicella: case report of a serologically proved long term survivor, *Pediatrics*, 59, 110, 1977.
60. Bai, P. V. A. and John, R. J., Congenital skin ulcers following varicella in late pregnancy, *J. Pediatr.*, 94, 95, 1979.
61. Kotchmar, G. S., Grose, C., and Brunell, P. A., Complete spectrum of the varicella congenital defects syndrome in 5-year-old child, *Pediatr. Infect. Dis.*, 3, 142, 1984.
62. Peterson, P. H. and Black, S. A. B., Varicella herpetiformis, *Br. Med. J.*, 2, 762, 1946.
63. Gordon, J. E., Chickenpox: an epidemiologic review, *Am. J. Med. Sci.*, 244, 362, 1962.
64. Hope-Simpson, R. E., The nature of herpes zoster: a long-term study and a new hypothesis, *Proc. R. Soc. Med.*, 58, 9, 1965.
65. London, W. P. and Yorke, J. A., Recurrent outbreaks of measles, chickenpox, and mumps. I. Seasonal variation in contact rates, *Am. J. Epidemiol.*, 98, 453, 1973.
66. Yorke, J. A. and London, W. P., Recurrent outbreaks of measles, chickenpox, and mumps. II. Systematic differences in contact rates and stochastic effects, *Am. J. Epidemiol.*, 98, 469, 1973.
67. Rifkind, D., The activation of varicella-zoster virus infections by immunosuppressive therapy, *J. Lab. Clin. Med.*, 68, 463, 1966.
68. Mazur, M. H. and Dolin, R., Herpes zoster at the NIH: a 20-year experience, *Am. J. Med.*, 65, 738, 1978.

69. Moutsopoulos, H. M., Gallagher, J. D., Decker, J. L., and Steinberg, A. D., Herpes zoster in patients with systemic lupus erythematosis, *Arthritis Rheum.*, 21, 798, 1978.
70. Tausch, M., Beitrag Zum Vorkommen eines mit auf die Welt gebrachten Herpes zoster am Neugeborenen, *Zentralbl. Gynakol.*, 56, 1629, 1932.
71. Lomer, D., Herpes zoster bei einem 4 Tage alten Kinde, *Zentralbl. Gynakol*, 13, 778, 1889.
72. Feldman, G. V., Herpes zoster neonatorum, *Arch. Dis. Child.*, 27, 126, 1952.
73. Brunell, P. A., Miller, L. H., and Lovejoy, F., Zoster in children, *Am. J. Dis. Child.*, 115, 432, 1968.
74. Kouvalainen, K., Salmi, A., and Salmi, T. T., Infantile herpes zoster, *Scand. J. Infect. Dis.*, 4, 91, 1972.
75. McKendrick, G. D. W. and Raychoudhury, S. C., Herpes zoster in childhood, *Scand. J. Infect. Dis.*, 4, 23, 1972.
76. Almeida, J. D., Howatson, A. F., and Williams, M. G., Morphology of varicella (chickenpox) virus, *Virology*, 16, 353, 1962.
77. Becker, P., Melnick, J. L., and Mayor, H. D., A morphologic comparison between the developmental stages of herpes zoster and human cytomegalovirus, *Exp. Mol. Pathol.*, 4, 11, 1965.
78. Cook, M. L. and Stevens, J. G., Replication of varicella-zoster virus in cell culture: an ultrastructural study, *J. Ultrastruct. Res.*, 32, 334, 1970.
79. Grose, C., Perrotta, D. M., Brunell, P. A., and Smith, G. C., Cell-free varicella-zoster virus in cultured human melanoma cells, *J. Gen. Virol.*, 43, 15, 1979.
80. Weller, T. H., Witton, H. M., and Bell, E. J., The etiologic agents of varicella and herpes zoster. Isolation, propagation, and cultural characteristics *in vitro*, *J. Exp. Med.*, 108, 843, 1958.
81. Taylor-Robinson, D. and Caunt, A. E., *Varicella Virus*, Springer-Verlag, New York, 1972, 1.
82. Grose, C. and Brunell, P. A., Varicella-zoster virus: isolation and propagation in human melanoma cells at 36 and 32°C, *Infect. Immun.*, 19, 199, 1978.
83. Caunt, A. E. and Taylor-Robinson, D., Cell-free varicella-zoster virus in tissue culture, *J. Hyg. Cambridge*, 62, 413, 1964.
84. Brunell, P. A., Separation of infectious varicella-zoster virus from human embryonic lung fibroblasts, *Virology*, 31, 732, 1967.
85. Schmidt, N. J. and Lennette, E. H., Improved yields of cell-free varicella-zoster virus, *Infect. Immun.*, 14, 709, 1976.
86. Grose, C., Friedrichs, W. E., and Smith, K. O., Cryopreservation of varicella-zoster virions without loss of structural integrity or infectivity, *Interviriology*, 15, 154, 1981.
87. Ludwig, H., Haines, H. G., Biswal, N., and Benyesh-Melnick, M., The characterization of varicella-zoster virus DNA, *J. Gen. Virol.*, 14, 111, 1972.
88. Grose, C., Friedrichs, W. E., and Smith, G. C., Purification and molecular anatomy of the varicella-zoster virion, *Biken J.*, 26, 1, 1983.
89. Shemer, Y., Leventon-Kriss, A., and Sarov, I., Isolation and polypeptide characterization of varicella-zoster virus, *Virology*, 106, 133, 1980.
90. Shiraki, K., Okuno, T., Yamanishi, K., and Takahashi, M., Polypeptides of varicella-zoster virus (VZV) and immunological relationship of VZV and herpes simplex virus (HSV), *J. Gen. Virol.*, 61, 255, 1982.
91. Shiraki, K. and Takahashi, M., Virus particles and glycoproteins excreted from cultured cells infected with varicella-zoster virus (VZV), *J. Gen. Virol.*, 61, 271, 1982.
92. Killington, R. A., Yeo, L., Honess, R. W., Watson, D. H., Duncan, B. E., Halliburton, I. W., and Mumford, J., Comparative analyses of the proteins and antigens of five herpesviruses, *J. Gen. Virol.*, 37, 297, 1977.
93. Hondo, R., Shibuta, H., and Matumoto, M., An improved plaque assay for varicella virus, *Arch. Virol.*, 51, 355, 1976.
94. Friedrichs, W. E. and Grose, C., Varicella-zoster virus p32/p36 complex is present in both the viral capsid and the nuclear matrix of the infected cell, *J. Virol.*, 57, 155, 1986.
95. Montalvo, E. A., Parmley, R. T., and Grose, C., Varicella-zoster viral glycoprotein envelopment: ultrastructural cytochemical localization, *J. Histochem. Cytochem.*, 34, 281, 1986.
96. Rapp, F. and Vanderslice, D., Spread of zoster virus in human embryonic lung cells and the inhibitory effect of iododeoxyuridine, *Virology*, 22, 321, 1964.
97. Yamanishi, K., Matsunaga, Y., Ogino, T., Takahashi, M., and Takamizawa, A., Virus replication and localization of varicella-zoster virus antigens in human embryonic fibroblast cells infected with cell-free virus, *Infect. Immun.*, 28, 536, 1980.
98. Spear, P. G. and Roizman, B., Herpes simplex viruses, in *DNA Tumor Viruses*, 2nd ed., Tooze, J., Ed., Cold Spring Harbor Laboratory, Cold Spring Harbor, New York, 1981, chap. 12.
99. Elbein, A. D., The role of lipid-linked saccharides in the biosynthesis of complex carbohydrates, *Annu. Rev. Plant Physiol.*, 30, 239, 1979.

100. Kornfeld, R. and Kornfeld, S., Assembly of asparagine-linked oligosaccharides, *Annu. Rev. Biochem.*, 54, 631, 1985.
101. Feizi, T., Gooi, H. C., Childs, R. A., Picard, J. K., Uemura, K., Loomes, L. M., Thorpe, S. J., and Hounsell, E. F., Tumor-associated and differentiation antigens on the carbohydrate moieties of mucin-type glycoproteins, *Biochem. Soc. Trans.*, 12, 591, 1984.
102. Davison, A. J., Edson, C. M., Ellis, R. W., Forghani, G., Gilden, D., Grose, C., Keller, P. M., Vafai, A., Wroblewska, Z., and Yamanishi, K., A new common nomenclature for the glycoprotein genes of varicella-zoster virus and their glycosylated products, *J. Virol.*, 57, 1195, 1986.
103. Grose, C., The synthesis of glycoproteins in human melanoma cells infected with varicella-zoster virus, *Virology*, 101, 1, 1980.
104. Grose, C., Edmond, B. J., and Friedrichs, W. E., Immunogenic glycoproteins of laboratory and vaccine strains of varicella-zoster virus, *Infect. Immun.*, 31, 1044, 1981.
105. Grose, C. and Friedrichs, W. E., Immunoprecipitable polypeptides specified by varicella-zoster virus, *Virology*, 118, 86, 1982.
106. Weigle, K. A. and Grose, C., Common expression of varicella-zoster viral glycoproteins antigens *in vitro* and in chickenpox and zoster vesicles, *J. Infect. Dis.*, 148, 630, 1983.
107. Grose, C., Edwards, D. P., Friedrichs, W. E., Weigel, K. A., and McGuire, W. L., Monoclonal antibodies against three major glycoproteins of varicella-zoster virus, *Infect. Immun.*, 40, 381, 1983.
108. Montalvo, E. A., Parmley, R. T., and Grose, C., Structural analysis of the varicella-zoster virus gp98-gp62 complex: posttranslational addition of N-linked and O-linked oligosaccharide moieties, *J. Virol.*, 53, 761, 1985.
109. Asano, Y. and Takahashi, M., Studies on the polypeptides of varicella-zoster (V-Z) virus. I. Detection of varicella-zoster virus polypeptides in infected cells, *Biken J.*, 22, 81, 1979.
110. Asano, Y. and Takahashi, M., Studies on the polypeptides of varicella-zoster (V-Z) virus. II. Synthesis of viral polypeptides in infected cells, *Biken J.*, 23, 95, 1980.
111. Okuno, T., Yaminishi, K., Shiraki, K., and Takahashi, M., Synthesis and processing of glycoproteins of varicella-zoster virus as studied with monoclonal antibodies to VZV antigens, *Virology*, 129, 357, 1983.
112. Namazue, J., Campo-Vera, H., Kitamura, K., Okuno, T., and Yamanishi, N., Processing of virus-specific glycoproteins of varicella-zoster virus, *Virology*, 143, 252, 1985.
113. Keller, P. M., Neff, B. J., and Ellis, R. W., Three major glycoprotein genes of varicella-zoster virus whose products have neutralization epitopes, *J. Virol.*, 52, 293, 1984.
114. Ellis, R. W., Keller, P. M., Lowe, R. S., and Zivin, R. A., Use of a bacterial expression vector to map the varicella-zoster virus major glycoprotein gene, gC, *J. Virol.*, 53, 81, 1985.
115. Forghani, B., Schmidt, N. J., Myoraku, C. K., and Gallo, D., Serological reactivity of some monoclonal antibodies to varicella-zoster virus, *Arch. Virol.*, 73, 311, 1982.
116. Forghani, B., DuPuis, V. W., and Schmidt, N. J., Varicella-zoster viral glycoproteins analysed with monoclonal antibodies, *J. Virol.*, 52, 55, 1984.
117. Edson, C. A., Hosler, B. A., Poodry, C. A., Schooley, R. T., Waters, D. J., and Thorley-Lawson, D. A., Varicella-zoster virus envelope glycoproteins: biochemical characterization and identification in clinical material, *Virology*, 145, 62, 1985.
118. Weigle, K. A. and Grose, C., Molecular dissection of the humoral immune response to individual varicella-zoster viral proteins during chickenpox, quiescence, reinfection, and reactivation, *J. Infect. Dis.*, 149, 741, 1984.
119. Grose, C., Edwards, D. P., Friedrichs, W. E., Weigle, K. A., and McGuire, W. L., Varicella-zoster virus specific gp140: a highly immunogenic and disulfide-linked structural glycoprotein, *Virology*, 132, 138, 1984.
120. Ito, M., Ihara, T., Grose, C., and Starr, S., Human leukocytes kill varicella-zoster virus infection fibroblasts in the presence of murine monoclonal antibodies to virus-specific glycoproteins, *J. Virol.*, 54, 98, 1985.
121. Vafai, A., Wroblewska, Z., Wellish, M., Green, M., and Gilden, D., Analysis of three late varicella-zoster virus proteins, a 125,000 molecular-weight protein and gp1 and gp3, *J. Virol.*, 52, 953, 1984.
122. Wroblewska, Z., Gilden, D., Green, M., Devlin, M., and Vafai, A., Affinity-purified varicella-zoster virus glycoprotein gp1/gp3 stimulates the production of neutralizing antibody, *J. Gen. Virol.*, 66, 1795, 1985.
123. Edson, C. M., Hosler, B. A., Repress, R. A., Waters, D. J., and Thorley-Lawson, D. A., Cross-reactivity between herpes simplex virus glycoprotein B and a 63,000-dalton varicella-zoster virus envelope glycoprotein, *J. Virol.*, 56, 333, 1985.
124. Friedrichs, W. E. and Grose, C., Glycoprotein gp118 of varicella-zoster virus: purification by serial affinity chromatography, *J. Virol.*, 49, 992, 1984.
125. Montalvo, E. A. and Grose, C., Neutralization epitope of varicella zoster virus on native viral glycoprotein gp118 (VZV glycoprotein gpIII), *Virology*, 149, 900, 1986.

126. Davison, A. J., Waters, D. J., and Edson, C. M., Identification of the products of a varicella-zoster virus glycoprotein gene, *J. Gen. Virol.*, 66, 2237, 1986.
127. Berezney, R. and Coffey, D. S., Nuclear matrix: isolation and characterization of a framework structure from rat liver nuclei, *J. Cell Biol.*, 73, 616, 1977.
128. Hackstadt, T. and Mallavia, L. P., Deoxypyrimidine nucleoside metabolism in varicella-zoster virus-infected cells, *J. Virol.*, 25, 510, 1978.
129. Lopetegui, P., Matsunaga, Y., Okuno, T., Ogino, T., and Yamanishi, K., Expression of varicella-zoster virus-related antigens in biochemically transformed cells, *J. Gen. Virol.*, 64, 1181, 1983.
130. Shiraki, K., Ogino, T., Yamanishi, K., and Takahashi, M., Immunochemical characterization of pyrimidine kinase induced by varicella-zoster virus, *J. Gen. Virol.*, 66, 221—229, 1985.
131. Grose, C., Zoster in children with cancer: radioimmune precipitation profiles of sera before and after illness, *J. Infect. Dis.*, 147, 47, 1983.
132. Rubin, C. S. and Rosen, O. M., Protein phosphorylation, *Annu. Rev. Biochem.*, 44, 831, 1975.
133. Strand, M. and August, J. T., Protein kinase and phosphate acceptor proteins in Rauscher murine leukemia virus, *Nature (London) New Biol.*, 233, 137, 1971.
134. Randall, C. G., Rodgers, H. W., Downer, D. N., and Gentry, G. A., Protein kinase activity in equine herpesvirus, *J. Virol.*, 9, 216, 1972.
135. Rubenstein, A. S., Gravell, M., and Darlington, R., Protein kinases in enveloped herpes simplex virions, *Virology*, 50, 287, 1972.
136. Lemaster, S. and Roizman, B., Herpes simplex virus phosphoproteins. II. Characterization of the virion protein kinase and the polypeptides phosphorylated in the virion, *J. Virol.*, 35, 798, 1980.
137. Blue, W. T. and Stobbs, D. G., Isolation of a protein kinase induced by herpes simplex virus type 1, *J. Virol.*, 38, 383, 1981.
138. Flugel, R. M. and Darai, G., Protein kinase and specific phosphate acceptor proteins associated with tupaia herpesvirus, *J. Virol.*, 43, 410, 1982.
139. Michelson, S., Tardy-Panit, M., and Barzu, O., Properties of a human cytomegalovirus-induced protein kinase, *Virology*, 134, 259, 1984.
140. Katan, M., Stevely, W. S., and Leader, D. P., Partial purification and characterization of a new phosphoprotein kinase from cells infected with pseudorabies virus, *Eur. J. Biochem.*, 152, 57, 1985.
141. Stevely, W. S., Katan, M., Stirling, V., Smith, G., and Leader, D. P., Protein kinase activities associated with the virions of pseudorabies and herpes simplex virus, *J. Gen. Virol.*, 66, 661, 1985.
142. Montalvo, E. A. and Grose, C., Varicella-zoster virus glycoprotein gpI is selectively phosphorylated by a virus-induced protein kinase, *Proc. Natl. Acad. Sci. U.S.A.*, in press.

Chapter 2

CLINICAL MANIFESTATIONS OF VARICELLA AND HERPES ZOSTER AND THE IMMUNE RESPONSE TO VARICELLA-ZOSTER VIRUS

Ann M. Arvin

TABLE OF CONTENTS

I. CLINICAL MANIFESTATIONS OF VARICELLA AND HERPES ZOSTER

A. Varicella

1. Introduction

Fifty years ago, Bullow and Wishik observed that chickenpox is "the disease of childhood which is taken least seriously by the laity and by the medical profession."[1] Their description of the complications of primary infection with varicella-zoster virus (VZV) which was published in 1935 does not differ substantially from the analysis of the morbidity and mortality of varicella among previously healthy children reported by Fleisher et al. in 1981.[2] Varicella was and is usually a benign exanthem of childhood but its clinical consequences must now be considered in a population which is largely protected against other systemic viral infections such as smallpox and measles.[3]

2. Usual Clinical Manifestations of Varicella in Healthy Children

Primary VZV infection causes fever and a generalized vesicular exanthem.[4-7] Prodromal symptoms may precede the appearance of the first lesions which usually occur on the scalp, face, or trunk. Each lesion evolves within a few hours from an erythematous macule to the formation of a vesicle surrounded by erythema, the classic "dew drop on a rose petal" (Chapter 1, Figure 1). The vesicle fluid becomes cloudy and may develop an umbilicated appearance as the lesion begins to crust. As the initial lesions progress through these stages, crops of new lesions continue to form, usually in a "centrifugal" pattern, i.e., from the face and trunk to the extremities. The lesions which appear later in the course of the illness are distributed less densely and may not evolve to vesicle formation. Occasionally, the cutaneous rash will exhibit a more pronounced eruption in one dermatome.[8] Vesicles can occur on the mucous membranes of the oropharynx and conjunctivae and also of the vagina and rectum. Hypopigmentation of the underlying skin may be seen as the crusts resolve, but extensive scarring is unusual. Faint, shallow scars often persist at the sites of the initial larger lesions along the scalp hairline or eyebrows. In an African population, only 2.4% of patients had five or more facial scars 1 year after varicella.[9]

The clinical course of varicella in healthy children was documented in exceptional detail by Ross.[10] Some children had prodromal symptoms of headache, malaise, and abdominal pain just before the appearance of the exanthem, but many patients had no symptoms during the incubation period (Table 1). Fever, pruritis, throat pain, irritability, headache, listlessness, and anorexia were the prominent symptoms associated with the acute phase of the illness which was defined as the 6 days following the appearance of the exanthem. Respiratory symptoms, with the exception of mild rhinitis, were absent during the prodromal and acute phases of varicella and vomiting was also unusual (Table 1). The average temperature elevation was below 101.5°F, but the range was up to 106°F. New cutaneous lesions continued to emerge for a period of 1 to 6 days. More than half of the children developed new lesions for 3 to 6 days. Only 1 of 521 children had new lesions for 7 days. Most children had fewer than 300 lesions, but children who acquired varicella by household exposure were more likely to develop an extensive cutaneous eruption than the index case (Table 1). The total number of varicella lesions in healthy children ranged from 12 to 1968.[10]

The white blood cell counts of children with varicella were investigated by Holbrook.[11] The total white cell count fell during the last few days of the incubation period, reaching a nadir on the second and third day of rash. The decrease in the total white cell count reflected lower polymorphonuclear leukocyte and lymphocyte counts. Beginning on the second day of the rash, the lymphocyte count rose rapidly, resulting in both a relative and an absolute lymphocytosis by the fifth day (Figure 1). Lymphoblasts and prolymphocytes were evident in the peripheral blood during the acute and convalescent phases of varicella. Mitchell and Fletcher found no difference between the leukocyte response of adults with varicella and that of children of various ages.[12]

3. Complications of Varicella in Healthy Children and Adults

a. *Skin Complications*

The vesicular rash of varicella can be severe if the skin is already damaged or inflamed.[13] Varicella lesions in infants may be concentrated in the diaper area, and children with eczema are subject to extensive cutaneous varicella. A dramatic "photodistribution" of the varicella exanthem has been described in children who acquired sunburn during the late incubation period of varicella.[14,15] The rash resembled smallpox in the density of the eruption and the simultaneous occurrence of lesions exhibiting the same stage of development.

Secondary bacterial infection of the cutaneous lesions is the most frequent complication of varicella. Bacterial superinfection was the predominant reason for hospitalization of children under the age of 5 years who had varicella.[16] *Staphylococcus aureus*

Table 1
PRODROMAL AND ACUTE SYMPTOMS OF VARICELLA IN OTHERWISE HEALTHY CHILDREN[a]

Symptom	Primary Cases	Secondary Cases
Prodrome (%)		
Headache	10	6
Malaise	20	13
Abdominal pain	5	4
Anorexia	3	1
Vomiting	1	1
Respiratory Symptoms	8	4
Acute illness		
Av. max. temp., °F		
Day 1	101.2	100.7
Day 3	100.6	100.9
Day 6	99.3	99.4
Av. no. of lesions		
10 — 100	30[b]	14[b]
100 — 200	24	22
200 — 300	21	14
300 — 400	9	9
400 — 500	6	10
500 — 1000	9	27
>1000	1	4
Symptoms (%)		
Pruritis	96[c]	86[c]
Throat pain	58	49
Irritability	76	79
Headache	46	35
Listlessness	48	45
Anorexia	57	56
Vomiting	11	15
Abdominal pain	31	20
Rhinitis	27	21

[a] Abstracted from Reference 10.
[b] Estimated from Reference 10.
[c] Average duration 3.7 days (primary) and 3.6 days (secondary).

Reprinted by permission of the *New England Journal of Medicine*, 267, 369, 1962.

and *Streptococcus pyogenes* (group A beta-hemolytic streptococcus) are the usual bacterial pathogens. "Impetigo" associated with varicella lesions is diagnosed commonly but may be difficult to differentiate from larger lesions caused by the virus alone. *S. aureus* and *S. pyogenes* are normal skin flora, so the isolation of these bacteria from varicella lesions may be incidental. Nevertheless, the cutaneous lesions provide a portal of entry for these organisms in some patients. Soft tissue invasion by these organisms produces cellulitis or subcutaneous abscesses. *S. aureus* is now the predominant pathogen, whereas *S. pyogenes* was common in the early decades of the century.[1,2] In the preantibiotic era, abscess formation represented effective restriction of the bacterial infection by the host which was treatable by surgical drainage, whereas 6 of 16 patients with cellulitis developed sepsis.[17] Sepsis remains a possible consequence of cellulitis.

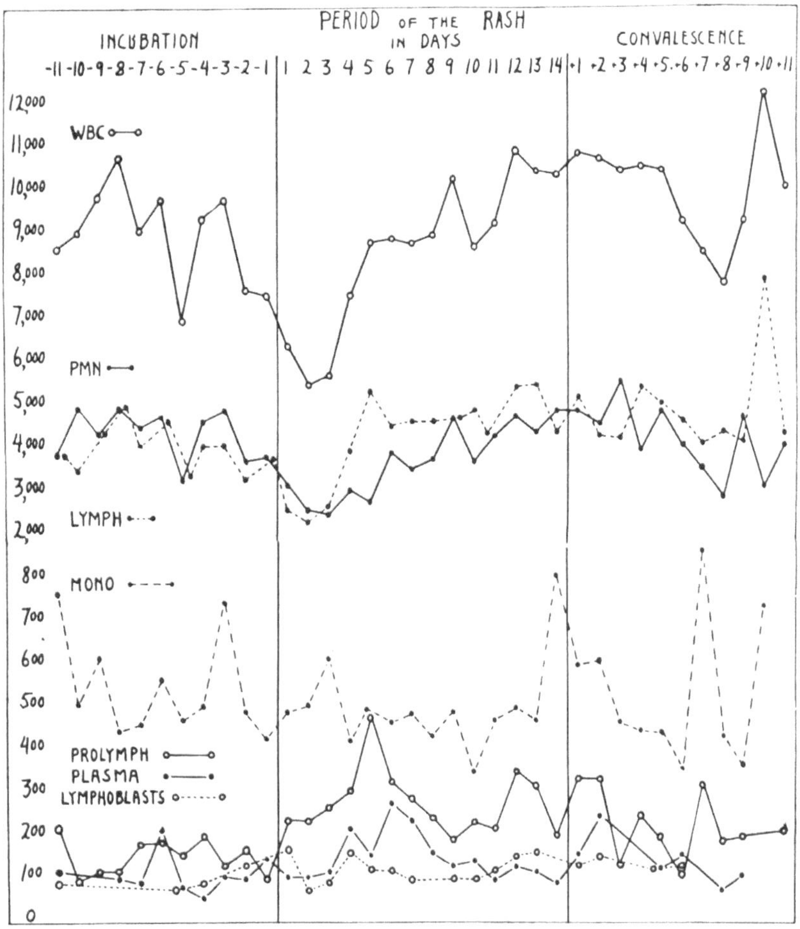

FIGURE 1. Changes in the peripheral blood white cells of otherwise healthy children with varicella during the incubation period, the acute phase, and convalescence. (From Holbrook, A. A., The blood picture in chickenpox, *Arch. Intern. Med.*, 97, 294, 1941. With permission.)

Rarely, cellulitis of the head and neck causes a marked "brawny" induration which may compromise the airway.

Pyogenic cutaneous infection can produce bacterial lymphadenitis of the adjacent lymph nodes which is also usually caused by *S. aureus* or *S. pyogenes.* Lymphadenitis can occur without an obvious cutaneous infection in the region of drainage of the involved node and can appear at various times in the course of varicella, including after the crusting of the cutaneous lesions. Adenitis, like abscess formation, was rarely accompanied by sepsis in the pre-antibiotic era and is usually a well-localized infection of the cervical or inguinal nodes.

Four cases of "bullous varicella" were described by Pye-Smith in 1897 and occasional cases have been reported subsequently.[18-23] Whether this entity represents an unusual pathogenesis of the cutaneous lesion caused directly by VZV or whether it must be attributed to bacterial superinfection is unclear. Wald et al.[24] and Melish[25] isolated *S. aureus,* phage group II, from two children with bullous varicella. This phage type has also been implicated in the staphylococcal scalded skin syndrome.[24,25] In one patient, bullous lesions occurred at cutaneous sites not involved by varicella lesions. Regardless of their etiology, the bullous lesions are superficial and do not represent a serious complication of varicella.

"Varicella gangrenosa" is a cutaneous complication of varicella which is very rare but remains potentially life-threatening.[26-28] Since *S. pyogenes* was isolated from the gangrenous lesion and the blood of a child with varicella gangrenosa,[29] this descriptive term has implied a secondary bacterial pathogenesis. Barenberg and Lewis estimated its frequency at 0.05 to 0.16% of varicella cases in 1927.[30] *S. pyogenes* is still recognized as the most common cause of varicella gangrenosa.[31-33] In its classical presentation, a single varicella lesion, usually on the trunk or an extremity, manifests an unusual appearance early in the course of the varicella eruption. The surrounding skin becomes erythematous, warm, and painful to touch. The erythematous area enlarges rapidly over a few hours, often turning a dusky, dark-red color, and extensive edema develops in the soft tissue around the lesion. Bullae may appear as the infection progresses but are clinically distinct from the lesions of bullous varicella because of the other cutaneous signs. The gross pathology of varicella gangrenosa consists of necrosis of the soft tissue down to the fascia with extension along the fascial planes and myonecrosis in the most severe cases. In current terminology or in advanced cases, this entity is referred to as *necrotizing fasciitis*.[34] As in other forms of necrotizing fasciitis, organisms other than streptococcus or staphylococcus may be isolated, including pseudomonas and anerobic bacteria. The tissue pathology of varicella gangrenosa exhibits thrombosis of the subcutaneous vessels and coagulation necrosis of the subcutaneous tissue. Smith et al. demonstrated Gram-positive cocci within the adventitia of the thrombosed vessels and between adipose cells.[33] Immediate, extensive surgical debridement to the deep fascia is required because the response to antibiotic therapy is poor. *S. pyogenes* was isolated from devitalized tissue debrided 11 days after starting intravenous penicillin.[33] In 1905, Kieffer noted the association of varicella gangrenosa with other disorders such as severe malnutrition, tuberculosis, and congenital syphilis, which may explain why it is rarely encountered now.[28] Clinically, the manifestations of this "infectious gangrene" may be difficult to distinguish from the tissue necrosis which follows arterial thrombosis or the purpuric lesions caused by disseminated intravascular coagulopathy.[35]

Erythema multiforme has been described in two patients with varicella.[36] Its pathogenesis in relation to VZV infection is not known. The cases could represent the coincidental occurrence of the vesicular variant of idiopathic erythema multiforme,[37] although this explanation seems less likely because it also occurs in association with herpes simplex virus infection.

b. Bacterial Sepsis and Focal Bacterial Infections

Children who have cutaneous bacterial infection associated with varicella may develop bacteremia, usually caused by *S. aureus* or *S. pyogenes*. The clinical manifestations of acute bacterial sepsis, including high fever, hypotension, and disseminated intravascular coagulopathy, can occur. Alternatively, the hematogenous spread of the bacterial pathogen may not be detected until signs of a focal infection appear. Staphylococcal or streptococcal pneumonia, arthritis, and osteomyelitis following varicella exemplify this process (Table 2). Obvious bacterial infection of the cutaneous varicella lesions need not have been present in these patients. The secondary bacterial complications of varicella led to death in 11 of 57 patients (19.3%) described by Bullowa and Wishik.[1] The mortality from secondary bacterial sepsis remains significant despite antibiotic therapy.[2] Life-threatening hypotension has also been described in an adult who had the staphylococcal toxic shock syndrome associated with varicella.[38] Bacterial meningitis is associated with varicella so infrequently that its occurrence is probably

Table 2
BACTERIAL INFECTIONS IN PREVIOUSLY HEALTHY CHILDREN HOSPITALIZED WITH VARICELLA

	Reference 2 (n = 81)			Reference 1 (n = 191)	
Infection	Number of patients	%	Infection	Number of patients	%
Cellulitis	11	13.6	Cellulitis/erysipelas	15/7	11.5
Sepsis	2	2.5	Sepsis	14	7.3
Pneumonia	1	1.2	Pneumonia[a]/empyema	21/3	12.6
Adenitis	2	1.2	Abscess/adenitis	24/16	20.9
Arthritis	2	2.5	Osteomyelitis	2	1.0
Impetigo	5	6.2	Impetigo	4	2.1
			Pericarditis/endocarditis	2	1.0
			Meningitis	3	1.6
Total	22	27.4	Total	111	57.0

[a] No distinction between viral and probable bacterial pneumonia.

Column 1 from Fleisher, G., Henry, W., McSorely, M., et al., *Am. J. Dis. Child.*, 135, 896, 1981. Column 2 from Bollowa, J. G. M. and Wishik, S. M., *Am. J. Dis. Child.*, 49, 923, 1935. Copyright 1981 and 1935, respectively, by the American Medical Association. With permission.

coincidental. The reported pathogens are *Streptococcus pneumoniae* and *Neisseria meningitidis* which are the common pathogens in childhood meningitis.[39]

c. Varicella Pneumonia

Varicella pneumonia is most often recognized as a manifestation of disseminated varicella in the immunocompromised host. Although it is very rare in children without underlying disease, the susceptibility of otherwise healthy adults to varicella pneumonia accounts for most of the increased morbidity and mortality caused by primary VZV infection among adults.[16] The cough and dyspnea associated with varicella pneumonia usually begin within 1 to 6 days after the appearance of the initial cutaneous lesions.[40-44] Among 251 cases reviewed by Triebwasser et al.[44] respiratory symptoms developed within 1 to 3 days after the onset of the rash in 70% of the cases.[44] Of adults with varicella pneumonia, 42% had cyanosis; pleuritic chest pain and hemoptysis were common. Although rales, rhonchi, and wheezes may be detected, the physical examination of patients with varicella pneumonia is often unremarkable. The specific findings may include only fever, tachypnea, and intercostal retractions. Fever was above 103°F in one third of the cases compiled by Triebwasser et al.[44] Hypoxemia is often much more severe than is suggested by the physical examination or by the chest X-ray. The chest X-ray may be normal or, more often, demonstrates diffuse bilateral infiltrates with small nodular densities especially in the perihilar area (Figure 2). Pleural effusions may be present. Triebwasser et al. found that the total leukocyte count was above 10,000 cells/mℓ in 38% of the cases; 43% of the patients had more than 70% polymorphonuclear cells.[44] Varicella pneumonia is often transient, resolving completely within 24 to 72 hours. In severe cases, the interstitial pneumonia progresses with equal rapidity to cause respiratory failure. Impaired oxygenation from varicella pneumonia is due to ventilation/perfusion abnormalities and poor alveolar-capillary diffusion. The epithelial cells of the pulmonary alveoli become actively infected with VZV (Figure 3). The damaged cells are sloughed, producing atelectasis secondary to obstruc-

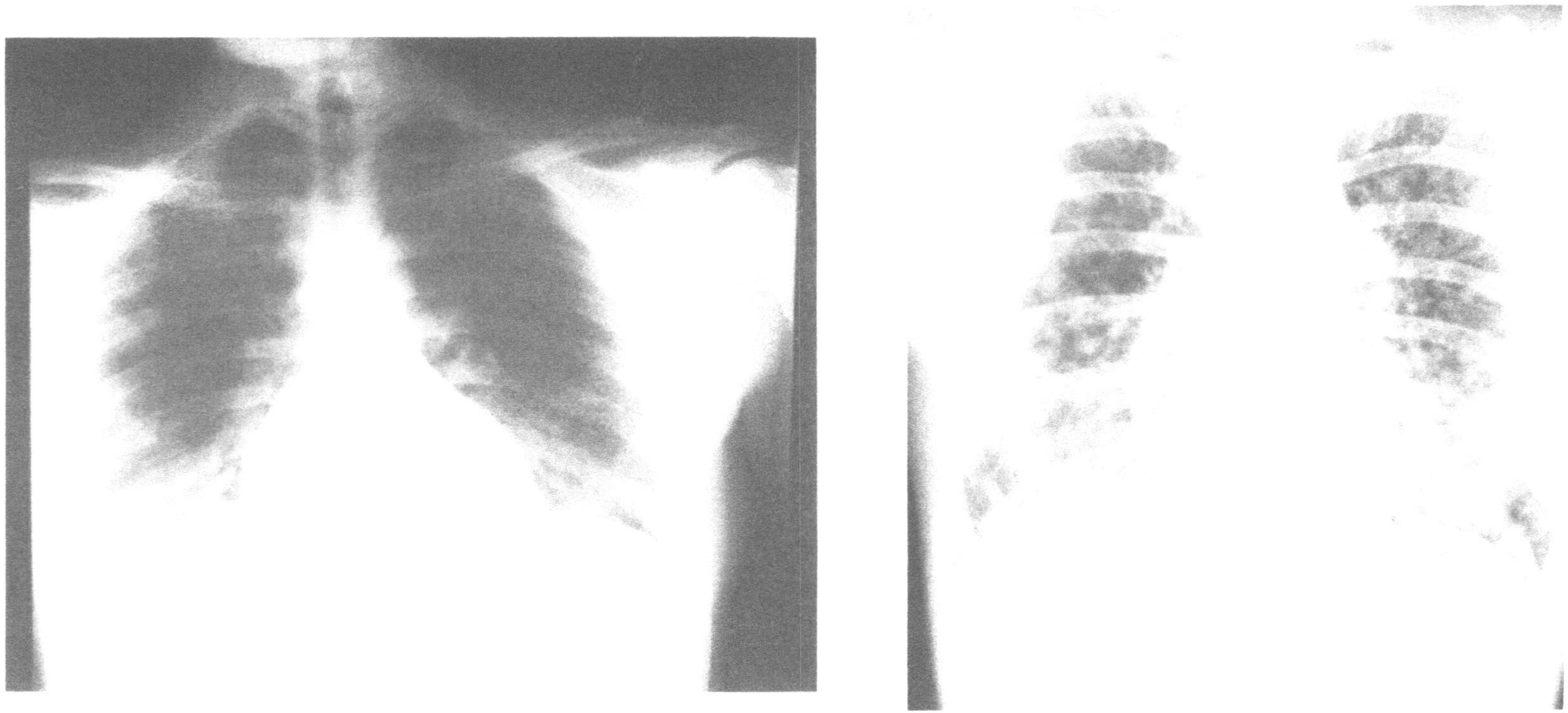

A

B

FIGURE 2. The chest X-ray findings of varicella pneumonia in a healthy adult (A) and in an immunocompromised patient (B).

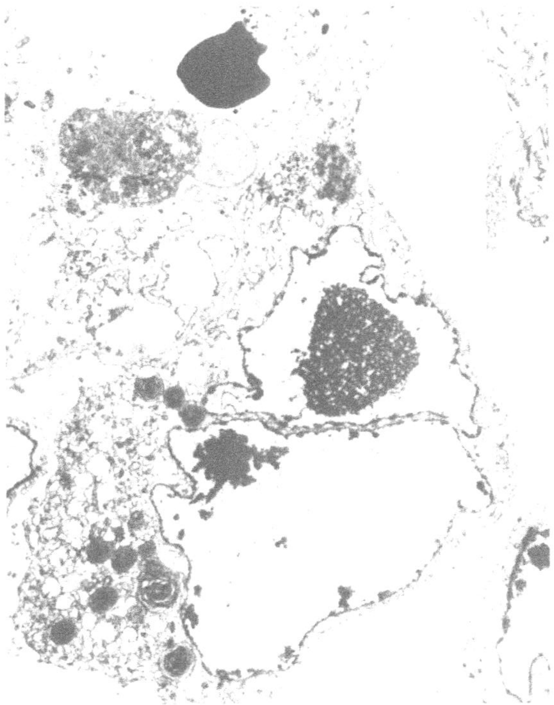

A

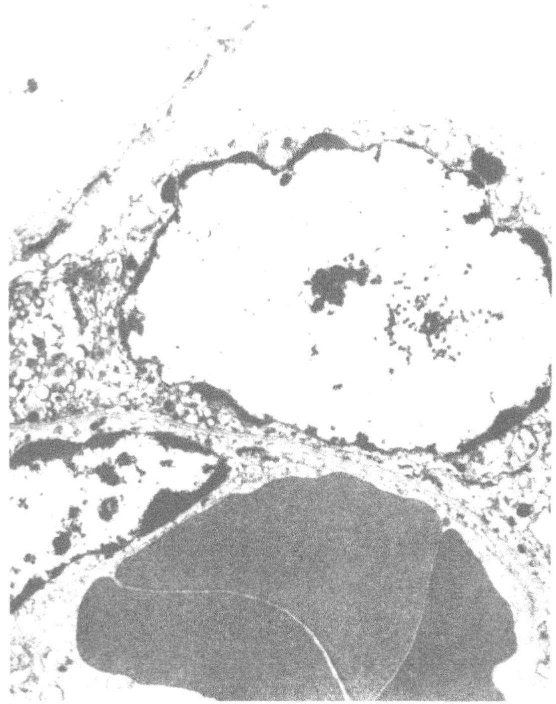

B

FIGURE 3. The electronmicroscopic detection of herpesviral infection of the lungs in a child with leukemia who had fatal varicella: the presence of an inclusion and cytoplasmic virions in a type II pneumocyte (A) and intranuclear virions in a virus-infected alveolar lining cell.

tion of the terminal bronchioles by necrotic debris. Edema and a mononuclear cell infiltration of the alveolar septae occurs along with septal cell proliferation. These pathologic changes may make it impossible to achieve adequate arterial oxygenation despite maximum ventilatory support. Focal necrosis of the lung is apparent at post-mortem. Varicella vesicles have been also observed along the tracheobronchial mucosa and on the pleura in fatal cases.

d. Varicella Hepatitis

Severe varicella in immunocompromised patients and in adults is often associated with clinical symptoms of hepatitis. Abnormal liver function tests have also been reported in otherwise healthy children with varicella.[45,47] In a prospective study of SGOT values among children with varicella, 23% of patients had normal SGOT (<31 IU/ℓ), 49% had mild elevations (32 to 50 IU/ℓ), and 28% had SGOT above 50 IU/ℓ. Other liver functions, including prothrombin time, ammonia, and glucose levels, were normal. Varicella hepatitis was subclinical in most cases, although some children with the highest elevations of SGOT (range 200 to 800 IU/ℓ) had severe vomiting. Vomiting was not associated with the progressive neurologic deterioration which is characteristic of Reye syndrome.[46] The occurrence of vomiting and abnormal SGOT levels in children with varicella creates a clinical question about the difference between varicella hepatitis and stage I Reye syndrome, which is defined as vomiting, lethargy, evidence of liver dysfunction, and a type I EEG. By definition, the diagnosis of Reye syndrome requires an elevation of transaminases by at least threefold, a situation which occurred in fewer than 10% of children with varicella hepatitis. Regardless of their clinical classification, most patients with these findings have rapid resolution of symptoms so that definitive diagnosis by liver biopsy is not warranted.

e. Neurologic Complications of Varicella

Neurologic complications of varicella have been described since the late 19th century.[48-50] With the prevention of measles by immunization, varicella has emerged as the most common childhood exanthem associated with neurologic disease. Of encephalitis cases of known etiology which were reported to the Center for Disease Control between 1972 and 1977, 13% were associated with varicella.[51] Encephalitis was the second most frequent indication for hospitalization of nonimmunocompromised children with varicella.[2] Gibel et al.[52] and Preblud[53] demonstrated a disproportionate incidence of neurologic complications of varicella among patients under 5 years and those older than 20 years. Nevertheless, in the 5- to 9-year age group in which most cases of varicella occur, Guess et al. found that encephalitis and ataxia were the most common indications for hospitalization.[16]

Meningoencephalitis and cerebellar ataxia are the major neurologic complications of varicella. The pathogenesis of these disorders is not well defined, despite a series of clinical descriptions of varicella encephalitis and cerebellar ataxia, beginning with the report of Underwood.[54-61] Varicella encephalitis is one of the complications associated with persistent viremia in the immunocompromised host which suggests a role for direct viral invasion. VZV was isolated from the brain tissue of a leukemic child with fatal varicella.[62] Elevated titers of VZV IgG and VZV IgM antibodies in the cerebrospinal fluid of immunocompromised patients with varicella encephalitis and the detection of herpesviral particles in cerebrospinal fluid by electron microscopy are evidence for local viral replication in varicella encephalitis.[63,64] VZV immunofluorescence staining of cerebrospinal fluid cells from patients with cerebellar ataxia[65] is difficult to interpret because activated mononuclear cells may bind immunoglobulins nonspecifically. Immunofluorescence staining of brain tissue has not demonstrated VZV antigen expression in neural cells.

The histopathology of brain tissue from the few reported cases of fatal varicella encephalitis in otherwise healthy children and adults demonstrates demyelination, axonal damage, perivascular infiltration with mononuclear cells, microglial proliferation, and neuronal degeneration, but the histologic findings of encephalitis are absent in many cases despite clinical signs of encephalitis pre-mortem.[55] Nicolaides observed inclusion-containing cells in brain tissue from one previously healthy adult with varicella encephalitis, but this pathologic marker of herpesviral infection is also usually absent.[66] The histopathologic changes of cerebellar ataxia have not been described because this complication is not fatal.

Some neurologic complications of varicella have been attributed to "postinfectious encephalitis" by analogy with experimental allergic encephalitis and the postinfectious encephalitis of measles and vaccinia. The occasional observation of encephalitis or cerebellar signs which begin after the resolution of cutaneous viral replication has suggested an immune-mediated pathogenesis. This pathogenic mechanism has been invoked to explain the failure to detect VZV in neural tissue. Griffith et al. proposed such an etiology for cases which showed perivenous demyelination, mononuclear cell infiltration, and microglial proliferation, as opposed to the focal necrosis which is considered more characteristic of direct viral infection,[60] but pointed out that only a few cases of varicella-related CNS disease had these histopathologic findings. McKendall and Klawans found six cases which demonstrated the demyelination and microgliosis of an immune-mediated process, but stated that similar histopathologic changes may be seen with direct invasion.[67] Bauman and Bergman reported a probable case of postinfectious encephalitis which began 6 days after the appearance of varicella lesions.[68] The computerized tomographic brain scan showed extensive areas of low absorption in the white matter consistent with demyelination, which finally resolved after 4 months.

Clinically, patients with neurologic complications of varicella may present with symptoms of both cerebral and cerebellar disease. The clinical analysis of varicella-related CNS disorders is also difficult because patients with early stages of Reye syndrome are not easily distinguished from those with encephalitis. Johnson and Milbourne found that about half of the patients with neurologic complications had encephalitis and half had cerebellar disease (Table 3).[57] Patients with both types of CNS disease had vomiting, but fever was more common in those with encephalitis. By clinical definition, cerebellar disease is associated with ataxia, nystagmus, and slurred speech, while meningoencephalitis presents with seizures, diminished level of consciousness, nuchal rigidity, and extensor plantar reflexes. The symptoms are typically more dramatic among patients with encephalitis who have the sudden onset of seizures and altered sensorium, while those with cerebellar disease show a gradual progression of irritability and gait and speech disturbances. Some patients have the clinical symptoms of viral meningitis with fever, headache, and meningismus, but without altered consciousness.

Signs of varicella encephalitis and of cerebellar ataxia have appeared during the incubation period of varicella from 11 to 1 days before the exanthem.[69,70] Neurologic symptoms occurred in 51% of 84 cases between 2 and 6 days after the cutaneous lesions had appeared.[67] Appelbaum reported one case with symptoms which developed 21 days after the varicella rash.[55]

The cerebrospinal fluid in patients with varicella and CNS disease usually shows a mild lymphocytic pleocytosis with fewer than 100 cells/mℓ in most cases and a slight to moderate elevation of protein (<200 mg). The glucose is usually normal. Johnson and Milbourne found higher cell counts and CSF protein concentrations among patients with encephalitis than among those with cerebellar disease.[59] As in other viral infections of the central nervous system, the cerebrospinal fluid is often normal.

Table 3
THE CLINICAL MANIFESTATIONS OF CNS COMPLICATIONS ASSOCIATED WITH VARICELLA (%)

Symptoms	Cerebral type (n = 23)	Cerebellar type (n = 29)
General		
Vomiting	57	76
Fever	48	21
Neurological		
Lethargy	43	45
Convulsions	39	0
Headache	30	35
Delirium	9	0
Ataxia	0	83
Dizziness	0	21
Slurred speech	0	10
Physical findings		
Semi-coma	52	0
Coma	17	0
Lethargy	0	28
Irritability	9	24
Altered pupillary response	22	3
Nystagmus	0	24
Nucal rigidity	39	17
Ataxia/incoordination	0	86
Hyperactive tendon reflexes	35	34
Extensor plantar responses	26	0

Adapted from Reference 59. (Originally published in *Canadian Medical Association Journal*, Vol. 102, April 25, 1970. With permission.)

The seizures associated with varicella encephalitis are generalized, brief, usually few in number, and easily controlled with anticonvulsant drugs. Diffuse but not focal electroencephalographic abnormalities are present.[57] Of 21 children with cerebellar ataxia, 4 also had excessive slow wave activity.[56] Seizures and abnormal electroencephalograms in children with varicella may also be related to fever rather than encephalitis.[71]

The resolution of encephalitic symptoms is usually rapid, often occurring within 24 to 48 hours, while ataxia may persist for days and sometimes weeks. The mortality rate from CNS complications of varicella among otherwise healthy children and adults is difficult to determine because some series included immunocompromised patients who also had varicella pneumonia[55] or patients with Reye syndrome.[59] The range is from 5 to 18%.[55,56,58,59] When cases of encephalitis were distinguished from cerebellar ataxia, the mortality rates were 35% and 0%. Only a few of these patients have been evaluated for long-term sequelae. Appelbaum found abnormalities in 3 of 24 patients who had no neurologic findings at discharge and in 7 of 9 patients who had deficits immediately after the acute illness (Table 4).[55] Recurrent seizures are rare, but persistent electroencephalographic abnormalities were noted in 44% of patients evaluated 1 year after varicella encephalitis.[57] However, some of these children had changes which are now considered normal variations, e.g., 14 and 6 positive spikes per second. One child who had cerebellar ataxia developed sudden deafness, but hearing loss is a very rare consequence of varicella.[72]

Table 4
SEQUELAE AFTER VARICELLA-RELATED CNS DISEASE

Symptoms	Cerebral type	Cerebellar type
No abnormality	23	70.0
Retardation	3	13.0
Personality change	2	8.7
Hemiparesis	2	8.7
Paresis of one extremity	1	4.3
Paralysis of both legs	1	4.3
Recurrent seizures	1	4.3
Blindness	1	4.3
Speech disturbance	1	4.3

Adapted from Appelbaum, E., Rachelson, M. H. et al., *Am. J. Med.*, 15, 223, 1953. With permission.

In addition to encephalitis and cerebellar ataxia, a few cases of transverse myelitis have been associated with varicella.[73] Evidence of spinal involvement including paraplegia, loss of sphincter control, and sensory deficits occur in some patients with encephalitic symptoms. McCarthy and Amer reviewed the eight cases of "pure" transverse myelitis described before 1978 and an additional case.[74] Paraplegia and sensory loss occurred 5 to 16 days after the onset of varicella lesions. One patient had residual bladder dysfunction but the symptoms in the other cases resolved completely. Optic neuritis has been described as a complication of varicella alone or along with acute transverse myelitis.[75,76] Optic neuritis with transverse myelitis is called neuromyelitis optica or Devic disease. The symptoms are blindness and paraplegia. Its histopathology when associated with other viral diseases consists of demyelinating lesions of the white matter of the optic nerves and the spinal cord. One of the two patients with this complication of varicella recovered completely, but the other had persistent paraplegia.[76] Facial palsy followed varicella in one reported case.[77] Only a few cases of Guillain-Barré syndrome have been associated with varicella.[78,79] These patients are distinguished from those with acute transverse myelitis by the clinical findings of polyneuritis, including an ascending pattern of lower motor nerve paralysis accompanied by paresthesias. Pain sensation and sphincter control are intact, in contrast to transverse myelitis where a clearly defined level of sensory and motor loss can be demonstrated. The cerebrospinal fluid in Guillain-Barre syndrome typically shows elevated protein without pleocytosis, whereas mild pleocytosis occurs with acute transverse myelitis. Gradual recovery occurred in all of the cases of varicella-associated polyneuritis summarized by McKendall and Klawans.[67]

f. Reye Syndrome

Reye syndrome is a clinicopathologic diagnosis based on findings of acute encephalopathy, elevated serum transaminases and ammonia, and fatty degeneration of the liver. Early cases of the syndrome were related to varicella.[80] In periods when influenza A and B infections are not prevalent, most cases of Reye syndrome are associated with varicella (Figure 4).[81] Hurwitz and Goodman estimated that Reye syndrome occurred in 2.5 per 10,000 cases of varicella.[82] These patients present with vomiting, signs of increased intracranial pressure, and progressive neurologic deterioration; cerebral edema without inflammatory changes is characteristic. The symptoms begin from 3 to 5 days after the appearance of the varicella exanthem[83] and have not been reported

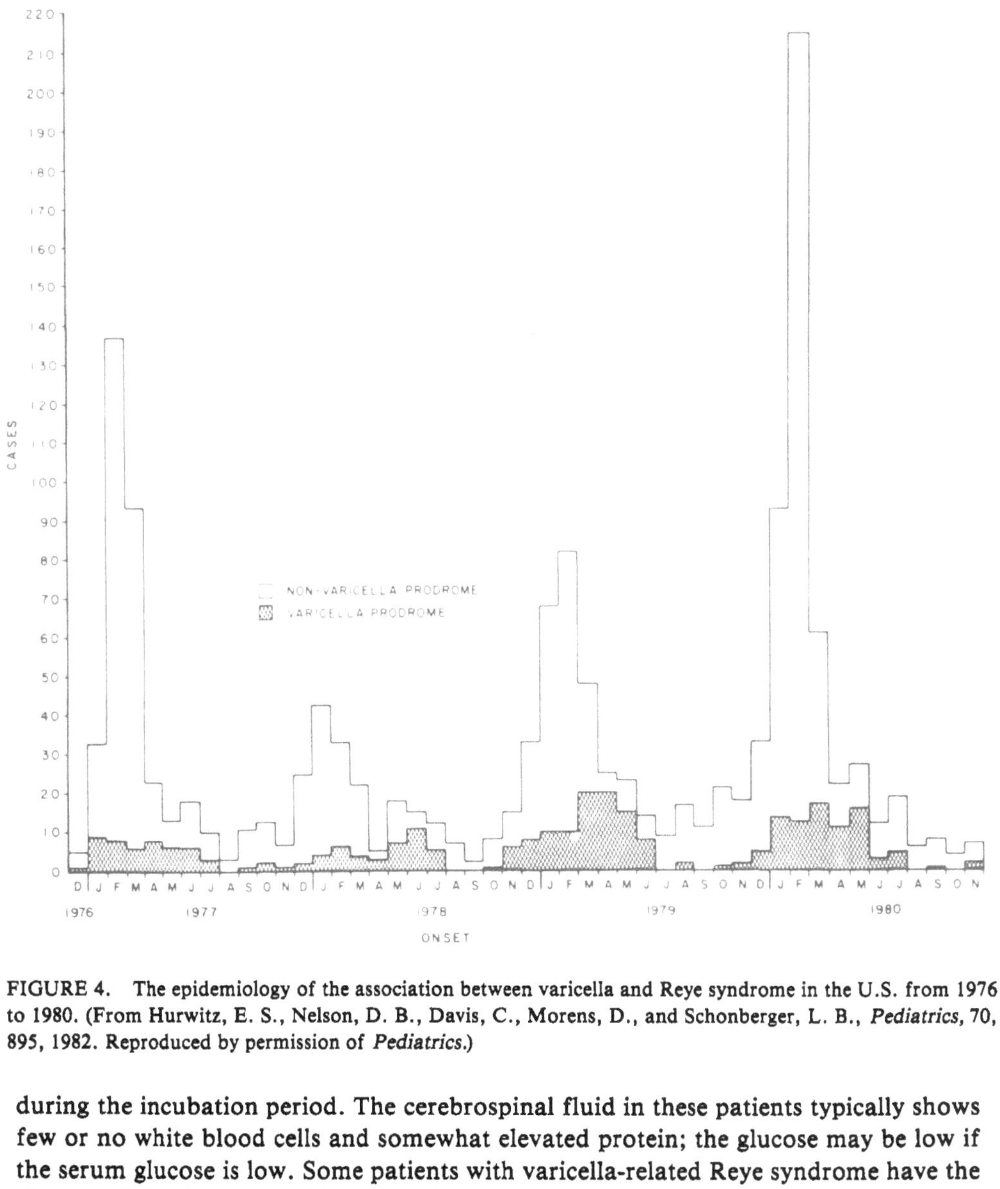

FIGURE 4. The epidemiology of the association between varicella and Reye syndrome in the U.S. from 1976 to 1980. (From Hurwitz, E. S., Nelson, D. B., Davis, C., Morens, D., and Schonberger, L. B., *Pediatrics*, 70, 895, 1982. Reproduced by permission of *Pediatrics*.)

during the incubation period. The cerebrospinal fluid in these patients typically shows few or no white blood cells and somewhat elevated protein; the glucose may be low if the serum glucose is low. Some patients with varicella-related Reye syndrome have the characteristic alteration of hepatic cell ultrastructure but do not progress to encephalopathic symptoms.[84] Although the pathogenesis of Reye syndrome remains uncertain, the epidemologic evidence substantiates a role for salicylates as one important factor.[85,86] Administration of aspirin to children with varicella is contraindicated for this reason. Reye syndrome associated with varicella is the major cause of fatal varicella among otherwise healthy children.[2]

g. Hemorrhage Complications

Bleeding as a complication of varicella in the otherwise healthy child is usually a consequence of thrombocytopenia. Thrombocytopenia during the acute phase of the illness has been called "infectious" thrombocytopenia, while its onset during a period of weeks after varicella is considered a "postinfectious" complication of varicella. Stroh described hemorrhagic varicella in approximately 2% of cases,[87] but Mitchell and Fletcher[12] and Bullowa and Wishik[1] do not mention this complication among 3309 patients with varicella. Patients who present with "hemorrhagic chickenpox", i.e., with bleeding complications in the first few days after the appearance of the rash, have toxic changes of the megakaryocytes in the bone marrow with vacuolization and degeneration of the nucleus.[88] The megakaryocytes are usually present in normal or increased numbers but may be decreased. One case of thrombocytopenia during the incubation period of varicella showed similar megakaryocyte abnormalities.[89] Pathogenesis by direct viral infection of the megakaryocytes is suggested by the detection of viral particles which have the morphology of herpesviruses within megakaryocytes from patients with varicella and purpura.[90] Platelet survival may also be reduced because of platelet destruction which is mediated by antiplatelet IgG or IgM antibodies,[91] vasculitis with endothelial cell damage, or transient hypersplenism.

Yeager and Zinkham summarized 18 cases of thrombocytopenia associated with acute varicella.[92] Fourteen children developed symptoms within 3 days after the appearance of varicella vesicles and all patients had hemorrhagic symptoms within 1 week. Thrombocytopenia was suggested by the presence of petechiae and purpura and by hemorrhage into the varicella vesicles. Most patients also had hematuria. Melanoma may occur as well as more significant gastrointestinal bleeding. Epistaxis is another common clinical symptom in these patients. The duration of bleeding manifestations is usually brief and self-limited, but the platelet count may remain low for a period of days to weeks after the resolution of symptoms. Among the patients described by Yeager and Zinkham, the platelet count was below 150,000 in four patients who were followed for at least 8 and up to 20 days.[92] Transfusion to replace blood loss may be necessary in some patients. Platelet transfusions are of little value because of the shortened platelet survival time. Steroid therapy is not indicated and may exacerbate the course of varicella. Fatal hemorrhagic complications such as cerebral hemorrhage occur rarely.[93]

Postinfectious thrombocytopenia occurs from 1 to 2 weeks or more after varicella. Charkes reviewed 17 cases of this syndrome and found that these patients characteristically had symptomatic bleeding for an average of 5 weeks, and that the thrombocytopenia persisted as long as 4 months.[94] Patients with this syndrome usually have an increase in bone marrow megakaryocytes but a markedly shortened platelet survival time. In some cases, platelet production is also arrested. Antiplatelet antibodies have been demonstrated in a few patients with the late onset of thrombocytopenia following varicella.[95,96]

Purpura fulminans is a very rare but life-threatening complication of varicella.[97-100] In this disorder, the clinical course of varicella is typical until a rapidly enlarging ecchymotic area appears, usually on the leg or trunk. The pathology of this lesion is arterial thrombosis with hemorrhagic gangrene. The immediate consequence of these events is shock which may be irreversible. Survival of the patient requires intensive supportive care and often amputation of the limb.

h. Renal Complications

Clinical evidence of renal involvement associated with varicella is rare.[101] Glomerulonephritis has been reported in patients with and without recent group A streptococcal infection,[101-105] but a poststreptococcal etiology can be difficult to exclude. Glomerulonephritis was noted at autopsy in several cases of fatal adult varicella.[106-108] The clinical symptoms of varicella nephritis include hematuria and diffuse edema with or without hypertension. The symptoms appear within 3 weeks after the varicella exanthem in most cases. One hospitalized child was noted to have microscopic hematuria 4 days before the rash.[103] Laboratory abnormalities include proteinuria, microscopic hematuria, elevated BUN and creatinine, and a decreased creatinine clearance. The histopathology of varicella nephritis resembles poststreptococcal glomerulonephritis even in patients without serologic evidence of recent streptococcal infection. The renal biopsy typically shows hypercellularity of the glomeruli, crescent formation, epithelial proliferation, focal necrosis, and an interstitial inflammatory infiltrate.[108] Some patients progress to anuria which requires dialysis for weeks or months. Singhal et al. reported a case in which the symptoms of renal failure resolved within 2 months but the creatinine clearance remained 45 mℓ/min/m^2.[105] The mortality rate of varicella accompanied by acute glomerulonephritis is high in reported cases but may be accounted for by its occurrence in adult patients; its association with other visceral involvement, e.g., pneumonia and encephalitis; and the availability of pathologic specimens from fatal cases.

Hematuria caused by varicella lesions in the bladder has been described.[109] A few cases of transient nephrotic syndrome have been reported in children with varicella.[110] Varicella may also exacerbate the underlying disease in children who have idiopathic nephrotic syndrome. The hemolytic uremia syndrome has occurred following varicella.[111]

i. Varicella Arthritis

Arthritis is an infrequent complication of varicella.[112] VZV was isolated from the joint fluid of a child with monoarticular arthritis.[113] This case and varicella-associated arthritis cases in which the joint fluid inflammatory response was lymphocytic and no bacterial pathogen was isolated confirm that VZV alone can cause arthritis.[114-118] Varicella arthritis occurs in otherwise healthy children who have no evidence of visceral dissemination. Bacterial infection of the joint also occurs in patients with varicella.[119,120] Pyogenic arthritis associated with varicella is usually caused by *S. aureus* or *S. pyogenes*. Both primary varicella arthritis and bacterial arthritis as a secondary complication of varicella are typically monoarticular and usually involve the large joints. Patients with varicella arthritis develop symptoms within a few days after the appearance of the rash. The joint is swollen, slightly or markedly painful with movement or weight bearing, and there may be associated erythema and warmth of the skin. Occasionally, the joint symptoms are polyarticular.[114] The cell counts in joint fluid range from 2500 to 25,000 cells/mℓ, with more than 85% lymphocytes; these patients may also have an elevated erythrocyte sedimentation rate (range 7 to 107 mm/m). Because of the similarity of these clinical findings to those of pyogenic arthritis, varicella arthritis must be diagnosed by aspirating the joint effusion. If the inflammatory response is predominantly lymphocytic and the Gram stain is negative, a presumptive diagnosis of varicella arthritis can be made. Antibiotic therapy is not indicated in such cases unless bacterial cultures of the joint fluid yield a pathogen. Varicella arthritis resolves spontaneously within 3 to 5 days and has not been associated with residual joint disease.

j. Varicella Orchitis

A few cases of orchitis associated with varicella have been reported.[121,123] Symptoms occurred 1 to 5 days after the appearance of the varicella rash. One of two postadolescent males had evidence of visceral dissemination with varicella pneumonia and developed testicular atrophy.[122]

k. Varicella Myocarditis

Myocarditis is a very rare complication of varicella. Hackel reported seven cases of varicella myocarditis which were noted at autopsy.[124] The characteristic finding was focal necrosis of the myocardium with a perivascular inflammatory infiltrate. Myocarditis was not the clinical cause of death in these patients, and these focal myocardial lesions were considered incidental. Most of the patients had other evidence of visceral dissemination, including pneumonia, hepatitis, hemorrhage, and nephritis. Moore et al. described two children who developed myocarditis and clinical symptoms of congestive heart failure associated with varicella.[125] Cardiac symptoms occurred 1 and 3 days after the appearance of the cutaneous lesions. Both patients recovered after a short course of digitalis and diuretics. Noren et al. identified unsuspected interstitial myocarditis in 11 of 17 cases of fatal varicella.[126] Four of six children with varicella and Reye syndrome had clinical signs of acute congestive cardiomyopathy.[126] One adult and one child with varicella pericarditis have also been described.[127,128]

l. Ocular Complications

Serious ocular complications of varicella are rare despite the frequent occurrence of vesicular lesions on the eyelids and conjuctivae. These lesions alone may cause tearing, injection, photophobia and erythema and edema of the eyelids. Such lesions were noted in 4 of 125 patients (3.2%) examined by Griffin and Searle.[129] Occasionally, children with varicella develop unilateral anterior uveitis[130-134] or corneal lesions.[135] Uveitis may be associated with internal ophthalmoplegia and postinflammatory mydriasis. Jordan et al. diagnosed anterior uveitis in 6 of 24 patients referred for ophthalmologic complications of varicella and corneal involvement in 2 of 24 patients.[136] Corneal lesions usually resolve spontaneously without sequelae, although formation of dendritic lesions and corneal scars can occur.[137] One case of optic disc pigmentation with a persistent field defect has been described.[138] Optic neuritis and ocular motor palsies produce disturbances of vision, but can be classified as neurologic complications of varicella.[139,140] Whether varicella uveitis and corneal disease are caused by direct viral infection of ocular tissue or nerves or are immunologically mediated, as has been suggested for some neurologic complications of varicella, is not known.

m. Reinfection with VZV

Physicians occasionally observe apparent second episodes of varicella in otherwise healthy children, but their attribution to reinfection with VZV is difficult because vesicular rashes have other etiologies. Talbot et al. reported an adult with preexisting antibodies to VZV who developed varicella 15 days after exposure to VZV.[141] Such serologic evidence of reinfection with VZV is likely to provide the only documentation for this phenomenon in an otherwise healthy subject with naturally induced immunity to VZV since the DNA from VZV isolates will not be available for restriction endonuclease analysis.

4. Varicella in the Immunocompromised Patient

a. Immunosuppression Associated with Malignancy, Organ Transplantation, and Other Diseases

The advent of immunosuppressive chemotherapy and radiation for the treatment of acute leukemia and other malignancies in childhood was soon followed by reports of fatal varicella in these patients.[142-145] VZV was isolated from tissue in two early cases.[146] In a comprehensive review of varicella in children with lymphoproliferative malignancy and solid tumors, Feldman et al. found that 32% of these patients developed visceral dissemination of VZV, varicella pneumonia occurred in 20% of cases, and the mortality rate was 7%.[147] Children at greatest risk were those who had acute lymphoblastoid leukemia or lymphoma. The likelihood of progression and severe illness was increased if chemotherapy was given during the incubation period, especially within 5 days before the appearance of the varicella exanthem and if the patient had

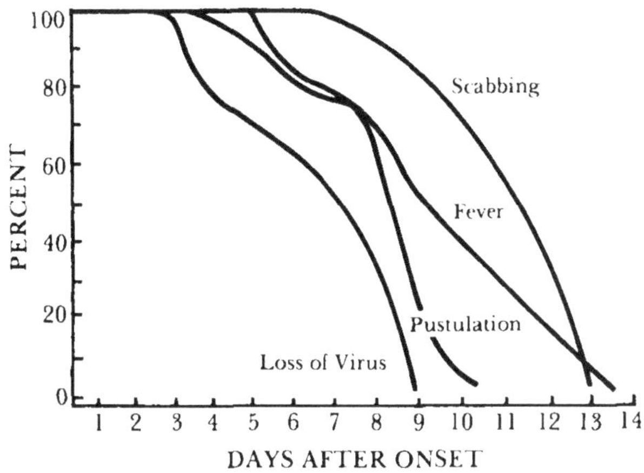

FIGURE 5. The clinical course of varicella among immunocompromised children. (From Whitley, R. J., in *Advances in Therapy against Herpes Virus Infections in Immunocompromised Hosts*, Balfour, H. H., Ed., Park Row Publications, New York, 1985, 25. With permission.)

an absolute lymphopenia of less than 500 cells. Half of the placebo recipients in the early trials of antiviral therapy for varicella in high-risk children had life-threatening primary VZV infection; and 17% of these patients died.[148]

The clinical differences between varicella in immunocompromised children and varicella in otherwise healthy children include a prolonged period of new lesion formation and visceral dissemination manifesting as varicella pneumonia, hepatitis, encephalitis, and disseminated intravascular coagulopathy. The pathogenesis of the visceral complications and the histopathologic changes resemble their description in otherwise healthy children and adults, but the clinical progression of these complications is often accelerated in immunocompromised children. Immunocompromised children fail to restrict viral replication at the cutaneous lesion sites and, most important, are susceptible to viremia which can be detected during the incubation period and persists during the acute phase among children who develop progressive varicella.[149-152]

Varicella vesicles in these patients are usually more numerous and larger than the typical varicella lesions in healthy children. More than half of the immunocompromised children with varicella had new lesion formation for at least 7 days, and the average time to crusting of lesions was 14 days (Figure 5).[148] The diffuse interstitial nodular pneumonitis of varicella pneumonia develops within 3 to 7 days after the onset of skin lesions and progresses rapidly to respiratory failure (Figures 2 and 3). All of the deaths from varicella in the series of Feldman et al. occurred within 3 days following the diagnosis of varicella pneumonia.[147] Varicella hepatitis is accompanied by jaundice, elevated serum transaminases, and coagulopathy caused by decreased hepatic production of clotting factors. Thrombocytopenia is common and occurs in association with disseminated intravascular coagulopathy. Hemorrhage into the cutaneous lesions is a clinical sign of severe varicella in the immunocompromised child. Severe abdominal or back pain is another marker of potentially life-threatening varicella although its pathogenesis is uncertain.[153] Varicella encephalitis is diagnosed by symptoms of lethargy, confusion, and seizures. While some patients develop progressive encephalitis with coma, varicella encephalitis is rarely the immediate cause of death among immunocompromised patients. The syndrome of inappropriate antidiuretic hormone secretion may accompany disseminated varicella with or without clinical encephalitis.[154] Myocarditis

Table 5
VZV INFECTION AFTER BONE MARROW TRANSPLANTATION

	Varicella	Herpes zoster
No. of patients	36	195
Pretransplant history of varicella	39%	41%
Interval to onset after transplant	Av 4.5 mo	Av 5.0 mo
Dissemination		
Cutaneous only	—	23%
Visceral	44%	13.3% $p < .05$
Mortality	27.8%	6.7% $p < .05$

From Locksley, R. M., Flournoy, N., Sullivan, K. M., and Myers, J. D., *J. Infect. Dis.*, 152, 1172, 1985. Copyright by University of Chicago Press, 1986. With permission.

and severe hypertension, possibly related to nephritis, are less frequent complications. Disseminated VZV infection in children with cancer has also produced pancreatitis, necrotizing splenitis, esophagitis, and enterocolitis.[155] The leukopenia associated with chemotherapy and radiation also makes these children highly susceptible to secondary bacterial infection, usually caused by *S. aureus* or *S. pyogenes.*

Children who acquire varicella following renal transplantation are also at risk for progressive VZV infection because of the immunosuppression used to maintain the renal allograft, e.g., azothioprine, prednisone, antilymphocyte globulin, and radiation. Feldhoff et al. observed one fatal infection (5%) and severe varicella in 8 of 19 children (42%) who had renal transplants.[156] Thrombocytopenia and hepatitis were the major clinical complications; varicella pneumonia did not occur. Continuation of azothioprine therapy after the appearance of the varicella exanthem placed these patients at increased risk for serious VZV infection.

Bone marrow transplantation has provided a new approach to the treatment of aplastic anemia, hematologic malignancies, and other childhood cancers. These patients are immunosuppressed with combinations of cyclophosphamide and supralethal total body irradiation, methotrexate, cyclosporine, azothioprine, prednisone, and antithymocyte globulin. Bone marrow transplant recipients constitute a unique population with respect to the definition of primary and recurrent VZV infection. The eradication of previous immunity to VZV means that a generalized VZV exanthem in these patients may result from the reactivation of latent VZV as well as from reexposure to the virus. Locksley et al. defined varicella in these patients as a vesicular eruption which had a generlized onset and no dermatomal localization at any time.[157] By this definition, "varicella" in these patients had a high risk of visceral dissemination (44.4%) and a high mortality rate (27.8%) (Table 5). Of 26 patients with the early onset of varicella (i.e., within 9 months after transplant), 16 had visceral dissemination compared to 2 of 10 patients with late onset. Almost all patients with fatal VZV infection had diffuse interstitial pneumonia. Of special note was the observation that two patients had widespread tissue infection with VZV at autopsy that was not associated with cutaneous lesions, and a third patient had VZV viremia without lesions.

Reinfection with exogenous VZV probably occurs more frequently among immunocompromised patients than among healthy individuals who have had primary VZV infection. This assumption is difficult to confirm because the distinction between generalized reactivation of latent VZV and reinfection with VZV requires a comparison of restriction endonuclease patterns of the DNAs from the VZV isolates from the first and second episodes. Gershon et al. reported three leukemic children who developed clinical varicella despite serologic evidence of past infection with VZV.[158] Feldman et al. found a recurrence rate of 8% but could not relate all of these episodes to reexposure.[147] Two of three renal transplant recipients who developed second episodes of varicella had definite reexposure.[156] We have observed episodes of varicella in two patients with leukemia who had VZV IgG antibody at the time of household exposure to varicella and whose chemotherapy was not interrupted because of known past VZV infection. The first episode of varicella in one of these children had occurred just after reinduction therapy for a relapse of leukemia. Both patients had extensive cutaneous infection during the second episode of varicella with new lesion formation for 7 and 9 days, respectively, and one patient had transient pulmonary infiltrates.

Children who develop varicella after the initiation of antileukemic therapy are also subject to prolonged, atypical infection. Weel-Sipman described two patients with such a syndrome who had not developed VZV antibody after the initial primary VZV infection.[159]

Corticosteroid treatment for indications other than malignancy or transplantation has not been definitely associated with an increase in the mortality of primary VZV infection. In two early series, Falliers and Ellis[160] and Finkel[161] concluded that the underlying disease was the primary risk factor. Children on long-term, low-dose steroid therapy for asthma were not at risk for serious varicella. However, Gershon et al. reported fatal varicella in a child with rheumatic fever who received prednisone at 2 mg/kg/day during the incubation period and in a second child whose high-dose steroid therapy had been tapered off just before the onset of the exanthem.[162] There are occasional reports of severe varicella in children receiving steroid therapy for other diseases, e.g., rheumatoid arthritis, nephrotic syndrome, and ulcerative colitis.[163] Most patients with these diseases are treated with less than 2 mg/kg of prednisone per day. The cutaneous eruption may be more severe and prolonged, but visceral dissemination is unusual (Plate 1)*.

b. Congenital Immunodeficiency

Fatal varicella has been associated with congenital immunodeficiency diseases. The congenital disorders of T-lymphocyte function often occur along with partial or severe deficiencies of B-lymphocyte function.[164] Varicella is severe or fatal in children with the severe combined immunodeficiency disorders and in those with relatively specific defects of T-cell function, including adenosine deaminase deficiency, nucleoside phosphorylase deficiency, and cartilage hair hypoplasia/short-limbed dwarfism. The critical defect in the host response to VZV is probably the T-lymphocyte deficiency because varicella is usually uncomplicated in patients who have deficient antibody production only. The original description of cartilage hair hypoplasia by McKusick noted two deaths from varicella and three very severe cases among 77 affected children.[165] Two cases described by Lux et al. had new lesion formation for 9 and 14 days.[166] Children with cartilage hair hypoplasia and the enzyme deficiencies may not have had serious complications from other viral infections but are at risk for life-threatening varicella nonetheless. Varicella can also be severe in children with Wiskott-Aldrich syndrome[167] and ataxia telangiectasia (Plate 2). *Hook et al. described a family in which two of five children died from varicella, but fatal varicella in a previously healthy child does not suggest that siblings are at risk of severe VZV in the absence of an identifiable immunodeficiency disorder.[168]

* Plates 1 through 4 follow page 90.

Table 6
CLINICAL FINDINGS OF VARICELLA EMBRYOPATHY

Cutaneous scarring
Limb hypoplasia with rudimentary digits
Paralysis with limb atrophy
Cataracts and other eye pathology
Microcephaly
Chorioretinitis
Cortical atrophy
Hydroureter, hydronephrosis
Gastroesophageal reflux

5. Varicella During Pregnancy and the Newborn Period

a. Maternal Morbidity

The risk of maternal varicella during pregnancy is low in the U.S. because most women of childbearing age had had varicella. In 1968, Sever and White found an incidence of 0.7 per 1000 pregnancies among 30,000 pregnancies.[169] Many cases of varicella pneumonia and other manifestations of visceral dissemination have been reported in otherwise healthy pregnant women.[170-178] One of the few reported cases of varicella myocarditis occurred in a pregnant woman. Whether the predisposition to pneumonia and other complications of varicella relates to pregnancy or to adult age cannot be determined from the literature. In a retrospective study of Harris and Rhodes, the maternal mortality rate was 41%.[173] Siegel et al. found one maternal death, which was due to varicella pneumonia, among 150 pregnancies (0.7%) complicated by varicella.[179] Among 44 pregnancies which were evaluated, 22% of women with varicella had associated morbidity including pneumonia (9.3%), fatal pneumonia (2.4%), premature labor (9.3%), and premature delivery (4.7%).[180]

b. Consequences of Maternal Varicella for the Fetus

Maternal varicella, especially varicella pneumonia, is associated with a risk of fetal death even when varicella is not fatal for the mother.[175] Siegel et al. did not note an increase in fetal loss when maternal varicella was uncomplicated,[179] but Enders reported spontaneous abortion in 4 of 35 pregnancies in which the mother had varicella during the first trimester.[181] Spontaneous abortion did not occur in our series of pregnancies complicated by first trimester varicella, but several women chose therapeutic abortion because of the lack of a method to detect fetal involvement; this was also noted by Enders.[180,181] The progression of premature labor to the delivery of two infants at 31 and 35 weeks gestation resulted in some morbidity from prematurity but both infants survived.[180]

The most serious fetal complication of maternal varicella is the congenital varicella syndrome (Table 6). This term describes the consequences of intrauterine infection with VZV which Brunell has called varicella embryopathy.[178] Based upon a series of case reports beginning with that of Laforet and Lynch,[182,183] it is clear that VZV can reach the fetus during maternal varicella. The most striking anomalies of the syndrome which distinguish it from other viral embryopathies are the unusual cutaneous defects, limb atrophy, and evidence of autonomic dysfunction (Plate 3).[184-200] Many of these infants have microcephaly and cortical atrophy secondary to probable intrauterine VZV encephalitis; seizures and mental retardation are common sequelae. Chorioretinitis, microophthalmia, and cataracts also occur. The involvement of the autonomic

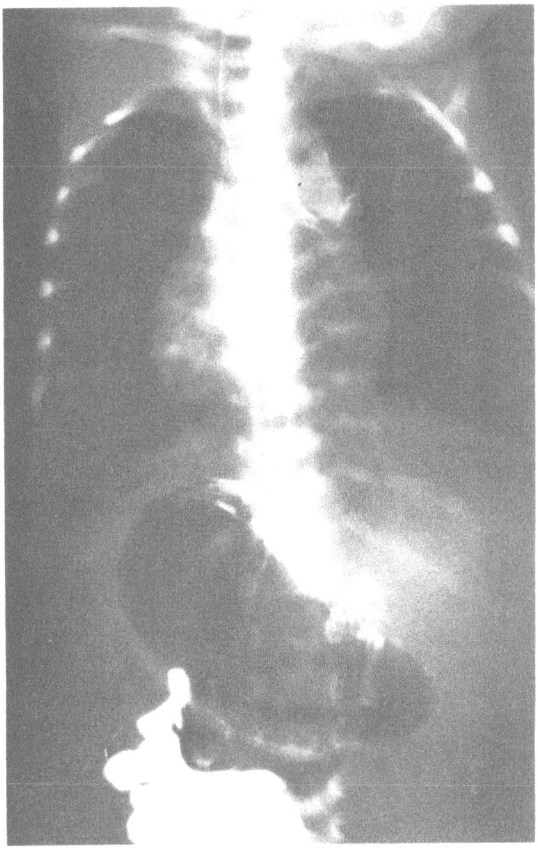

FIGURE 6. Gastroesophageal reflux in a child with congenital varicella syndrome which led to death at 6 months of age from recurrent aspiration pneumonia.

nervous system produces a neurogenic bladder, hydroureter, hydronephrosis, and severe gastroesophageal reflux. Recurrent aspiration pneumonia is a common reason for early postnatal death among infants with the congenital varicella syndrome (Figure 6).

Based upon the limited data available, the highest risk of varicella embryopathy seems to correlate with maternal varicella during the first trimester. The risk of symptomatic intrauterine VZV infection following first trimester maternal varicella is estimated to be 4.9% (Table 7).[180,181,201] One case of varicella embryopathy has been reported after maternal varicella later in pregnancy.[202]

As is true of other viral infections during pregnancy, VZV may infect the fetus without causing obvious anomalies. Infants born to mothers who have had varicella during pregnancy may develop herpes zoster early in life (Plate 4).[203]

By immunologic criteria, 12.5% of asymptomatic infants had had subclinical intrauterine VZV associated with maternal varicella during pregnancy.[180] Although symptomatic fetal infection is rare, herpes zoster in infancy and the presence of VZV-specific immunity in asymptomatic infants indicate that the virus does reach the fetus during second and third trimester maternal varicella with approximately the same frequency as in the first trimester.[180]

c. *Neonatal Varicella*

Neonatal varicella is distinguished clinically from the congenital varicella syndrome,

Table 7
PHYSICAL ANOMALIES IN INFANTS BORN TO MOTHERS WITH VARICELLA DURING PREGNANCY

	First Trimester			Second Trimester			Third Trimester			Total		
Series	No.	Defects	(%)	No.	Defects	(%)	No.	Defects	(%)	No.	Defects	(%)
Siegel, 1966	27	2	(7.4)	32	0	(0)	76	2	(2.6)	135	4	(3.0)
Enders, 1984	23	0	(0)	8	0	(0)[a]	—	—	—	31	0	(0)
Paryani and Arvin, 1986	11	1	(9.0)	11	0	(0)	16	0	(0)	38	1	(3.6)
Total	61	3	(4.9)	51	0	(0)	92	2	(2.2)	204	5	(2.5)

[a] Analyzed as pregnancies complicated by varicella in the 4 to 7th months.

Adapted from References 180, 181, and 201.

even though varicella in the newborn period is often due to the acquisition of VZV from the mother prior to delivery. These infants are born with the typical varicella exanthem in various stages of evolution (vesicles, crusted lesions, hypopigmented skin areas) or develop varicella during the first few weeks of life.[204-210] Those infants whose lesions are present at birth or within the first 5 days of life are not at risk for severe varicella, but those who are born within 4 days after or 2 days before the onset of maternal varicella are at risk for fatal varicella.[200,210] The infants in the latter group are probably born after intrauterine VZV infection but before the transplacental transmission of VZV antibodies (Figure 7).[177,211] Some of these infants may be exposed to the virus during or after delivery from contact with maternal lesions. The attack rate for neonatal varicella is approximately 20%. The estimated risk of severe or fatal neonatal varicella is based upon the cases reported in the literature before VZIG was available for passive antibody prophylaxis and probably reflects a bias towards serious disease. Based upon the summaries of Myers[210] and of DeNicola and Hanshaw,[195] the mortality rate was about 30% in infants whose mothers had symptoms from 4 days before to 2 days after delivery, while none of the infants whose mothers had varicella from 5 to 21 days before delivery had serious disease.

Whether infants in the first few weeks of life are more susceptible to serious varicella from any exposure or whether this risk is specifically related to exposure to maternal varicella just before or after delivery is not known. Preblud found a varicella death rate of 6.9% among 51 infants who had varicella during the first year of life but the number who were infected during early infancy was unknown (Chapter 1, Figure 8).[212] The chance that an infant will lack transplacental antibodies, i.e., was born to a varicella-susceptible mother, and is also coincidentally exposed to varicella by close contact during the first few weeks of life is slight. Neonatal varicella was reported following household exposure of two infants whose mothers were apparently immune to VZV.[213] Secondary cases of varicella have occurred among infants exposed in a newborn nursery, but the attack rate in this circumstance is very low.[214,215] Gustafson et al. described an outbreak of varicella in a newborn intensive care unit in which two infants developed varicella after direct, prolonged contact with personnel who had varicella.[216] Both of these infants had passively acquired VZV antibodies and one received VZIG.

B. Herpes Zoster

1. Introduction

Clinical evidence suggesting a relationship between varicella and herpes zoster[217] was

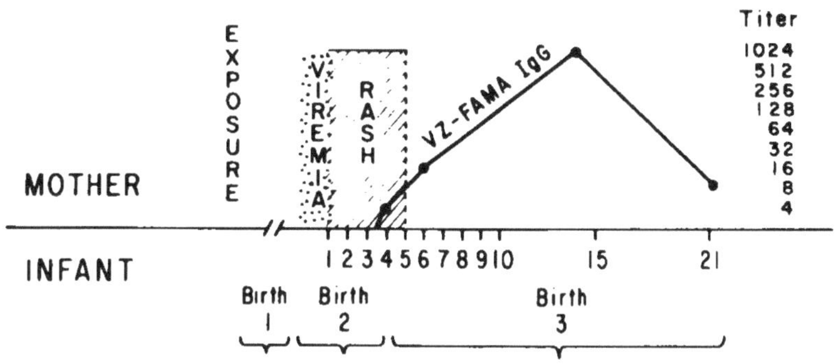

FIGURE 7. Events in VZV infection and varicella-zoster IgG antibody production which affect maternal to fetal transmission of the virus late in gestation. Birth 1: Delivery prior to maternal viremia with no transmission to the fetus; Birth 2: Delivery after transplacental infection of the fetus but before acquisition of passive antibody with a high risk of serious neonatal varicella; Birth 3: Delivery after transplacental infection and after acquisition of passive antibody with a low risk of serious neonatal varicella. (From Gershon, A. A., *Infections of the Fetus and Newborn. Progress in Clinical and Biological Research*, Vol. 3, Krugman, S. and Gershon, A. A., Eds., Alan R. Liss, New York, 1975, 79. With permission.)

confirmed by early in vitro investigations which demonstrated the antigenic similarity of VZV isolates from patients with both types of infection.[218,219] Although it has not been possible to demonstrate the identity of varicella and herpes zoster isolates from a healthy subject because of the long interval between primary and recurrent infection, Straus et al. have used restriction endonuclease analysis of VZV DNA to confirm the hypothesis that herpes zoster resulted from the reactivation of latent VZV in an immunocompromised patient.[220] Cutaneous and blood isolates of VZV-DNA from patients with herpes zoster and viremia also have indistinguishable restriction endonuclease patterns.[221]

The presence of VZV in the dorsal root ganglia of patients with herpes zoster has been documented by the detection of inclusion-containing cells histopathologically, by immunofluorescence, by viral culture, and by *in situ* hybridization.[222-227]

Although herpes zoster is even less likely to be fatal than varicella, this infection is the cause of significant morbidity for otherwise healthy as well as immunocompromised patients. The large epidemiologic studies of McGregor and Hope-Simpson in Britain showed an annual rate of 3.4 to 4.8 cases per 1000 persons.[228,229] McGregor found that 75% of patients with herpes zoster were more than 45 years old, and Hope-Simpson observed that the incidence rose to more than 10 cases per 1000 persons during the ninth decade (Figure 8). The age-adjusted average annual incidence was 1.3 per 1000 persons in a U.S population, with an increase to more than 4.0 cases per 1000 persons by the age of 75.[230] The risk of herpes zoster is very low in chidren, but Guess et al. demonstrated that varicella in the first year of life was a risk factor for childhood zoster,[231] as had been suggested by Latif and Shope (Chapter 1, Figure 12).[232] Maternal varicella during pregnancy also predisposes to herpes zoster in early childhood.[203,233]

2. Usual Clinical Manifestations of Herpes Zoster in Healthy Children and Adults

The unique signs of herpes zoster have made it a well-recognized clinical entity for centuries. Its classic manifestation is the occurrence of a cluster of vesicular lesions with surrounding erythema which appear unilaterally in the dermatomal distribution of one or adjacent dorsal nerves (Chapter 1, Figure 2). The cutaneous lesions are associated with or preceded by the pain of acute neuritis and by pruritis. The localized dermatomal infection is occasionally accompanied by disseminated cutaneous lesions

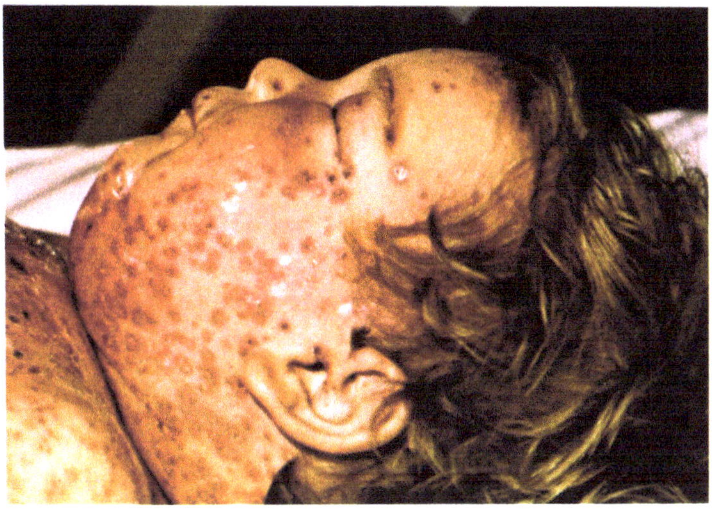

PLATE 1. The extensive cutaneous eruption of varicella in a child with nephrotic syndrome.

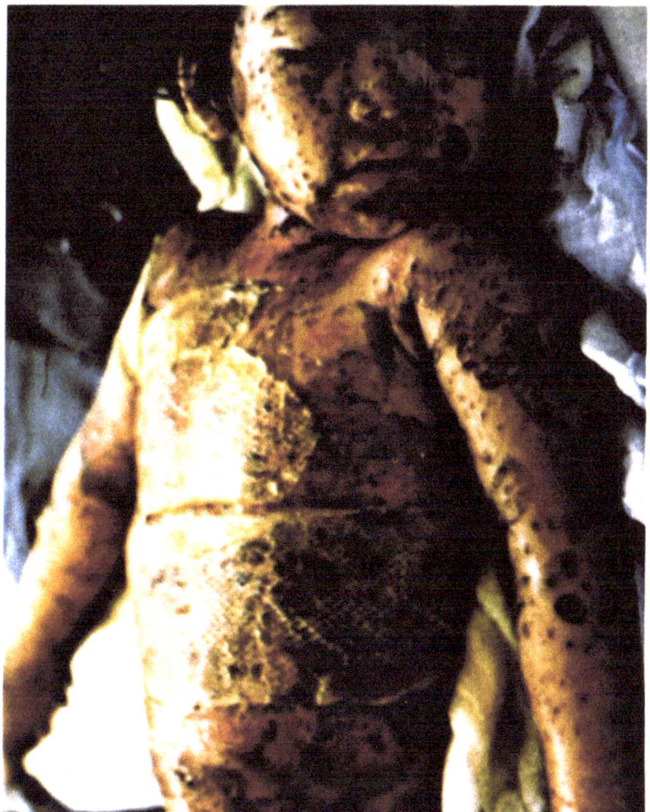

PLATE 2. Severe varicella in a child with ataxia telangiectasia.

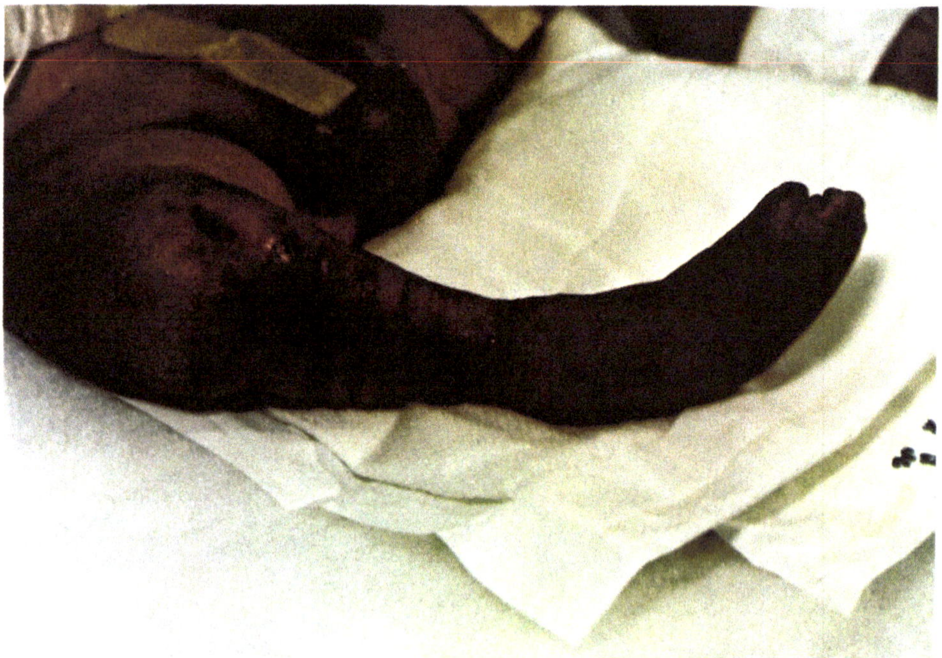

PLATE 3. Limb atrophy and cutaneous scarring in an infant exposed to maternal varicella at 13 weeks gestation.

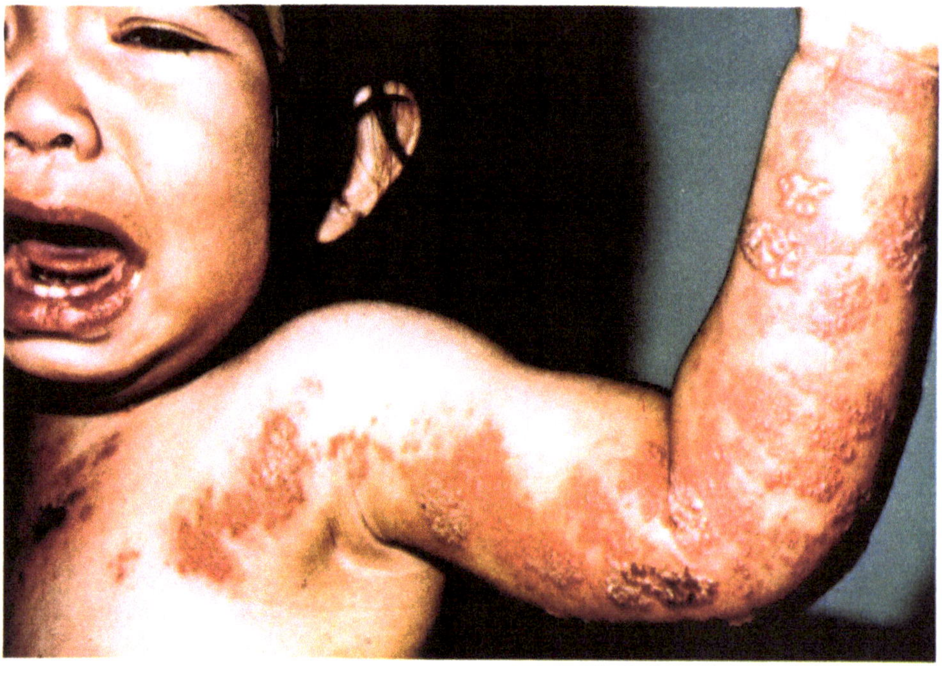

PLATE 4. Herpes zoster in a 15-month old child who was exposed to maternal varicella during the third trimester.

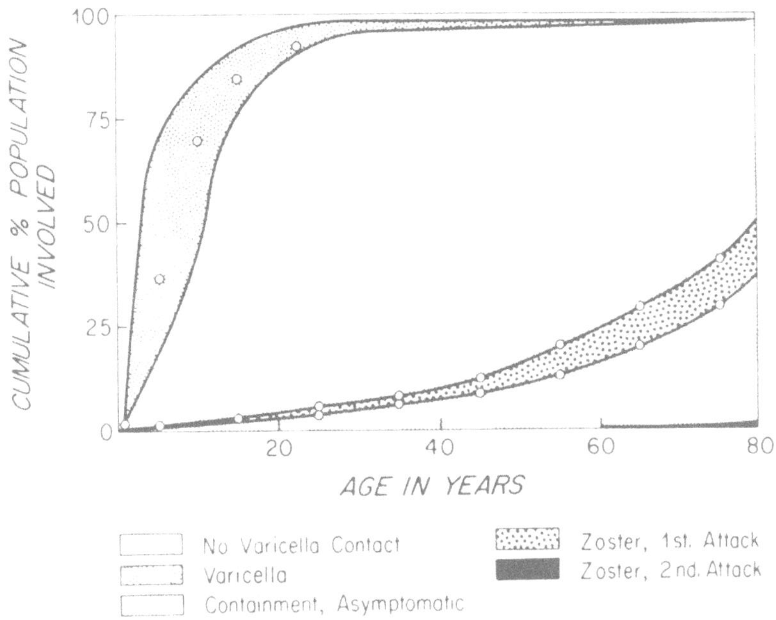

FIGURE 8. The acquisition of varicella and the occurrence of herpes zoster in the general population in relation to age. (From Weller, T. H., reprinted by permission of the *N. Engl. J. Med.*, 309, 1362, 1983.)

in the normal host. About half of the cases involve the thoracic dermatomes and a cranial nerve distribution occurs in 13 to 20% of all patients[228-230] (Chapter 1, Figure 11). The cutaneous lesions often appear as clusters of vesicles in one or several sites anteriorly and posteriorly along the dermatome. The infection in the normal host usually subsides rapidly with new lesion formation in the primary dermatome for 3 to 7 days followed by pustulation and crust formation.[234-236] In some otherwise healthy subjects, the cutaneous process may progress to involve the whole dermatome before it begins to resolve. Complete healing is expected within 2 weeks but may require as long as 4 to 6 weeks. Hypersensitivity of the skin to touch or temperature changes and intermittent pain may persist for a variable period, as long as 6 months among 5 to 10% of patients.[230] When herpes zoster does occur in children, the symptoms are milder with less acute neuritis and the lesions resolve quickly.[232,237,238] The possibility of herpes simplex virus infection must be considered in cases of apparent herpes zoster which involve the lumbosacral dermatomes.

The histopathology of VZV reactivation has been described in a few cases of patients who died coincidentally during an episode of herpes zoster.[222] These patients had inflammatory necrosis of spinal ganglia, motor and sensory root degeneration, neuritis and segmental myelitis, and localized leptomeningitis.

Second episodes of herpes zoster are rare among otherwise healthy individuals. The incidence of 5 to 8% for second episodes does not differ from the age-adjusted frequency of first episodes.[229,230]

3. Usual Clinical Manifestations of Herpes Zoster in Immunocompromised Patients

As malignant diseases have been treated with more intensive immunosuppressive therapy, including combinations of chemotherapy and radiation, herpes zoster has become a common complication.[239-247] Mazur and Dolin found that 15% of patients treated for cancer over a 20 year period developed herpes zoster.[242] Herpes zoster has

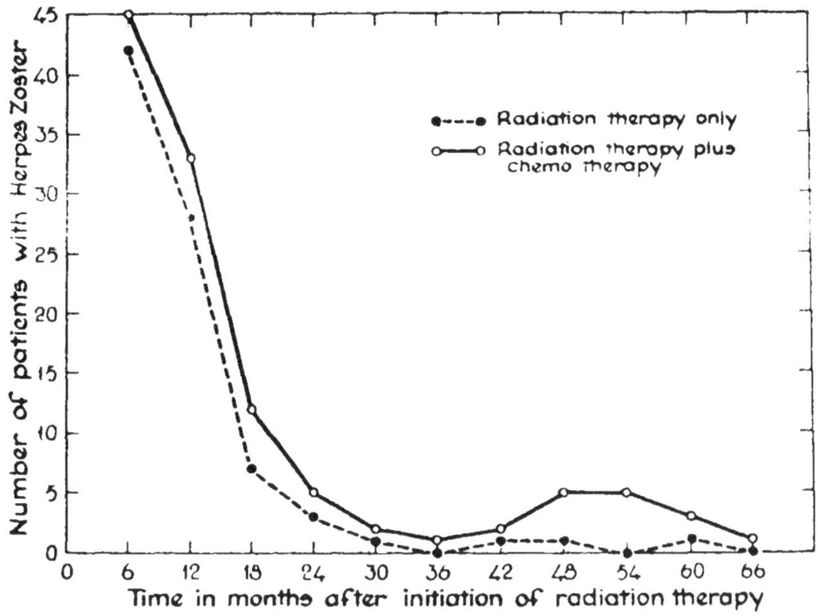

FIGURE 9. The occurrence of herpes zoster in patients with lymphoma in relation to the time in months after initiation of therapy and the use of radiation therapy only or radiation combined with chemotherapy. (From Goffinet, D. R., Glatstein, E. J., and Merigan, T. C., *Ann. Int. Med.*, 76, 235, 1972. With permission.)

been particularly common among patients with Hodgkin's disease and other lymphomas and in patients with oat cell carcinoma of the lung. Herpes zoster was associated with Hodgkin's disease before modern therapy,[248] but the use of lymphoid radiation along with cytotoxic drugs has increased its incidence to more than 25% of patients in some series (Figure 9). Patients with Hodgkin's disease often undergo splenectomy, but a splenectomy alone does not appear to predispose patients without malignancy to VZV reactivation.[249] Children with cancer who have had varicella also develop herpes zoster during the course of treatment of the underlying disease.[250,251] This complication is also especially common in children with Hodgkin's disease.[252-254] Although previously diagnosed cancer is correlated with a higher risk of herpes zoster, herpes zoster in otherwise healthy individuals was not associated with an increased risk of malignancy in the years immediately after its occurrence.[255]

Patients who are immunocompromised because of renal, heart, or bone marrow transplantation are very likely to develop herpes zoster.[157,256-258] Patients who are receiving steroid therapy for chronic diseases such as systemic lupus erythematosis are at somewhat increased risk for herpes zoster.[259] Herpes zoster is also recognized as a complication of the acquired immunodeficiency syndrome.[260]

The clinical manifestations of herpes zoster in immunocompromised patients differ from the findings in healthy individuals in that the local dermatomal infection is more severe and the risk of visceral dissemination is significantly higher. On the basis of data from patients who did not receive antiviral therapy, Whitley estimated that the duration of local symptoms was extended by approximately one week.[236] The average time to cessation of new lesion formation was 8 days and complete crusting did not occur until an average of 18 days. Occasionally, immunocompromised patients develop a chronic reactivation of VZV which can persist for months unless immunosuppressive therapy is reduced or antiviral therapy is given.[261]

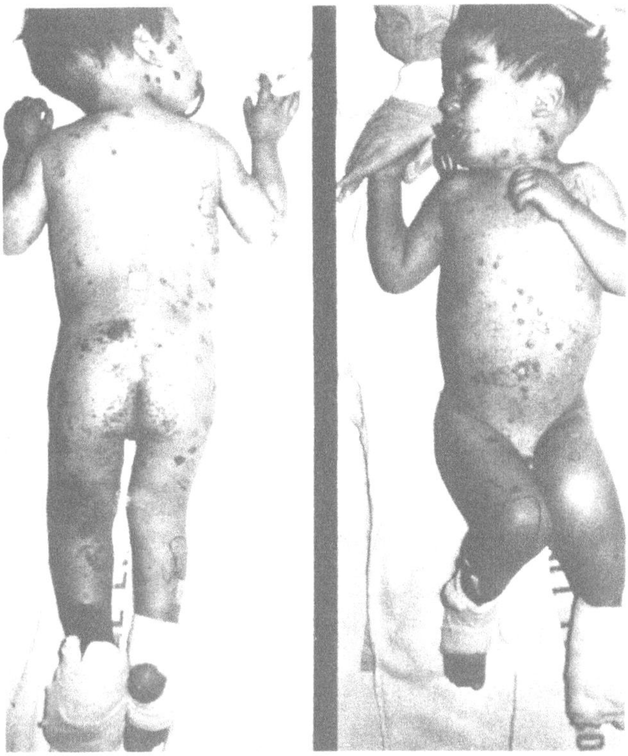

FIGURE 10. Disseminated herpes zoster in a child with malignancy.

The incidence of cutaneous dissemination in immunocompromised patients with herpes zoster, which includes primarily patients who have lymphoproliferative malignancies, ranges from 10 to 40% (Figure 10). Cutaneous lesions outside of the primary dermatome substantiate the presence of VZV viremia and provide an important marker for the risk of life-threatening visceral dissemination.[262] VZV viremia secondary to herpes zoster can produce all of the complications described with primary VZV infection, including pneumonia, hepatitis, encephalitis, and disseminated intravascular coagulopathy.[263] The mortality which accompanies herpes zoster in immunocompromised patients is primarily due to VZV pneumonia. The risk of fatal VZV infection after reactivation of the virus in immunocompromised patients is estimated to be less than 1%, which is significantly lower than that associated with varicella in such patients. Unfortunately, there are no precise criteria to predict the risk of serious infection at the time of onset of localized herpes zoster, except the diagnosis of lymphoproliferative malignancy[236] or recent bone marrow transplantation (Table 5).

4. Neurologic and Ophthalmologic Complications of Herpes Zoster in Normal and Immunocompromised Patients

a. Postherpetic Neuralgia

The most common complication of herpes zoster is postherpetic neuralgia. This complication can result in severe, debilitating, chronic pain. Ragozzino found postherpetic neuralgia ranging in duration from 4 weeks to more than 10 years in 9% of patients with herpes zoster in a general population.[230] Among individuals who had postherpetic neuralgia, symptoms persisted for more than 1 year in 22%. Interestingly, postherpetic neuralgia was much less common in patients with lumbar herpes zoster.

Table 8
CLINICAL FINDINGS IN 39 PATIENTS WITH HERPES ZOSTER ENCEPHALITIS

CNS signs/symptoms	No.	%
Mild confusion	7	17.9
Delirium/hallucinations	18	46.2
Stupor	6	15.4
Meningismus	4	10.3
Photophobia	3	7.7
Headache	5	12.8
Ataxia	4	10.3
Seizure	4	10.3
Cranial nerve palsies	4	10.3
Extracranial nerve palsies	8	20.5

From Jemsek, J., Greenberg, S., and Taber, L., *Medicine*, 62, 81, 1982. Copyright permission by the University of Chicago Press, 1986. With permission.

Its frequency increased with the age of the patient as has also been observed among immunocompromised patients. The average age of patients with postherpetic neuralgia was 67 years, while the remainder of patients were 46 years old on average. Increased susceptibility to postherpetic neuralgia among the elderly was also described by Rogers and Tindall.[264] Postherpetic neuralgia is very unusual in children with herpes zoster, including those with malignancy.

b. CNS Infection

The true incidence of CNS infection during herpes zoster is not known. Gold et al. found a lymphocytic pleocytosis and elevated protein in the cerebrospinal fluid of 40% of patients with herpes zoster who did not have clinical signs of meningoencephalitis.[265] VZV has been demonstrated by electron microscopy[266] and can be isolated from the cerebrospinal fluid of patients with herpes zoster.[265] Local VZV antibody production is also common[63,267,268] Symptomatic meningoencephalitis is rare, even in immunocompromised patients with herpes zoster.[269-272] Ragozzino et al.[230] found an incidence of 0.2% for zoster-associated encephalitis which was comparable to the incidence of 0.5% observed by Burgoon.[234] Of 1210 patients hospitalized for herpes zoster, 2 had encephalitis.[273] Jemsek et al. analyzed 12 cases of encephalitis associated with herpes zoster and 27 cases from the published literature in which patients with recent herpes zoster had CNS symptoms and an abnormal cerebrospinal fluid.[274] The risk factors for herpes zoster encephalitis include an older age, cranial nerve involvement, and immunosuppression. Jemsek identified encephalitis in 9 of 31 immunocompromised patients who had disseminated herpes zoster.[274]

Zoster-related encephalitis causes the symptoms which are expected with viral encephalitis, i.e., altered level of consciousness, headache, photophobia, and meningismus. Fever is mild or absent. Seizures and evidence of increased intracranial pressure are rare. Patients with CNS complications of herpes zoster may also have associated cranial or peripheral nerve paresis (Table 8). The EEG may be abnormal, with diffuse slowing or epileptiform activity. In contrast to herpes simplex virus, focal abnormalities of the EEG are not characteristic. The average time from the occurrence of lesions in the initial dermatome to the onset of symptoms of encephalitis was 9 days.[67] The reported range is up to 6 weeks. McCormick et al. reported a patient whose encephalitic symptoms preceded the herpes zoster rash by 30 days.[62] The mean duration of symptoms was 16 days; prolonged symptoms were more common among patients with associated paresis.[67] Patients with encephalitis secondary to disseminated herpes zoster usually develop symptoms within a few days after cutaneous dissemination is noted.[272] The mortality rate is higher among immunocompromised patients who have herpes zoster encephalitis and VZV pneumonia than it is for immunocompromised patients without other evidence of visceral dissemination. Only one death from herpes zoster encephalitis has been reported in a patient without underlying disease.[275] The pathology of the brain in patients with fatal herpes zoster encephalitis is as variable as that seen in varicella encephalitis. Most patients recover from herpes zoster encephalitis without CNS sequelae. One patient with recurrent herpes zoster encephalitis has been reported.[276] Two patients who had subacute encephalitis and acquired immunodeficiency syndrome had VZV isolated from brain tissue,[277] but progressive panencephalitis is not otherwise described with VZV reactivation.

c. Cerebral Angiitis

Patients with herpes zoster of the cranial nerves are at risk for the complication of cerebral angiitis. This entity consists of vascular inflammation with thrombosis and microinfarcts, and is particularly associated with herpes zoster ophthalmicus. If the thrombosis is extensive, these patients may develop a contralateral hemiparesis.[279,283] The interval between symptoms of herpes zoster and those of hemiparesis averaged 7 weeks and the mortality rate was 20% among cases reviewed by Reshef et al.[282] Cerebral angiography demonstrates segmental constriction of the ipsilateral cerebral arteries and a computerized tomographic brain scan usually shows an infarct in the distribution of the middle cerebral artery.[282,284] The pathology of this complication in its original description as well as in a recent report has been demonstrated in patients with lymphoma, although most cases actually occur in patients with no underlying disease.[285,286] In addition to the characteristic granulomatous inflammatory process with mononuclear cell infiltration of the arterial walls, vascular necrosis, and multiple small infarcts, Linneman and Alvira demonstrated viral particles within smooth muscle cells from cerebral arteries which showed the histopathology of granulomatous angiitis.[286] The CNS symptoms in some cases of cerebral angiitis resemble herpes zoster encephalitis. The observation that herpes zoster encephalitis is more common in patients with involvement of the cranial nerves and that its symptoms are more prolonged in patients who develop associated paresis suggests that some cases of apparent encephalitis are due to cerebral angiitis.

d. Herpes Zoster Ophthalmicus and Optic Neuritis

The reactivation of herpes zoster in the trigeminal ganglion is associated most often with clinical manifestations in the first division of the trigeminal nerve, i.e., the ophthalmic branch.[287] The presence of vesicles on the tip of the nose is a classic clinical sign of the involvement of the nasociliary branch and therefore, of risk for ocular disease. The clinical manifestations of this infection include conjunctivitis, dendritic keratitis, anterior uveitis, iridocyclitis with secondary glaucoma, and panophthalmitis.[288-290] Corneal disease was noted in 47 of 86 patients referred for ophthalmologic evaluation.[290] Blindness following herpes zoster ophthalmicus is very uncommon. Loss of vision associated with herpes zoster is predominantly caused by retrobulbar neuritis and optic atrophy. Optic neuritis may sometimes occur along with herpes zoster of other divisions of the trigeminal nerve.

e. Other Complications of Herpes Zoster Involving the Cranial Nerves

Herpes zoster of the seventh cranial nerve produces facial palsy on the same side as the cutaneous lesions.[291-293] Facial palsy may accompany manifestations of VZV reactivation affecting other cranial nerves, e.g., trigeminal herpes zoster, ophthalmic zoster, ophthalmoplegia as evidence of involvement of the third, fourth, and sixth cranial nerves, and herpes zoster oticus.[294] Herpes zoster oticus[295] is associated with VZV reactivation in the geniculate ganglion of the seventh cranial nerve and of the eighth nerve and produces the Ramsey-Hunt syndrome.[296] Unilateral deafness and vestibular symptoms can result. Idiopathic facial palsy without obvious cutaneous herpes zoster has also been attributed to VZV reactivation based upon serologic evidence of recent primary or recurrent VZV infection.[297,298] This has been considered to be a manifestation of "zoster sine herpetae".[299] Herpes zoster of the second branch of the fifth cranial nerve produces oral lesions, primarily on the palate. Lesions on the tongue indicate seventh nerve involvement and are associated with loss of taste.

f. Motor Paralysis as a Complication of Herpes Zoster

Although the pathology of VZV reactivation primarily involves the dorsal root ganglion and the posterior portion of the spinal cord, the inflammation and necrosis sometimes extend into the anterior horn cells, producing deficits of motor function. Partial paralysis after herpes zoster was first described by Broadbent.[300] Ragozzino et al. reported a motor deficit in 1% of cases, but 5 of the 6 cases were facial palsy.[230] One patient had cervical zoster followed by permanent atrophy of muscles of the hand. Thomas and Howard identified motor paralysis in 11.5% of patients with cervical zoster, 0.3% with thoracic zoster, and 6.8% of those with lumbosacral zoster.[274] As pointed out by McKendall and Klawans, motor deficits may be recognized less frequently with thoracic zoster because intercostal weakness is less obvious than weakness of an extremity.[67] In general, the prognosis for recovery from the motor deficits of herpes zoster is good, but significant residual loss of function may occur in 10 to 15% of patients,[301,302] and can be a debilitating complication of herpes zoster in the elderly.

Neurogenic bladder dysfunction can occur with lumbosacral herpes zoster,[303-305] and rare cases of ileus with intestinal obstruction have been reported.[306-307] One case of cervical herpes zoster associated with diaphragmatic paralysis has been described.[308]

g. Transverse Myelitis and Ascending Paralysis in Herpes Zoster

A few cases of transverse myelitis with paraplegia have been described in patients with herpes zoster.[309-311] Hogan and Krigman isolated VZV from the spinal cord of such a patient and demonstrated herpes particles in the glial cells of the spinal cord by electron microscopy.[312] Extensive necrosis of the affected segments of spinal cord was observed with destruction of gray and white matter, mononuclear cell infiltration, and replacement of neural tissue with fibrin. This process produces an ascending paralysis along with an associated sensory deficit in some patients and can be fatal.[313,314] The mortality rate for this complication was 22 of 41 cases as reviewed by McKendall and Klawans.[67] Recurrent symptoms have also been described.[315] Guillain-Barré syndrome as a cause of ascending paralysis is very rare among patients with herpes zoster.[316,317]

h. Herpes Zoster in Pregnancy

Herpes zoster in pregnancy is not associated with the maternal morbidity of varicella.[180,181,318] Eyal et al. found only 13 reported cases but the author's experience with identification of 14 cases over a few years suggests that herpes zoster in pregnancy is not as unusual as the literature suggests. Most of the cases identified occurred in the third trimester and followed the mild clinical course which is typical of herpes zoster in the young adult.[180] One woman had cutaneous lesions outside the primary dermatome. There are a few case reports of fetal anomalies associated with herpes zoster, but none of the infants have had the typical syndrome of congenital varicella.[319,320] Disseminated cutaneous lesions indicate viremia which could theoretically cause transplacental transmission of VZV, but the fetus of the mother with recurrent VZV can be expected to have preexisting passive antibodies, especially if most maternal reactivation occurs after the first trimester. Maternal herpes zoster in the perinatal period has not been associated with neonatal varicella; the infants of these mothers probably have high titers of VZV antibodies. Herpes simplex virus must be ruled out in maternal "herpes zoster" which involves the lumbosacral dermatomes because of the risks associated with neonatal exposure to herpes simplex virus.

II. THE IMMUNE RESPONSE TO VZV

Viral replication terminates rapidly in the normal host with primary VZV infection. Viremia can be detected during the late incubation period and for 24 to 48 hours after the appearance of cutaneous lesions in some otherwise healthy subjects.[321,322] VZV can be cultured from the cutaneous lesions of healthy children for only 24 to 96 hours.[322,323] The cessation of viral replication is accompanied by an immediate humoral and cell-mediated immune response which is VZV specific and by the activation of the interferon system. In most healthy individuals, VZV immunity provides lasting protection against symptomatic reinfection with VZV and the reactivation of latent VZV.

A. Humoral Immunity to VZV

1. VZV IgG, IgM, and IgA Antibody Production during Primary VZV Infection

IgG, IgM, and IgA antibodies against VZV appear in serum within the first 5 days after the onset of varicella.[323-329] Antibodies of all of the immunoglobulin classes bind to VZV antigens which are expressed on the surfaces of VZV-infected cells, as demonstrated by the fluorescence antibody membrane antigen assay.[324]

VZV IgG antibodies during varicella are predominantly of the immunoglobulin G subclass 3.[330] IgG antibodies to VZV which have complement fixing-activity are present in serum by 3 to 7 days after the appearance of the varicella exanthem.[323] Functionally, in the presence of complement, VZV IgG antibodies can produce lysis of VZV infected cells.[331] VZV IgG antibodies which neutralize cell-free preparations of the virus efficiently are also detected in serum within a few days after the onset of the vesicular rash.[332-334] Viral neutralization by VZV IgG is enhanced significantly by complement.

VZV IgG also mediates antibody-dependent cellular cytotoxicity.[331,335-339] The lysis of VZV infected cells by peripheral blood mononuclear cells and leukocytes occurs in the presence of very low concentrations of VZV IgG antibodies. Kamiya et al. found that 12 of 25 sera diluted ≥1:1280 still had activity in an antibody-dependent cellular cytotoxicity assay, while the highest titer was 1:1024 when measured by fluorescence antibody membrane antigen assay.[336] Inactivation of 50 PFU of VZV was achieved with only 100 nonadherent lymphoid cells.[335] Lymphocytes provide the most efficient lysis of VZV infected cells in the presence of VZV IgG when compared to monocytes and leukocytes (Figure 11).[340] Lymphocytes capable of mediating antibody-dependent cellular cytotoxicity circulate in the peripheral blood of nonimmune subjects,[336] which suggests that this host response mechanism could be activated as soon as low concentrations of VZV IgG antibodies have been produced in the course of primary VZV infection. VZV IgG antibodies do not prevent the cell-to-cell spread of the virus,[341] but virions which are released by antibody-dependent complement or cell-mediated lysis of VZV-infected cells become accessible for neutralization by VZV antibodies which

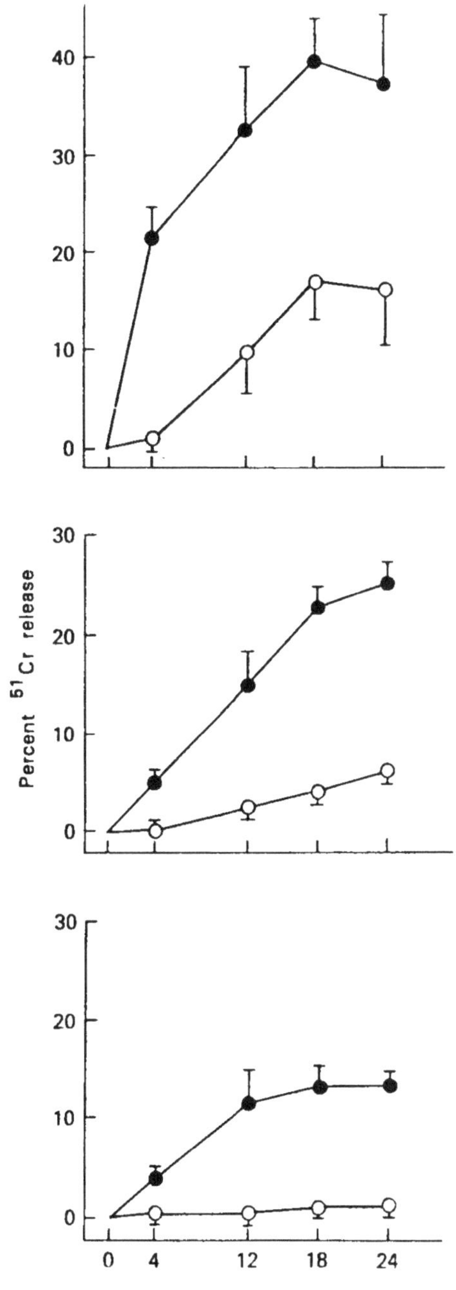

FIGURE 11. Kinetics of antibody-dependent cellular cytotoxicity measured by ^{51}Cr release using VZV-infected fibroblasts as targets and human lymphocytes (upper section), monocytes (middle section), and polymorphonuclear leukocytes (bottom section) a effector cells. Effector and target cells were incubated in serum with (•) and without (O) VZV IgG antibodies. Each point represents the mean ± SEM of these experiments. (From Ihara, T., Ito, M., and Starr, S. E., *Clin. Exp. Immunol.*, 63, 179, 1983. With permission.)

could help to terminate VZV replication. Leukocyte-mediated antibody-dependent cellular cytotoxicity may be important in restricting VZV replication in cutaneous lesions since the appearance of these cells in the base of the lesions correlated with resolution.[342]

The analysis of the IgG response to the major VZV proteins in subjects with primary VZV infection has been hampered by the extreme cell-associated nature of viral replication which interferes with the preparation of purified VZV proteins as reagents. Complement-fixing antibodies have been detected during the convalescent phase of varicella which are directed primarily against viral nucleocapsid antigens.[343] IgG antibodies against VZV "early" antigens, i.e., nonstructural proteins which are synthesized before viral DNA replication, also appear with primary VZV infection.[344,345] By using radioimmunoprecipitation to demonstrate IgG reactivity against VZV-infected cell proteins, Zweerink and Neff[346] found IgG antibodies to an average of 14 VZV-specific polypeptides in sera from subjects with acute varicella. Binding to the major glycoproteins and to a nonglycosylated protein of p155 predominated. Weigle and Grose described the appearance of IgG antibodies to VZV envelope glycoproteins of 66, 118, 98, and 62,000 mol wt in rapid succession after the onset of varicella.[347] Antibodies to the p155 protein appeared within 2 to 4 weeks after the onset of the rash while antibodies to the other nonglycosylated proteins were not present until 4 to 8 weeks later (Figure 12). Palumbo et al. found IgG antibodies to an average of three VZV infected cell proteins using the immune transfer method to test sera obtained within the first week after primary VZV infection.[348] During convalescence, IgG reactivity with an average of seven VZV-infected cell proteins was present. The immune transfer analysis of VZV IgG production during varicella showed prominent binding to a low molecular weight protein of 35,000, which is the approximate molecular weight of the viral thymidine kinase. Kollander et al. found antibody to the VZV thymidine kinase by an enzyme-blocking antibody method in sera from patients with herpes zoster, but not in sera from patients with recent varicella.[349] The immune transfer analysis of sera from subjects with varicella which had equivalent VZV IgG antibody titers by radioimmunoassay did not always show the same pattern of VZV IgG reactivity with VZV-infected cell proteins. Normal subjects with primary VZV infection had uncomplicated infection and recovered without the production of VZV IgG antibodies to all of the major VZV-infected cell proteins.[348]

When the IgG antibody response to VZV proteins was measured in a solid phase radioimmunoassay using antigenic "capture" by murine monoclonal antibodies to VZV, 63% of healthy subjects had detectable IgG to gpI (gp 90/58,000), 59% had IgG to gpIII (gp118) and 60% had IgG to a nonglycosylated p170 within 6 weeks after the onset of primary VZV infection (Figure 13).[350] The functional antiviral activity of IgG antibodies to specific VZV proteins is not certain. Based upon experiments using murine monoclonal antibodies, some IgG antibodies to the gpI and gpIII can mediate the neutralization of cell-free VZV (see Chapter 1). Human monoclonal IgG antibodies which had neutralizing activity against the virus were produced using B-lymphocytes obtained from a healthy donor 6 days after the onset of varicella.[351] One of these antibodies bound to VZV-infected cell proteins of 60,000, 95,000, and 97,000 mol wt by radioimmunoprecipitation. Murine monoclonal antibodies to gpI and gpIII also have antibody-dependent cellular cytotoxicity activity.[352]

VZV IgM antibodies which appear during primary VZV infection, like VZV IgG antibodies, have neutralizing activity.[333] However, complement enhancement of neutralization by VZV IgM antibodies was minimal. The immune transfer analysis of the IgM antibody response during the first week after the onset of varicella showed IgM reactivity against an average of seven VZV-ICPs, which was a broader spectrum of

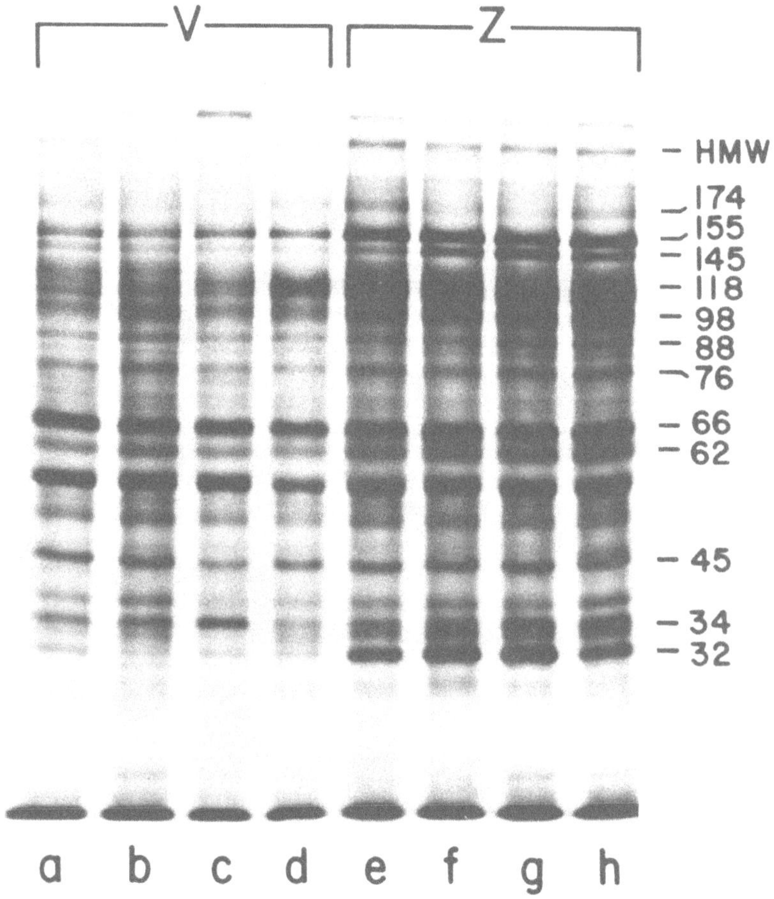

FIGURE 12. Radioimmune precipitation profiles of high FAMA titer (≥1:256) sera after varicella (V) and zoster (Z). Sera after chickenpox (lanes a through d) and after zoster (lanes e through h) from seven different patients were reacted with [^{35}S]methionine-labeled VZV proteins, and the immunoprecipitates were electrophoresed. One of the most noticeable differences between the "varicella" and the "zoster" profiles is the amount of polypeptide VZV p32. HMW = high-molecular-weight protein. (From Weigle, K. A. and Grose, C., *J. Infect. Dis.*, 149, 741, 1984. Copyright by the University of Chicago Press, 1986. With permission.)

reactivity than was seen for VZV IgG antibodies.[348] VZV IgM antibodies bound to the same VZV-infected cell proteins against which VZV IgG reacted, but VZV IgM antibodies also showed prominent binding to VZV-infected cell proteins of molecular weight ≥180,000.

IgA antibodies to VZV have been measured using several serologic methods.[324,327,353-356] In general, IgA antibodies have appeared a few days later than VZV IgG and IgM antibodies during acute varicella. The acute phase VZV IgA titers are also usually lower than the VZV IgG and IgM titers. VZV IgA antibodies can be detected in secretions from the nasopharynx during acute varicella.[327] By immune transfer, IgA antibodies were produced to an average of three VZV-infected cell proteins in subjects tested 6 to 49 days after the onset of varicella.[348] The pattern of IgA activity with VZV-infected cell proteins of specific molecular weights varied from subject to subject.

2. Humoral Immunity Following Primary VZV Infection

In the years after primary VZV infection, the VZV titers of all of the immunoglob-

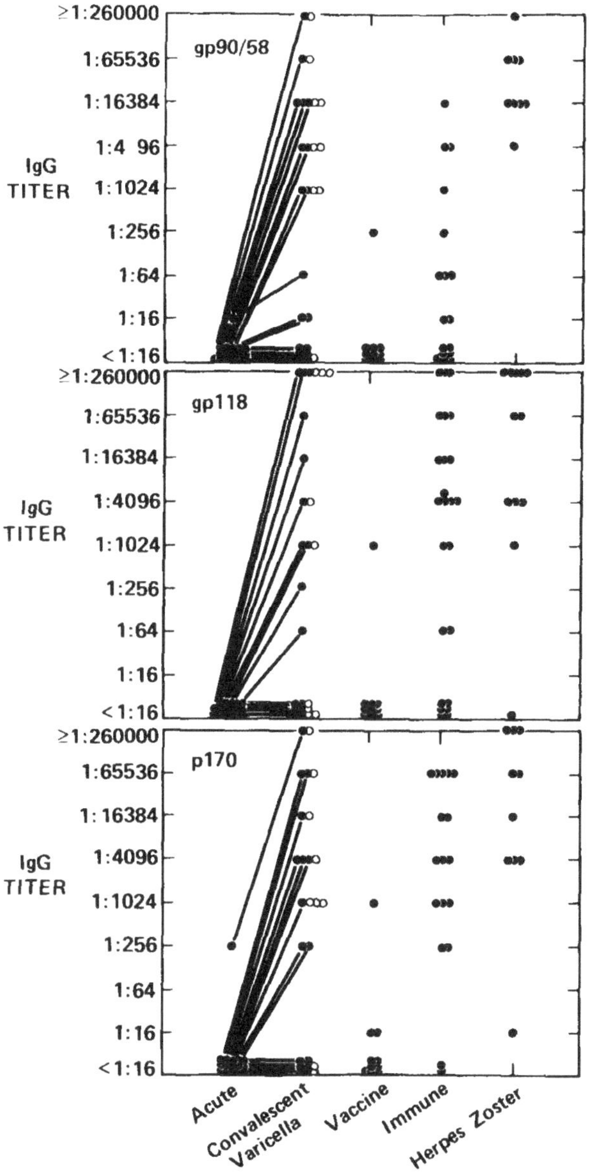

FIGURE 13. Antibody titers against varicella-zoster virus proteins gpI (gp90/58), gpIII (gp118), and p170 were measured by radioimmunoassay in sera from otherwise healthy subjects with acute primary varicella and during the convalescent phase, after varicella vaccine administration, in immune subjects with no recent clinical VZV infection, and in subjects convalescent from herpes zoster. The titer is expressed as the reciprocal of the serum dilution which gave a ratio of >2.5 mean counts per minute between VZV protein and control wells of the radioimmunoassay. Sera obtained from immunocompromised subjects are shown with open circles. Lines connect titers which were measured in paired sera from the same subject. (From Arvin, A. M., Kinney-Thomas, E., Shriver, K., Grose, C., Koropchak, C. M., Scranton, E., Wittek, A. E., and Diaz, P. S., *J. Immunol.*, 137, 1346, 1986. With permission.)

ulin subclasses decline. VZV IgM antibodies usually disappear from serum within 6 months,[324,326] as do VZV IgA antibodies as measured by most serologic methods.[354,356] Complement-fixing VZV IgG often becomes undetectable,[3] but the persistence of VZV IgG is demonstrated by other, more sensitive assays.[324-326,328,329,357-360] The VZV IgG antibodies which continue to be produced in the immune host have neutralizing activity which is enhanced by complement[332-334,357] and can mediate antibody-dependent cellular cytotoxicity.[335-340] As expected, the persistence of VZV IgG antibodies in serum is accompanied by the presence of circulating B-lymphocytes which can be stimulated to produce VZV IgG in vitro.[361] Weigle and Grose found persistent VZV IgG antibodies directed primarily against VZV gpII and gpIII with a decline in reactivity to VZV gpI over time.[347] Zweerink and Neff found IgG antibodies to all of the three major VZV glycoproteins and to VZV p155 in immune subjects without recent VZV infection.[346] Antibody to VZV thymidine kinase was detected in approximately 15% of VZV-immune subjects.[362] IgG antibodies to gpI, gpIII, and p170 were detected by radioimmunoassay in most immune subjects tested more than 20 years after primary VZV infection (Figure 17).[350]

3. Humoral Immunity with Reexposure to VZV

The reexposure of immune subjects to varicella by household contact resulted in an increase in VZV IgG antibody titers without clinical signs of infection.[324,363] This observation indicates that immunity after primary VZV infection does not protect against subclinical reinfection with the virus, but that symptomatic reinfection is usually prevented. Weigle and Grose observed the reemergence of prominent VZV IgG reactivity against the gpI following the reexposure of an immune adult.[347] Household reexposure to VZV was particularly associated with VZV IgA antibody production (Figure 14).[363] Although the relationship between serum IgA antibody titers and mucosal IgA is not certain, the production of serum IgA antibody to VZV in immune subjects exposed to varicella suggests a possible role for mucosal immunity in resistance to symptomatic reinfection with VZV. Baba et al. found an increase in nasopharyngeal VZV IgA among children after epidemic reexposure to varicella.[364]

4. Humoral Immunity with Reactivation of Latent VZV

Healthy immune subjects may have intermittent subclinical reactivation of latent VZV, as suggested by the detection of VZV IgM in some asymptomatic subjects.[365] In addition to boosts in VZV IgG from exogenous reexposure to this virus, episodes of silent VZV reactivation could also provide antigen stimulation to maintain VZV IgG in serum throughout adult life.

The occurrence of herpes zoster in the otherwise healthy subject is accompanied by a rapid rise in VZV IgG antibodies to very high titers as measured by many serologic assays. The dominant immunoglobulin G subclass is IgG1.[330] Complement-fixing and -neutralizing antibody titers increase by several logs.[333,366] VZV IgG antibodies to envelope and soluble antigens predominated in patients with herpes zoster.[343] Zweerink and Neff described the same pattern of antibody reactivity to VZV-infected cell proteins by radioimmunoprecipitation in subjects with herpes zoster as was observed after varicella.[346] Weigle and Grose detected IgG antibodies to all of the major VZV glycoproteins within 1 week after herpes zoster, compared to an interval of 5 weeks after varicella. The reappearance of IgG reactivity against gpI was very prominent after herpes zoster (Figure 15).[367] The rapid emergence of IgG antibody which bound a VZV protein of lower molecular weight, i.e., p32, was also striking in the IgG response to VZV reactivation (Figure 12 and Figure 15).

Depending upon the serologic method which is used, up to 80% of otherwise healthy

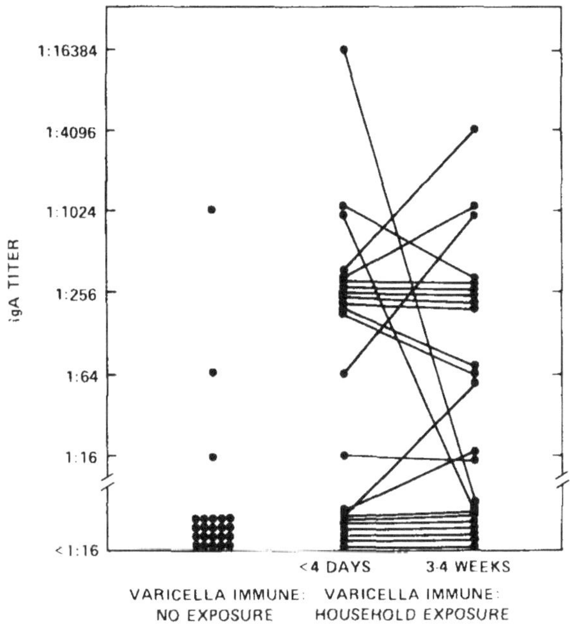

FIGURE 14. IgA antibody to VZV in immune subjects with no recent exposure and in immune subjects with household exposure to varicella. The lines connect IgA antibody titers in paired sera from immune subjects within 4 days after the onset of the varicella exanthem in the household contact and 3 to 4 weeks later. (From Arvin, A. A., Koropchak, C. M., and Wittek, A. E., *J. Infect. Dis.*, 48, 200, 1983. Copyright by the University of Chicago Press, 1986. With permission.)

subjects with herpes zoster and patients with lymphoma who have localized herpes zoster also have VZV IgM antibody production.[368-374] Complement-enhanced neutralizing antibody activity was detected in the IgM fraction of serum from 48% of patients with herpes zoster.[374] The VZV IgA response is also prominent in herpes zoster[353-356] and was detectable consistently by 5 days after the onset of recurrent VZV. Wittek et al. found that the initial geometric mean VZV IgA titer (1:512) was higher than the VZV IgM titer (1:64), and that VZV IgA remained elevated for a longer time after VZV reactivation that VZV IgM.[356]

5. Localized VZV Antibody Production in the CNS

Primary and recurrent VZV infections which are associated with CNS infection result in the localized production of VZV IgG and IgM antibodies.[63,267,268] VZV antibody concentrations in the cerebrospinal fluid are increased relative to serum VZV antibody concentrations in this circumstance.

6. Relationship of Humoral Immunity to the Clinical Manifestations of VZV in Normal and Immunocompromised Patients

Passive antibody prophylaxis with VZV IgG given early in the incubation period prevents or modifies the severity of primary VZV infection in the immunocompromised patient but only if it is given immediately after exposure.[375] (See Chapter 4.) Transplacentally acquired maternal VZV IgG or passive antibody prophylaxis also protects infants against severe varicella during the first few months of life.[376,377]

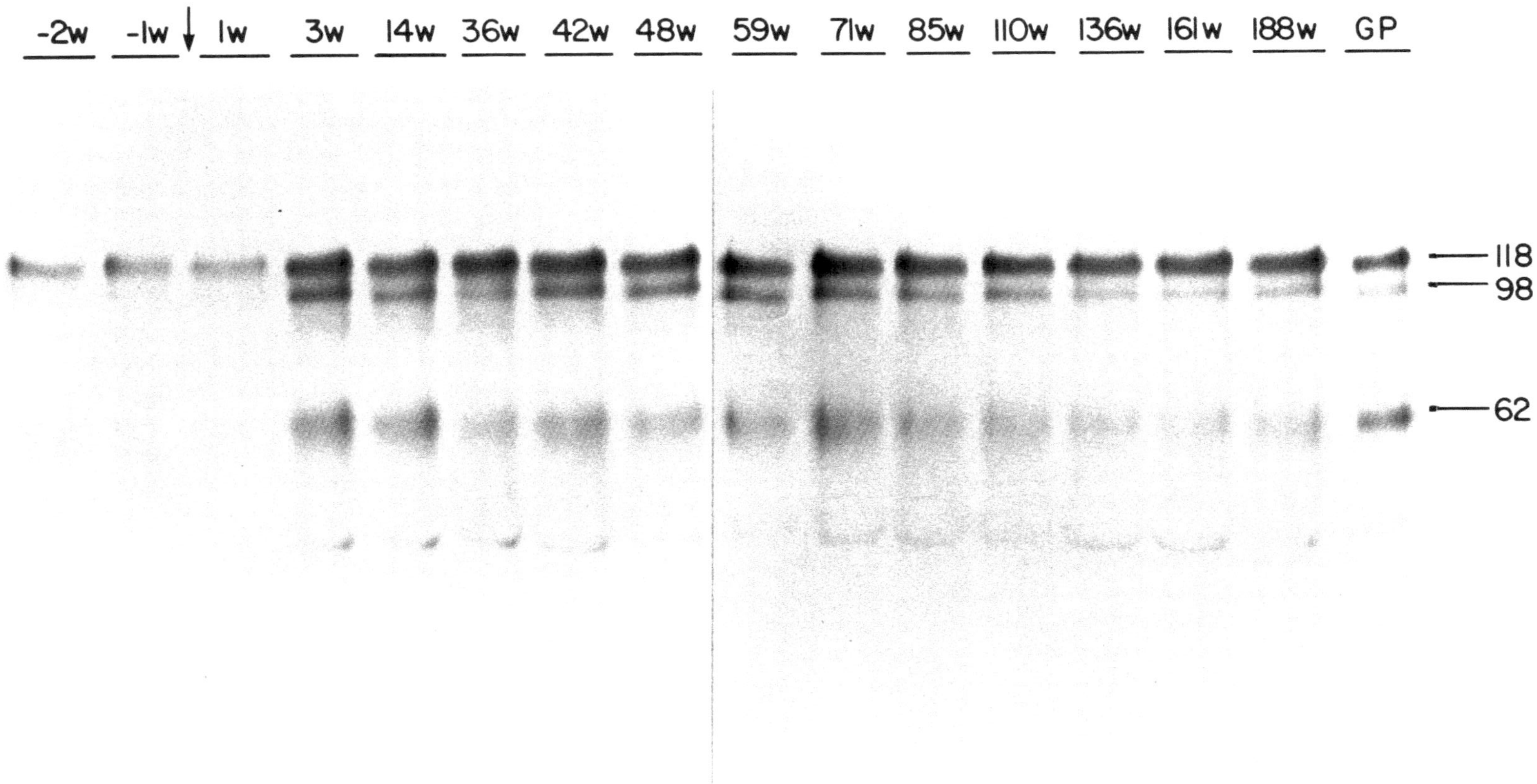

FIGURE 15. Varicella-zoster virus (VZV)-specific glycoproteins precipitated by sera from a child with cancer. Fifteen serum samples were collected from 2 weeks (w) before the onset of zoster (↓) until 188 weeks after zoster. These serum samples were reacted with [^{3}H]fucose-labeled antigen and the immunoprecipitates were fractioned in 10% acrylamide-*N,N'*-diallyltartardiamide slab gels. The lane on the far right represents the glycoproteins (GP) precipitated by a previously characterized caviid VZV immune serum. The apparent molecular weights (× 10^{-3}) of the VZV-specific immunogenic glycoproteins are enumerated in the right margin of the figure. Reprinted from Grose, C., Reference 367.

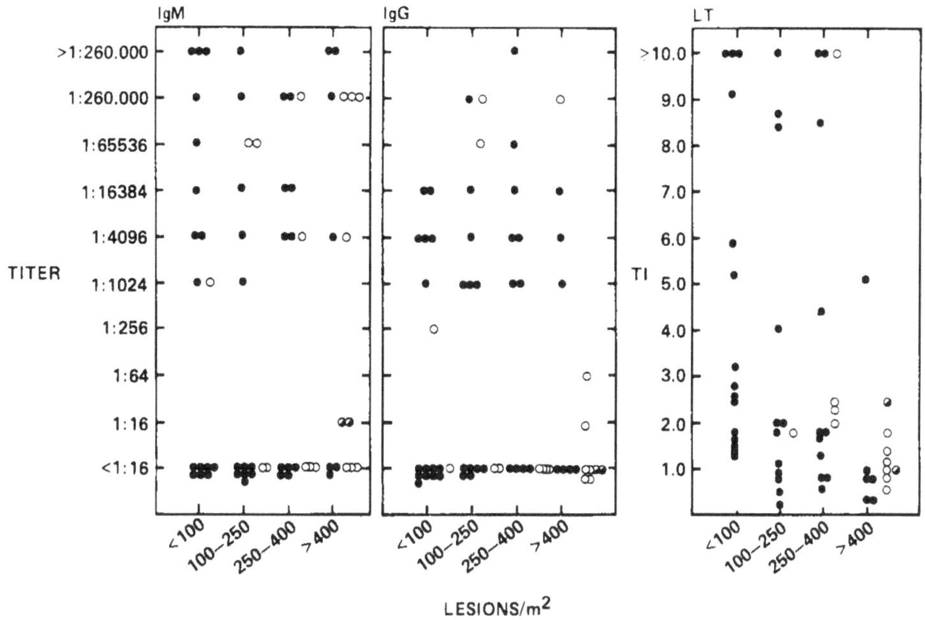

FIGURE 16. Humoral and cellular immunity to varicella-zoster virus in subjects tested during the early acute primary infection with VZV. IgM and IgG antibody titers against VZV antigen measured by radioimmunoassay and the lymphocyte transformation response of peripheral blood mononuclear cells stimulated in vitro with VZV antigen were determined in subjects tested within 3 days after the appearance of the varicella rash. Antibody titers are expressed as the reciprocal of the dilution of serum where the mean counts per minute for antigen wells was ≥2.5 compared to control wells stimulated with uninfected cell antigen. Lymphocyte transformation is expressed as the ratio of the mean counts per minute in antigen-stimulated wells to uninfected control wells (transformation index, TI). Laboratory assay results are plotted against lesions per square meter as a measure of the clinical severity of varicella. Lesions were counted 48 hours after no new lesions had appeared; lesions/m^2 was calculated by body surface area based on height and weight of the subject. The dark circles indicate subjects with normal immune function (●); the open circles indicate immunocompromised subjects (o); the half dark circles indicate immunocompromised subjects who had fatal VZV infection (◑). (From Arvin, A. A., Koropchak, C. M., Williams, B. R. G., Grumet, F. C., and Foung, S. K. H., The early immune response in healthy and immunocompromised subjects with primary varicella-zoster virus infection, *J. Infect. Dis.*, 154, 422, 1986. Copyright by the University of Chicago Press, 1986. With permission.)

The role of active humoral immunity in limiting the progression of primary VZV infection is uncertain aside from the observation that children with immunoglobulin deficiencies usually have uncomplicated varicella. VZV was isolated from the cutaneous lesions of children with varicella only when the VZV IgG complement fixation titer was ≤1:4.[323] However, the presence of early high titers of VZV IgG or IgM antibody titers by radioimmunoassay did not correlate with diminished viral replication in otherwise healthy subjects with varicella as judged by the number of cutaneous lesions (Figure 16).[322] Severe or fatal varicella occurs in immunocompromised children despite the presence of VZV IgG and IgM antibodies in concentrations comparable to those measured in healthy subjects with varicella (Figure 20).[322,323,378] Little is known about antibody production to specific VZV proteins in immunocompromised children with primary VZV infection. Subtle variations in humoral immunity to viral proteins could influence the relative severity of varicella. Two immunocompromised children with varicella had relatively diminished reactivity to gp98 and gp62 compared to normal children, and one infant with hypogammaglobulinemia and progressive VZV for 3 weeks failed to develop VZV IgG antibodies to the major viral proteins.[347]

Declining VZV IgG titers cannot be demonstrated with aging in normal adults despite their increased susceptibility to the reactivation of latent VZV.[379] Luby et al. observed equivalent VZV IgG titers among renal transplant recipients who developed VZV reactivation as compared with those who did not,[256] as was also demonstrated in prospective studies of heart transplant recipients and lymphoma patients.[380,381] There was no relationship between the clinical severity of herpes zoster and diminished production of VZV IgG measured by fluorescence antibody membrane antigen assay; VZV IgG titers were slightly higher among immunocompromised patients with herpes zoster who had disseminated infection compared to those with localized disease.[321,378] However, Mazur et al. found undetectable VZV IgG by immunoadherence, hemagglutination, and indirect fluorescence antibody titers <1:32 in 78% of immunocompromised patients with disseminated herpes zoster compared to 18% of patients with localized reactivation.[382] Children with cancer who had herpes zoster developed VZV IgG antibodies to all of the major VZV proteins by radioimmunoprecipitation.[367] Larkin et al. described less prominent reactivity against all of the VZV-infected cell proteins in immunocompromised patients with herpes zoster compared to normal subjects with recurrent VZV.[383] The administration of VZV IgG passive antibody after the onset of recurrent VZV infection did not alter its progression in patients with lymphoma.[384]

B. Cellular Immunity to VZV

Cellular immunity to viral pathogens can be considered in terms of cell-mediated responses which do not require prior exposure of the host to the specific pathogen, e.g., natural killer cell cytotoxicity, alpha-interferon production, and virus-specific responses which are acquired in the course of primary infection, e.g., T-lymphocyte proliferation, T-cell cytotoxicity, and gamma-interferon production.

1. Nonspecific Cellular Immunity to VZV

a. Natural Killer Cell Cytotoxicity

In vitro lysis of VZV-infected fibroblasts can be demonstrated using peripheral blood mononuclear cells from nonimmune subjects as effector cells.[385-387] Lysis occurs in the absence of VZV IgG antibodies and is mediated primarily by lymphocytes which do not express T-lymphocyte surface phenotypic markers. This cytotoxic activity against VZV-infected target cells is eliminated when peripheral blood mononuclear cells are incubated with monoclonal antibodies such as anti-Leu-11b which lyse natural killer cells in the presence of complement. Natural killer cell cytotoxicity provides the nonimmune host with an immediate cellular immune response to primary VZV infection. Natural killer cell cytotoxicity occurs in vitro without a requirement for a match between the major histocompatability (MHC) antigens of the effector cell and of the VZV-infected target cell. Natural killer cell cytotoxicity against VZV-infected fibroblasts was increased if effector cells were incubated with purified or recombinant interleukin-2 (Table 9).[387] Hayward et al. found that natural killer cell activity was enhanced in vitro by interleukin-2 release from T-lymphocytes which occurred with a partial match between the class I MHC antigens of the effector and VZV-infected target cells.[388]

b. Alpha-Interferon (Alpha-IFN) Production

Alpha-IFN is made in vitro by peripheral blood mononuclear cells, primarily B-lymphocytes, from nonimmune subjects stimulated with VZV antigen. Although not all natural killer cell activity is induced by interferons, alpha-IFN enhanced the in vitro lysis of VZV-infected fibroblasts by mononuclear cells.[385] In addition to its immunomodulating activity, alpha-IFN has a direct antiviral effect upon VZV which could be

Table 9
EFFECT OF RECOMBINANT INTERLEUKIN-2 ON NATURAL KILLER CELL ACTIVITY AGAINST VZV-INFECTED AND UNINFECTED FIBROBLASTS

	Concentration (U/ml)	Mean percent ^{51}Cr release ± SD	
		VZV infected	Uninfected
Control	0	26.7 ± 12.4	9.8 ± 5.5
IL-2	8	24.2 ± 18.2	43.2 ± 26.2
	16	30.0 ± 20.8	47.3 ± 26.5
	32	41.8 ± 29.4	54.3 ± 26.3

Abstracted from Ito, M., Bandyopadhyay, S., Matusumoto, M., Clark, S. C., Miller, D., and Starr, S., *Clin. Exp. Immunol.*, 65, 182, 1986. With permission from Blackwell Scientific Publications Limited.

important as an early nonspecific host response to limit VZV replication.[389] In vitro IFN production after VZV stimulation has not been assessed in nonimmune immunocompromised patients, but peripheral blood mononuclear cells from children with malignancy were deficient in mediating natural killer cell cytotoxicity by comparison with cells from nonimmune normal subjects.[386]

2. Cellular Immunity during Primary VZV Infection

The activation of the interferon system, along with an increase in the number of circulating large lymphocytes in the peripheral blood were the earliest immunologic changes detected in the investigation of the immune response to primary VZV infection.[322] Serum alpha-IFN, gamma-interferon (gamma-IFN), and 2′-5′A synthetase production by peripheral blood mononuclear cells were present during incubation in some subjects and during the early acute infection in most healthy subjects with varicella (Figure 17). The importance of the IFN response in primary VZV infection has been suggested by the efficacy of alpha-IFN administration to immunocompromised children immediately after the onset of varicella[390] and by the protective effect of alpha-IFN administration during the incubation period of simian varicella.[391] By comparison with normal subjects, immunocompromised children with varicella had diminished serum alpha-IFN and diminished in vitro alpha-IFN production by peripheral blood mononuclear cells.[322] Armstrong et al. found that three of seven immunocompromised patients with varicella had no detectable IFN in cutaneous vesicle fluid.[392] Low concentrations of IFN in vesicle fluid correlated with continued new lesion formation. Patel et al. observed that fatal varicella in immunocompromised children was associated with low serum and vesicle fluid IFN concentrations.[393] 2′-5′A synthetase in peripheral blood mononuclear cells is evidence that these cells have responded to circulating IFN. A low concentration of 2′-5′A synthetase in the peripheral blood mononuclear cells was associated with severe primary VZV infection in immunocompromised children.[394]

Circulating T-lymphocytes from normal subjects with varicella were activated in vitro, as evidenced by an increase in size and by the surface expression of HLA-DR antigen.[322] Primary VZV infection was accompanied by an inverted ratio of helper and suppressor/cytotoxic T-lymphocytes which resulted from a relative increase in the suppressor/cytotoxic subpopulation.[322,395,396] DR antigen expression occurred on cells of both helper and suppressor/cytotoxic T-lymphocyte subpopulations and developed before VZV-specific T-lymphocyte proliferation was detectable. Enhanced suppressor/

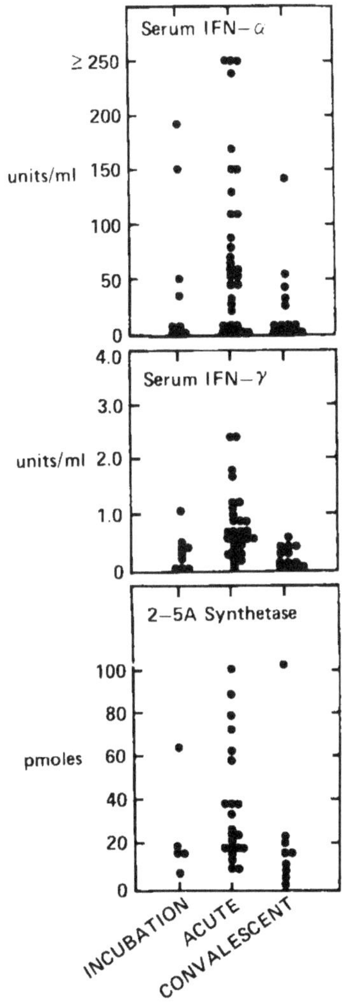

FIGURE 17. The interferon response in healthy subjects with primary varicella-zoster viral infection. Serum interferon alpha (IFN-α) (upper section), serum interferon gamma (IFN-γ) (middle section), and 2′-5′A synthetase production by peripheral blood mononuclear cells (lower section) were assayed in subjects during the incubation period, the acute phase, and the convalescent phase of primary VZV infection. IFN is expressed as units/mℓ with reference to a standard IFN-α or IFN-γ preparation. 2′-5′A synthetase production is expressed as pmoles of enzyme per hour per 10^5 cells; the mean value for control samples in this assay was 14.3 ± 9.0 SD. (From Arvin, A. A., Koropchak, C. M., Williams, B. R. G., Grumet, F. C., and Foung, S. K. H., The early immune response in healthy and immunocompromised subjects with primary varicella-zoster virus infection, *J. Infect. Dis.*, 154, 422, 1986. Copyright by the University of Chicago Press, 1986. With permission.)

Table 10
PERIPHERAL BLOOD LYMPHOCYTE COUNTS AND VZV LYMPHOCYTE PROLIFERATION IN NORMAL AND IMMUNOCOMPROMISED CHILDREN WITH VARICELLA

Subject group	No. tested	Peak cell counts ± SD: Lymphocytes	Peak cell counts ± SD: T-Lymphocytes	Peak Δ cpm ± SD to VZV antigen
Normal	(11)	2906 ± 774[a]	1590 ± 290[a]	37,170 ± 1.4[a]
Leukemia/lymphoma				
Relapse	(6)	1540 ± 280[a]	290 ± 101[a]	1800 ± 2.6[a]
Remission	(7)	1753 ± 302[a]	747 ± 285[a]	4500 ± 5.0[a]
Leukemia/lymphoma				
Fatal varicella	(5)	1346 ± 323	200 ± 60[b]	500 ± 2.5[b]
Survived	(8)	1960 ± 392	946 ± 130[b]	4709 ± 3.6[b]

[a] Peak absolute lymphocyte, T-lymphocyte counts, and Δ cpm to VZV were lower among patients with leukemia or lymphoma, including those in remission, than normal subjects ($p < .02$).

[b] Immunocompromised patients who survived varicella had higher T-lymphocyte counts ($p < .01$) and Δ cpm to VZV ($p < .02$) than those with fatal infection ($p =$ number of patients tested).

From Patel, P. A., Yoonessi, S., O'Malley, Freeman, A., Gershon, A. A., and Ogra, P., *J. Pediatr.*, 94, 223, 1979. With permission.

cytotoxic cell activity may account for the association of varicella with a transient decrease in tuberculin skin reactivity.[397-399] Peripheral blood mononuclear cells from patients with varicella caused suppression of the phytohemagglutinin response of convalescent cells from the same donor.[396]

VZV-specific cellular immunity can be demonstrated in healthy subjects within the first few days after the appearance of the varicella exanthem by in vitro proliferation of T-lymphocytes upon stimulation with VZV antigen. Early VZV-specific T-lymphocyte proliferation was the only immune response in healthy subjects which correlated with disease severity as measured by the number of cutaneous lesions. (Figure 16) The acquisition of lymphocyte proliferation to VZV gpI, gpIII and p170 was demonstrated in healthy subjects with primary VZV infection (Figure 18).[350] Successful recovery from varicella did not require the acquisition of T-lymphocyte proliferation to each of these VZV proteins in the normal host.

Patel et al. found that normal children with varicella developed a VZV lymphocyte proliferation response which was 30 to 60 times higher than that of children with leukemia or lymphoma in relapse who had varicella (Table 10).[393] Although the absolute lymphocyte counts were comparable, immunocompromised children who survived varicella had significantly higher T-lymphocyte counts and VZV lymphocyte proliferation than those with fatal varicella. Hayes and Feldman found that children with malignancy in remission were more likely to develop VZV-specific lymphocyte proliferation during acute varicella than those with relapse; varicella was usually uncomplicated in patients in remission.[400] The acquisition of VZV lymphocyte proliferation was delayed in patients with varicella meningoencephalitis compared to those with cutaneous infection only.[401] Production of leukocyte inhibitory factor was not detected after stimulation of peripheral blood cells from immunocompromised children.[402] Several healthy infants exposed to VZV in utero had VZV-specific cellular immunity but one infant with congenital varicella syndrome did not.[180] Baba et al. found that infants with varicella during the first 2 months of life had diminished delayed hypersensitivity re-

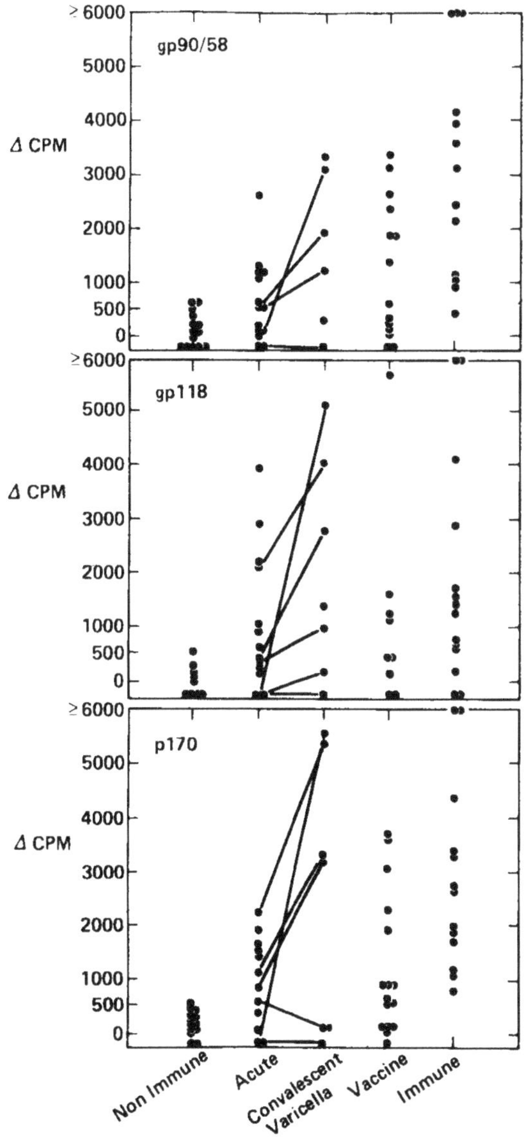

FIGURE 18. The lymphocyte proliferation responses of peripheral blood cells from VZV non-immune subjects, subjects with acute and convalescent varicella, healthy recipients of varicella vaccine and immune subjects with no recent clinical VZV infection when incubated with immunoaffintiy purified preparations of VZV proteins, gpI (gp 90/58), gpIII (gp118) and p170. The results are expressed as the Δ cpm, which is the difference between the mean cpm for VZV protein antigen-stimulated wells and wells incubated with the uninfected cell control preparation. Lines connect lymphocyte proliferation results for the same individual tested during the acute and convalescent phases of varicella. (From Arvin, A. M., Kinney-Thomas, E., Shriver, K., Grose, C., Koropchak, C. M., Scranton, E., Wittek, A. E., and Diaz, P. S., *J. Immunol.*, 137, 1346, 1986. With permission.)

Table 11
COMPARISON OF THE PEAK ANTIBODY AND CUTANEOUS REACTIVITY RESPONSE TO VZV IN DIFFERENT AGE GROUPS OF INFANTS AFTER VARICELLA

	Peak response ± SD		
Age	FAMA antibody	Neutralizing antibody	Skin test reaction
<2 mo	32.0 ± 1.0	16.0 ± 1.6	6.8 ± 2.5[a]
2.5 mo	55.7 ± 1.3	22.6 ± 1.3	19.3 ± 9.6
6.8 mo	55.7 ± 1.2	55.7 ± 1.2	21.0 ± 6.2

[a] Significantly less reactivity than infants 2 to 8 mo ($p < .001$).

Abstracted from Baba, K., Yabunchi, H., Takahashi, M., and Ogra, P. L., *J. Pediatr.*, 100, 881, 1982. With permission.

sponses to VZV antigen. These infants may be protected from severe varicella despite limited cellular immunity by maternal VZV IgG antibodies acquired transplacentally (Table 11).[376]

The presence of gamma-IFN which was detected in serum from healthy subjects during the incubation and acute phases of primary VZV infection can be considered an early marker of VZV-specific immunity since it is an immune-specific product of T-lymphocytes (Figure 17).[322] Gamma-IFN is also a product of cytotoxic T-lymphocytes and may be important for the recruitment of VZV-specific cytotoxic T-cells. Gershon et al. demonstrated the capacity of T-lymphocytes in the presence of monocytes to restrict VZV replication in vitro (Figure 19).[378,403,404] Two immunocompromised children with fatal varicella failed to develop this cellular immune response.[378] Cloned T-lymphocytes from VZV-immune subjects produced lysis of VZV-infected targets which shared MHC class II antigens (Figure 20).[405] The cloned T-lymphocytes which mediated this response had surface markers of helper T-cells. The interval to acquisition of T-lymphocyte cytotoxicity in the course of primary VZV infection has not been determined.

VZV viremia during primary VZV infection has been associated with mononuclear cells, but the specific cell types which contain the virus have not been identified.[321,322] Arbeit et al. observed that monocytes were resistant to VZV infection in vitro.[406] Baba and Shigeta observed VZV antigen expression in monocytes infected with VZV in vitro but complete viral replication did not occur.[407] The ability of monocytes to restrict VZV replication may be important in the host response to primary VZV infection. Gresser et al. observed that macrophages which synthesize 2′-5′A synthetase were non-permissive for viral infection.[408] As in other infections, monocytes are probably critical to the development of VZV-specific immunity through their function as antigen-presenting cells. In vitro VZV lymphocyte proliferation and VZV T-lymphocyte cytotoxicity require the presence of monocytes.[403,405]

3. Cellular Immunity Following Primary VZV Infection

The persistence of VZV lymphocyte proliferation by peripheral blood mononuclear cells can be demonstrated for years after primary VZV infection in healthy subjects

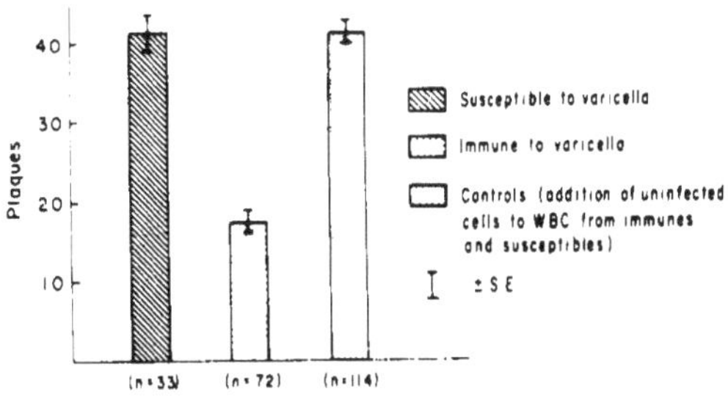

FIGURE 19. Summary of results of experiments comparing the ability of 500 to 1000 human WBC from persons immune to varicella and persons susceptible to varicella to inactivate VZV. The mean number of plaques is depicted for (1) WBC from susceptibles exposed to VZV antigen, (2) WBC from immunes exposed to VZV antigen, and (3) WBC from all donors exposed to control antigen. The ability of WBC from individuals immune to varicella to inactivate VZV is significantly greater than that of WBC from those susceptible to varicella. Total number of determinations is represented by "n". (From Gershon, A. A., Steinberg, S., and Smith, M., *Infect. Immun.*, 13, 1549, 1976. With permission.)

and correlates with the presence of IgG antibodies to VZV.[409-411] The number of T-lymphocytes in peripheral blood which can proliferate to whole VZV antigen in immune subjects ranges from 1:15,500 to 1:32,480.[426] Tonsillar lymphocytes from VZV-immune children also exhibit VZV-specific proliferation.[412] Lymphocyte proliferation to VZV gpI, gpIII, and p170 was demonstrated in VZV-immune subjects tested more than 20 years after primary VZV infection (Figure 18) but not all subjects had demonstrable VZV lymphocyte proliferation to all of these proteins.[350] Maintenance of VZV latency apparently does not require the host to have persistent detectable cellular immunity directed against each major viral protein. The number of T-lymphocytes which proliferate to different subunit antigens may fall below the threshold of detection, although clones of T-lymphocytes sensitized to the particular viral protein which can expand upon reexposure to the virus could persist.

The presence of lymphocytes which can mediate a delayed hypersensitivity reaction to VZV antigens is demonstrated by intradermal skin testing of VZV-immune subjects with VZV antigen.[413,414] Kamiya et al. noted positive responses by about 2 weeks after primary VZV infection. Delayed hypersensitivity reactions have been demonstrated in immune subjects given a soluble skin test antigen containing mainly glycoproteins of 115,000 and 45,000 mol wt.[415]

4. Cellular Immunity with Reexposure to VZV

VZV-immune healthy adults who had household exposure to varicella had enhanced cellular immunity to VZV documented by an increase in lymphocyte proliferation to VZV antigen (Figure 21).[363] The specificity of the response was demonstrated by comparison with lymphocyte proliferation to tetanus toxoid which remained unchanged. In vitro IFN production to VZV antigen did not increase. Children reexposed to varicella also had a significant increase in delayed hypersensitivity to VZV antigen.[364] The enhancement of VZV cellular and humoral immunity in these subjects suggested that subclinical reinfection may occur after close contact. Exposure to varicella may help to maintain persistent cellular immunity to VZV in immune subjects.

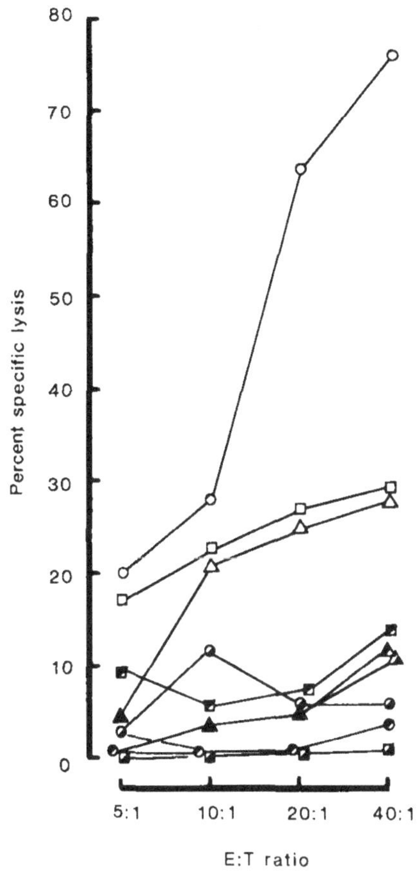

FIGURE 20. Specificity of VZV target cell lysis by cloned T-lymphocytes. Targets (with clones 4, 8, and 9, respectively, as the effector cells) were autologous VZV-infected B-cell line cells (□, ○, △), partially matched VZV infected fibroblasts, (◪, ◐, ▲) and autologous uninfected B-cell line cells (◪, ◑, ▲). E:T ratio = Effector/target ratio. (From Hayward, A. R., Pontesilli, O., Herberger, M., Laszlo, M., and Levin, M., *J. Virol.*, 58, 179, 1986. With permission.)

5. Cellular Immunity with Reactivation of VZV Infection

Studies of immunocompromised patients who have had previous primary VZV infection and of otherwise healthy elderly subjects suggest that a decrease in circulating T-lymphocytes capable of proliferating upon stimulation with VZV antigen is associated with an increased risk of recurrent VZV infection.[416-418] A selective impairment of VZV lymphocyte reactivity was observed among untreated patients with lymphoma.[419] Prolonged suppression of cellular immunity to VZV correlated with an increased susceptibility to VZV reactivation in lymphoma patients and in cardiac transplant recipients.[380,381] Ruckdeschel et al. observed that suppression of VZV lymphocyte proliferation was more frequent among patients with Hodgkin's disease who were receiving both radiation and chemotherapy, which is the patient population at greatest risk for herpes zoster.[420] Diminished VZV lymphocyte proliferation was also characteristic of bone marrow transplant recipients during the first 100 days after transplant and paralleled the period of increased susceptibility to VZV reactivation (Figure 22).[421] VZV-

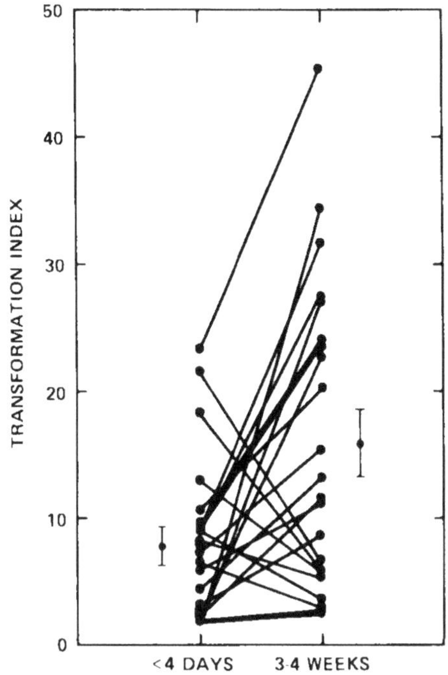

FIGURE 21. Lymphocyte transformation to VZV antigen in immune subjects with household exposure to varicella. The lines connect lymphocyte transformation to VZV antigen in paired samples from immune subjects within 4 days after the onset of the varicella exanthem in the household contact and 3 to 4 weeks later. (From Arvin, A. M., Koropchak, C. M., and Wittek, A. E., *J. Infect. Dis.*, 148, 200, 1983. Copyright by the University of Chicago Press, 1986. With permission.)

immune immunocompromised patients also have decreased in vitro IFN production by peripheral blood mononuclear cells stimulated with VZV antigen.

Healthy subjects with herpes zoster may demonstrate a low VZV lymphocyte proliferation response immediately after the onset of symptoms.[380,422,423] Recovery from VZV reactivation was associated with a marked increase in VZV lymphocyte proliferation in these subjects. A relationship between VZV lymphocyte proliferation and the ability of the host to localize recurrent VZV was suggested by the association of fever with low VZV lymphocyte proliferation observed by Sorensen et al. (Figure 23).[423] In vitro IFN production by peripheral blood mononuclear cells was present at the onset of herpes zoster and increases with recovery from herpes zoster in normal subjects.[380,423] Immunocompromised patients with herpes zoster who had uncomplicated infection also showed an increase in VZV lymphocyte proliferation and in vitro IFN production, but the response was delayed by comparison with normal subjects (Figure 24).[380] The resolution of herpes zoster was also accompanied by VZV-specific HLA-restricted cytotoxicity.[424] The cessation of dissemination of herpes zoster in immunocompromised patients occurred within 48 hours after the detection of vesicle fluid IFN.[425]

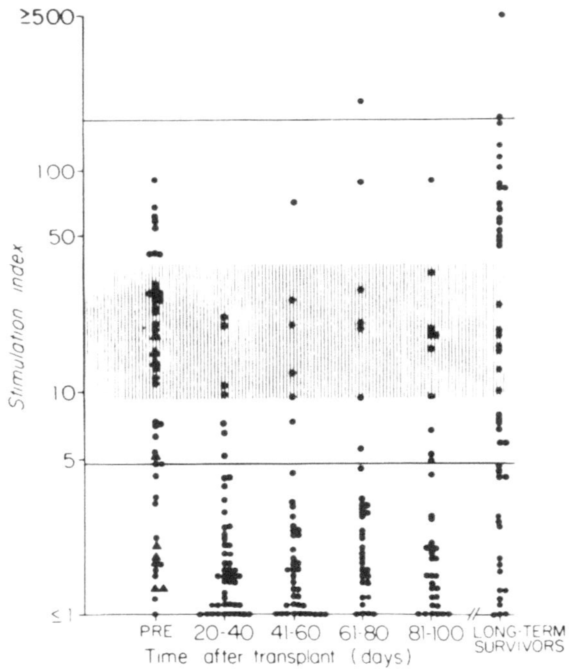

FIGURE 22. Lymphocyte transformation responses of marrow transplant recipients to varicella-zoster virus antigen. All of the patients and their donors had a history of VZV infection. The nine patients who had circulating myeloblasts or lymphoblasts are indicated by triangles. The 25 to 75th percentile of the normal response to VZV antigen is indicated by the hatched area, while the horizontal lines enclose a 95% range of normal responses. (From Myers, J. Flournoy, N., and Thomas, J. E., *J. Infect. Dis.*, 148, 200, 1983. Copyright by the University of Chicago Press, 1986. With permission.)

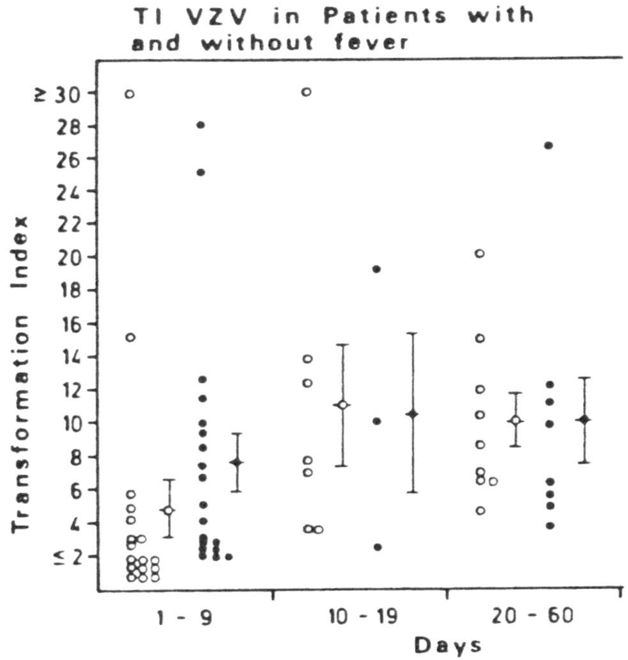

FIGURE 23. Lymphocyte transformation to VZV of mononuclear cells from blood samples of zoster patients with and without fever at various times after the start of the eruption. Symbols: (O) blood samples from patients with fever; (●) blood samples from patients without fever; (±) mean ± SE. (From Sorensen, O. S., Haahr, S., Moller-Larsen, A., and Wildenhoff, K., *Infect. Immun.*, 29, 369, 1980. With permission.)

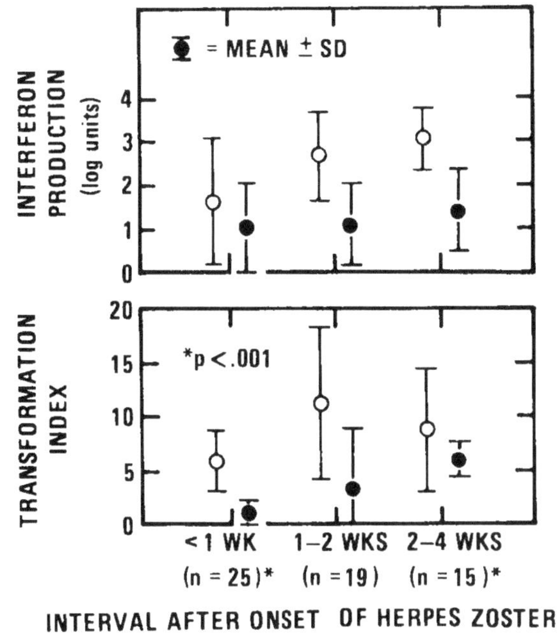

FIGURE 24. Cellular immune responses to VZV antigen in normal subjects and lymphoma patients with acute herpes zoster infection. The upper portion of the figure shows the mean log interferon production to VZV antigen ± SD in patients with lymphoma and acute zoster (●), compared with the responses seen in normal subjects (O) at intervals after the onset of infection. The lower portion of the figure shows the mean SI to VZV antigen ± SD; the SI is a ratio of counts per minute in antigen-stimulated cultures to counts per minute in control cultures. (From Reference 380. Reproduced from *J. Clin. Invest.*, 65, 869, 1980, by copyright permission of American Society of Clinical Investigation.)

ACKNOWLEDGMENT

The research from my laboratory which is reviewed in this chapter was supported by U.S. Public Health Service grants AI-17421 and AI-20459 and by a grant from The Thrasher Research Fund, Salt Lake City.

REFERENCES

1. Bullowa, J. G. M. and Wishik, S. M., Complications of varicella. I. Their occurrence among 2534 patients, *Am. J. Dis. Child.*, 49, 923, 1935.
2. Fleisher, G., Henry, W., McSorley, M., Arbeter, A., and Plotkin, S., Life-threatening complications of varicella, *Am. J. Dis. Child.*, 135, 896, 1981.
3. Weller, T. H., Varicella and herpes zoster: changing concepts of the natural history. Control and importance of a not-so-benign virus, *N. Engl. J. Med.*, 309, 1362, 1983.
4. Brunell, P. A., Varicella-zoster infections, in *Textbook of Pediatric Infectious Diseases*, Feigin, R. D., and Cherry, J. D., Eds., W. B. Saunders, Philadelphia, 1981, 1206.
5. Krugman, S., Katz, S., Gershon, A. A., and Wilfert, C. M., Varicella-zoster infections, in *Infectious Diseases of Children*, C. V. Mosby, St. Louis, 1985, 433.
6. Weller, T. H., Varicella-herpes zoster virus in *Viral Infections of Humans*, Evans, A. J., Ed., Plenum Press, New York, 1976, 462.
7. Grose, C., Varicella-zoster virus infections: Chickenpox (varicella) and shingles (zoster) in *Human Herpes Virus Infections: Clinical Aspects*, Glaser, R. and Gotlieb-Stematsky, T., Eds., Marcel Dekker, New York, 1982, 85.

8. Taylor-Robinson, D., Herpes-zoster occurring in a patient with chickenpox, *Br. Med. J.*, 1, 1713, 1960.
9. Jezek, Z., Hardjotanojo, W., and Rangaraj, A. G., Facial scarring after varicella, *Am. J. Epidemiol.*, 114, 798, 1981.
10. Ross, A., Modification of chickenpox in family contacts by administration of gamma globulin, *N. Engl. J. Med.*, 267, 369, 1962.
11. Holbrook, A. A., The blood picture in chickenpox, *Arch. Intern. Med.*, 60, 294, 1941.
12. Mitchell, A. G. and Fletcher, E. G., Studies on varicella: age and seasonal incidence, recurrences, complications and leukocyte counts, *JAMA*, 89, 279, 1927.
13. Krugman, S., Varicella and herpes virus infection, *Pediatr. Clin. N. Am.*, 7, 881, 1960.
14. Gilchrest, B. and Baden, H. P., Photodistribution of viral exanthems, *Pediatrics*, 54, 136, 1974.
15. Cupoli, J. M., Photodistribution of viral exanthems, *Pediatrics*, 59, 484, 1977.
16. Guess, H. A., Broughton, D. D., Melton, L. J., and Kurland, L. T., Chickenpox hospitalizations among residents of Olmsted County, Minnesota, 1961—1981. A population-based study, *Am. J. Dis. Child.*, 138, 1055, 1984.
17. Wishik, S. M., and Bullowa, J. G. M., Complications of varicella. II. Surface complications, *Am. J. Dis. Child.*, 49, 927, 1935.
18. Pye-Smith, P. H., Four cases of bullous varicella, *Br. J. Dermatol.*, 9, 148, 1897.
19. Elkridge, L. A. and Rivers, T. M., Bullous impetigo complicating varicella; report of a case, *Bull. Johns Hopkins Hosp.*, 41, 354, 1927.
20. Cardin, D. M. K., Two cases of bullous chickenpox, *Br. Med. J.*, 1, 266, 1958.
21. Saslow, S., Kluck, C., and Prior, J. A., Varicella bullosa, *JAMA*, 173, 118, 1960.
22. Seigerman, H. and Zorris, M. H., Varicella bullosa, a case report, *J. Pediatr.*, 62, 302, 1962.
23. Canby, J. P. and Blakely, L., Bullous chickenpox, *Clin. Pediatr.*, 2, 13, 1963.
24. Wald, E. R., Levine, M. R., and Togo, Y., Concomitant varicella and staphylococcal scalded skin syndrome, *J. Pediatr.*, 83, 1017, 1973.
25. Melish, M. E., Bullous varicella: its association with the staphylococcal scalded skin syndrome, *J. Pediatr.*, 83, 1019, 1973.
26. Stokes, W., An eruptive disease of chidren, *Dublin Med. Phys. Essays*, 1, 146, 1807.
27. Hutchinson, J., On gangrenous eruptions in connection with vaccination and chickenpox, *Med. Chir. Trans.*, 65, 1, 1882.
28. Kieffer, C. F., Varicella gangrenosa with report of a case, *N.Y. Med. J.*, 82, 1, 1905.
29. Banks, H. S. and McCartney, J. E., Varicella gangrenosa due to *Streptococcus pyogenes*, *Lancet*, 2, 311, 1937.
30. Barenberg, L. H. and Lewis, J. M., Gangrenous varicella, *Arch. Pediatr.*, 44, 653, 1927.
31. Illingworth, R. S. and Zachary, R. B., Superficial gangrene of the skin in chickenpox, *Arch. Dis. Child.*, 30, 177, 1955.
32. Crocker, A. M., and Middlecamp, J. N., Varicella gangrenosa, *J. Pediatr.*, 54, 104, 1959.
33. Smith, E. W. P., Garson, A., Boyleston, J. A., Katz, S. L., and Wilfert, C. M., Varicella gangrenosa due to group A Beta-hemolytic streptococcus, *Pediatrics*, 57, 306, 1976.
34. Kyong, C. U., Smith, C. D., Otherson, H. B., Necrotizing fasciitis of the abdominal wall as a complication of chickenpox, *Pediatr. Infect. Dis.*, 4, 420, 1985.
35. John, T. J., Varieties of varicella gangrenosa, *Pediatrics*, 60, 384, 1977.
36. Hosoya, R., Miwa, M., Nishimura, K., and Yamamoto, T., Erythema multiforme syndrome associated with varicella, *N. Engl. J. Med.*, 305, 582, 1981.
37. Brettle, R. P., Erythema multiforme syndrome associated with varicella, *N. Engl. J. Med.*, 305, 1472, 1981.
38. Jacobson, J. A., Burke, J. P., Benowitz, B. A., and Clark, P. V., Varicella-zoster and staphylococcal toxic shock syndrome in a young man, *J. Am. Med. Assoc.*, 249, 922, 1983.
39. Singer, J., Post-varicella suppurative meningitis: case reports and review of the literature, Am. J. Dis. Child., 133, 934, 1979.
40. Fitz, R. H., and Meiklejohn, G., Varicella pneumonia in adults, *Am. J. Med. Sci.*, 232, 489, 1956.
41. Weinstein, L., and Meade, R. H., Respiratory manifestations of chickenpox, *Arch. Int. Med.*, 98, 91, 1956.
42. Krugman, S., Goodrich, C. H., and Ward, R., Primary varicella pneumonia, *N. Engl. J. Med.*, 257, 843, 1957.
43. Weber, D. M. and Pellechia, J. A., Varicella pneumonia: study of prevalences in adult men, *J. Amer. Med. Assoc.*, 192, 572, 1965.
44. Triebwasser, J. H., Harris, R. E., Bryant, R. E., and Rhoades, E. R., Varicella pneumonia in adults: report of seven cases and a review of literature, *Medicine*, 46, 409, 1967.
45. Pitel, P. A., McCormick, K. L., Fitzgerald, E., and Orson, J. M., Subclinical hepatic changes in varicella infection, *Pediatrics*, 65, 631, 1980.

46. Ey, J. L., Smith, S. M., and Fulginiti, V. A., Varicella hepatitis without neurologic symptoms or findings, *Pediatrics*, 67, 285, 1981.
47. Meyers, M. G., Hepatic cellular injury during varicella, *Arch. Dis. Child.*, 57, 317, 1982.
48. Hunter, W., Case of varicella complicated with convulsions, *Lancet*, 1, 45, 1875.
49. Marfan, A. B., Paralysie spinale survenue a la site d'une varicelle, *Bull. et Mem. Soc. Med. Hop. Paris*, 10, 183, 1893.
50. Menko, M. L. H. S., Choreiforme Bewegungen nach Varicellen, *Dtsch. Med. Wochenschr.*, 25, 745, 1899.
51. Preblud, S. R. and D'Angelo, L. J., Chickenpox in the United States 1972—1977, *J. Infect. Dis.*, 140, 257, 1979.
52. Gibel, H., Kramer, B., and Naji, A. F., Encephalitis complicating chickenpox: report of a case with death two and a half years later and review of the literature, *AMA J. Dis. Child.*, 99, 137, 1960.
53. Preblud, S. R., Age specific risks of varicella complications, *Pediatrics*, 68, 14, 1981.
54. Underwood, E. A., The neurological complications of varicella. A clinical and epidemiological study, *Br. J. Child. Dis.*, 32, 83, 1935.
55. Appelbaum, E., Rachelson, M. H., and Dolgopol, V. B., Varicella encephalitis, *Am. J. Med.*, 15, 223, 1953.
56. Miller, H. G., Stanton, J. B., and Gibbon, J. L., Para-infectious encephalitis and related syndromes, *Quart. J. Med.*, 25, 427, 1956.
57. Gibbs, F. A., Gibbs, E. L., Spies, H. W., and Carpenter, P. R., Common types of childhood encephalitis, *Arch. Neurol.*, 10, 15, 1964.
58. Boughton, C. R., Varicella-zoster in Sydney. III. Neurological complications of varicella, *Med. J. Aust.*, 2, 444, 1966a.
59. Johnson, R. and Milbourne, P. S., Central nervous system manifestations of chickenpox, *Can. Med. Assoc. J.*, 102, 831, 1970.
60. Griffith, J. F., Salan, M. V., and Adams, R. D., The nervous system diseases associated with varicella, *Acta Neurol. Scand.*, 46, 279, 1970.
61. Shope, T. C., Chickenpox encephalitis and encephalopathy: evidence for differing pathogenesis, *Yale J. Biol. Med.*, 55, 321, 1982.
62. McCormick, W. F., Rodnitzky, R. E., Schochet, S. S., and McKee, A. P., Varicella-zoster encephalomyelitis, *Arch. Neurol.*, 21, 559, 1969.
63. Gershon, A., Steinberg, S., Greenberg, S., and Taber, L., Varicella-zoster associated encephalitis: Detection of specific antibody in cerebrospinal fluid, *J. Clin. Microb.*, 12, 764, 1980.
64. Steele, R. W., Keeney, R. E., Brodsky, R. W., Moses, E. B., and Soloff, B. L., Treatment of varicella-zoster meningoencephalitis with acyclovir: Demonstration of virus in cerebrospinal fluid by electron microscopy, *Am. J. Clin. Path.*, 80, 57, 1983.
65. Peters, A. C. B., Versteeg, J., Lindeman, J., Bots, G. T. A.M., Varicella and acute cerebellar ataxia, *Arch. Neurol.*, 35, 769, 1978.
66. Nicolaides, N. V., Fatal systemic varicella: a report of three cases, *Med. J. Aust.*, 2, 88, 1957.
67. McKendall, R. R. and Klawans, H. L., Nervous system complications of varicella-zoster virus, in *Handbook of Clinical Neurology*, Vinken, P. J. and Bruyn, G. W., Eds., Vol. 34, *Infections of the Nervous System*, North-Holland, Amsterdam, 1978, 161.
68. Bauman, M. L. and Bergman, I., Post-varicella encephalitis, *Arch. Neurol.*, 41, 556, 1984.
69. Goldston, A. S., Millichap, J. G., and Miller, R. H., Cerebellar ataxia with pre-eruptive varicella, *Am. J. Dis. Child.*, 106, 111, 1963.
70. Dreyfus, P. M. and Senter, T. P., Acute cerebellar ataxia of childhood: an unusual case of varicella, *West. J. Med.*, 120, 161, 1974.
71. Baird, H. W., and Garfunkel, J. M., Electroencephalographic changes in children with artificially induced hyperthermia, *J. Pediatr.*, 48, 28, 1956.
72. Bhandari, R. and Steinman, G. S., Sudden Deafness in chickenpox: a case report, *Ann. Neurol.*, 13, 347, 1983.
73. Gay, W., A case of peripheral paralysis following varicella, *Br. Med. J.*, 1, 679, 1894.
74. McCarthy, J. T. and Amer, J., Post-varicella acute transverse myelitis: a case presentation and review of the literature, *Pediatrics*, 62, 202, 1978.
75. Hatch, H. A., Bilateral optic neuritis following chickenpox: report of a case with apparently complete recovery, *J. Pediatr.*, 34, 758, 1949.
76. Chusid, M. J., Williamson, S. J., Murphy, J. V., and Ramey, L. S., Neuromyelitis optica (Devic's disease) following varicella, *J. Pediatr.*, 94, 737, 1979.
77. Shoji, H., Hirose, K., and Unono, M., A case of peripheral facial palsy following varicella, *J. Neurol.*, 209, 149, 1975.
78. Rab, S. M. and Choudhury, G. M., Landry-Guillain-Barre syndrome after chickenpox, *N. Engl. J. Med.*, 268, 200, 1963.

79. Zivin, I. and Schwager, V. A., The Guillain-Barre syndrome as a complication of varicella, *Dis. Nerv. Syst.*, 33, 742, 1972.
80. Wright, E. M., Lipson, M. J., and Mortimer, E. A., Varicella and severe hypoglycemia: a previously undescribed syndrome, *Am. J. Dis. Child.*, 92, 512, 1956.
81. Hurwitz, E. S., Nelson, D. B., Davis, C., Morens, D., and Schonberger, L. B., National surveillance for Reye syndrome: a five year review, *Pediatrics*, 70, 895, 1982.
82. Hurwitz, E. S. and Goodman, R. A., A cluster of cases of Reye syndrome associated with chickenpox, *Pediatrics*, 70, 901, 1982.
83. Glasgow, A. M. and Gold, M. B., The interval between varicella and Reye's syndrome, *Am. J. Dis. Child.*, 133, 653, 1979.
84. Lichtenstein, P. K., Heubi, J. E., and Daugherty, C. C., Farrell, M. K., Sokol, R. J., Rothbaum, R. J., Suchy, F. G., and Balistrevi, W. F., Grade I Reye's syndrome: a frequent cause of vomiting and liver dysfunction after varicella and upper respiratory tract infection, *N. Engl. J. Med.*, 309, 133, 1983.
85. Mortimer, E. A. and Lepow, M. L., Varicella with hypoglycemia possibly due to salicylates, *Am. J. Dis. Child.*, 103, 583, 1962.
86. Hurwitz, E. S., Barrett, M. J., and Bregman, D., Public health service study on Reye's syndrome and medications: report of the pilot phase, *N. Engl. J. Med.*, 313, 849, 1985.
87. Stroh, M., Zur Klinik der Varicellen mit besonderer Beruecksichtigung des Blutbefundes, *Z. Kinderheilkd.*, 26, 120, 1920.
88. Welch, R. G., Thrombocytopenic purpura and chickenpox, *Arch. Dis. Child.*, 31, 38, 1956.
89. Lopez, R. and Luby, A. L., Thrombocytopenic purpura and chickenpox, *Clin. Pediatr.*, 9, 100, 1970.
90. Espinoza, C. and Kuhn, C., Viral infection of megakaryocytes in varicella with purpura, *Am. J. Clin. Path.*, 61, 203, 1974.
91. Feusner, J. H., Slichter, S. J., and Harker, L. A., Mechanisms of thrombocytopenia in varicella, *Am. J. Hematol.*, 7, 255, 1979.
92. Yeager, A. M. and Zinkham, W. H., Varicella-associated thrombocytopenia: clues to the etiology of childhood idiopathic purpura, *Johns Hopkins Med. J.*, 146, 270, 1980.
93. Tobin, J. D. and ten Bensel, R. W., Varicella with thrombocytopenia causing fatal intracerebral hemorrhage, *Am. J. Dis. Child.*, 124, 577, 1972.
94. Charkes, N. D., Purpuric chickenpox: report of a case, review of the literature, and classification by clinical features, *Ann. Intern. Med.*, 54, 745, 1961.
95. Durard, C. H., Gavizzotti, B., and Pigney, P., Purpura thrombopenique aique apres un varicella commune, a propos d'un observation avec anticorps antiplaquettes dans le sang circulant, *Sem. Hop.*, Paris, 40, 1277, 1964.
96. Dixon, R., Rosse, W., and Ebbert, I., Quantitative determination of antibody in idiopathic thrombocytopenic purpura, *N. Engl. J. Med.*, 292, 230, 1975.
97. Batch, J. W. and Sepkowitz, S., Varicella complicated by gangrene of the lower extremities, *U.S. Armed Forces Med. J.*, 3, 759, 1952.
98. Stamey, C. C., Manley, J. H., and London, A. H., Purpura fulminans following chickenpox, *N.C. Med. J.*, 17, 115, 1956.
99. Hewes, W. J., Thrombocytopenia and massive gangrene following varicella, *Clin. Pediatr.*, 5, 161, 1966.
100. Becker, F. T., Purpura fulminans associated with varicella, *Arch. Derm.*, 94, 613, 1966.
101. Hensch, E., Nephritis nach varicella, *Klin. Wochenschr.*, 21, 17, 1884.
102. Denny, E. R. and Baker, B. M., Varicella complicated by acute nephritis: report of a case associated with streptococcal infection of the tonsils, *Bull. Johns Hopkins Hosp.*, 44, 201, 1929.
103. Yuceoglu, A. M., Berkovich, S., and Minkowitz, S., Acute glomerulonephritis as a complication of varicella, *JAMA*, 202, 113, 1967.
104. Minkowitz, S., Wenk, R., Friedman, E., Yuceoglu, A., and Berkovich, S., Acute glomerulonephritis associated with varicella infection, *Am. J. Med.*, 44, 489, 1968.
105. Singhal, P. C., Chugh, K. S., and Muthusethupathi, M. A., Rapidly progressive glomerulonephritis associated with varicella infection, *J. Pediatr.*, 91, 680, 1977.
106. Waring, J. J., Neuberger, K., and Geever, E. F., Severe forms of chickenpox in adults with autopsy observations in a case with associated penumonia and encephalitis, *Arch. Int. Med.*, 69, 384, 1942.
107. Clandy, W. O., Pneumonia associated with varicella: review of the literature and report of a fatal case with autopsy, *Arch. Int. Med.*, 80, 185, 1947.
108. Rigdon, R. H., Shojaii, S. A., and Garber, E. P., Fatal chickenpox: a review of the literature and a report of a case, *Am. Pract.*, 13, 292, 1962.
109. Amar, A. D., Hematuria caused by varicella lesions in the bladder, *JAMA*, 196, 450, 1966.
110. Krebs, R. A. and Burvant, M. U., Nephrotic syndrome in association with varicella, *JAMA*, 222, 325, 1972.

111. Sharman, V. L. and Goodwin, F. J., Hemolytic uremic syndrome following chickenpox, *Clin. Nephrol.*, 14, 49, 1980.
112. Ward, J. R. and Bishop, B., Varicella arthritis, *JAMA*, 212, 1954, 1970.
113. Priest, J. R., Urick, J. J., Groth, K. E., and Balfour, H. H., Varicella arthritis documented by isolation of virus from joint fluid, *J. Pediatr.*, 93, 990, 1978.
114. Friedman, A. and Naveh, Y., Polyarthritis associated with chickenpox, *Am. J. Dis. Child.*, 122, 179, 1971.
115. Mulhern, L. M., Friday, G. A., and Perri, J. A., Arthritis complicating varicella, *Pediatrics*, 48, 827, 1971.
116. DiLiberti, J. H., Bartel, S. J., Humphrey, T. R., and Pang, A. W., Acute monoarticular arthritis in association with varicella, *Clin. Pediatr.*, 16, 663, 1977.
117. Borgenicht, L., Varicella arthritis, *Pediatrics*, 66, 650, 1980.
118. Brook, I., Varicella arthritis in childhood, *Clin. Pediatr.*, 16, 1156, 1977.
119. Buck, R. E., Pyarthrosis of hip complicating chickenpox, *JAMA*, 206, 135, 1968.
120. Sethi, A. S. and Schloff, I., Purulent arthritis complicating chickenpox, *Clin. Pediatr.*, 13, 280, 1974.
121. Sabrazes, J., L'orchite de la varicelle, *Bull. Acad. Med. Paris*, 98, 122, 1927.
122. Wesselhoeft, C. and Person, C. M., Orchitis in the course of severe chickenpox with pneumonitis followed by testicular atrophy, *N. Engl. J. Med.*, 242, 651, 1950.
123. Ormiston, G., Orchitis as a complication of chickenpox, *Br. Med. J.*, 1, 1203, 1953.
124. Hackel, D. B., Myocarditis in association with varicella, *Am. J. Path.*, 29, 369, 1953.
125. Moore, C. M., Henry, J., Benzing, G., and Kaplan, S., Varicella myocarditis, *Am. J. Dis. Child.*, 118, 899, 1969.
126. Noren, G. R., Tobin, J. D., Staley, N. A., Asinger, R. W., and Einzig, S., Association of varicella, myocarditis, and congestive cardiomyopathy, *Pediatr. Cardiol.*, 3, 53, 1982.
127. Mandelbaum, T. and Terk, B. H., Pericarditis in association with chickenpox, *JAMA*, 170, 191, 1959.
128. Tatter, D., Fatal varicella pericarditis in a child, *Am. J. Dis. Child.*, 108, 88, 1964.
129. Griffin, W. P. and Searle, C. W. A., Ocular manifestations of varicella, *Lancet*, 2, 168, 1953.
130. Hallett, J. W., Report of a case of iridocyclitis associated with chickenpox, *Am. J. Ophthalmol.*, 29, 459, 1946.
131. Strachman, J., Uveitis associated with chickenpox, *J. Pediatr.*, 46, 327, 1955.
132. Edwards, T. S., Ophthalmic complications from varicella, *J. Pediatr. Ophthalmol.*, 2, 37, 1965.
133. Neil, L. P. and Watson, A. G., Internal ophthalmoplegia following chickenpox, *Can. J. Ophthalmol.*, 11, 267, 1976.
134. Appel, I., Frydman, M., Savin, H., and Elian, E., Uveitis and ophthalmoplegia complicating chickenpox, *J. Pediatr. Ophthalmol.*, 14, 346, 1977.
135. Robb, R. M., Cataracts acquired following varicella infection, *Arch. Ophthalmol.*, 87, 352, 1972.
136. Jordan, D. R., Noel, L. P., and Clarke, W. N., Ocular involvement in varicella, *Clin. Pediatr.*, 23, 435, 1983.
137. Uchida, Y., Kaneko, M., and Hayashi, K., Varicella dendritic keratitis, *Am. J. Ophthalmol.*, 89, 259, 1980.
138. Taylor, D. S. I., and Ffytche, T. J., Optic disc pigmentation associated with a field defect following chickenpox, *J. Pediatr. Ophthalmol.*, 13, 80, 1976.
139. Selbst, R. G., Selhorst, J. B., Harbison, J. W., and Myer, E. C., Parainfectious optic neuritis. Report and review following varicella, *Arch. Neurol.*, 40, 347, 1983.
140. Sharf, B., Hyams, S., Oculomotor palsy following varicella, *J. Pediatr. Ophthalmol.*, 9, 245, 1972.
141. Talbot, G. H., Skros, M., Fisher, M., and Friedman, H., Letter to the editor, *J. Infect. Dis.*, 149, 1035, 1984.
142. Toch, R., The interaction of varicella and acute leukemia, *Proc. Am. Assoc. Cancer. Res.*, 2, 255, 1957.
143. Pinkel, D., Chickenpox and leukemia, *J. Pediatr.*, 58, 729, 1961.
144. Bodey, G., McKelvey, E., and Karon, M., Chickenpox in leukemic patients: Factors in prognosis, *Pediatrics*, 34, 562, 1964.
145. Scheiman, J. I. and Stamler, F. W., Cyclophosphamide and fatal varicella, *J. Pediatr.*, 74, 118, 1969.
146. Cheatham, W. T., Weller, T. H., Dolan, T. F., and Dower, J. C., Varicella, report of two fatal cases with necroscopy, virus isolation with serologic studies, *Am. J. Path.*, 32, 1015, 1956.
147. Feldman, S., Hughes, W. T., and Daniel, C. B., Varicella in children with cancer, *Pediatrics*, 56, 388, 1975.
148. Whitley, R. J., Chickenpox in the Immunocompromised Host, in *Advances in Therapy against Herpes Virus Infections in Immunocompormised Hosts*, Balfour, H. H., Ed., Park Row Publications, New York, 1985, 25.

149. Feldman, S. and Epp, E., Isolation of varicella-zoster virus from blood, *J. Pediatr.*, 88, 265, 1976.
150. Gershon, A. A., Steinberg, S. P., and Silber, R., Varicella-zoster viremia, *J. Pediatr.*, 92, 1033, 1978.
151. Myers, M., Viremia caused by varicella-zoster virus: association with malignant progressive varicella, *J. Infect. Dis.*, 140, 229, 1979.
152. Feldman, S. and Epp, E., Detection of viremia during incubation of varicella, *J. Pediatr.*, 94, 746, 1979.
153. Morgan, E. R. and Smalley, L. A., Varicella in immunocompromised children, *Am. J. Dis. Child.*, 137, 883, 1983.
154. Ingraham, J. E., Estes, N. A., Bern, M. M., and DeGirolami, P. C., Disseminated varicella-zoster virus infection with the syndrome of inappropriate antidiuretic hormone, *Arch. Intern. Med.*, 143, 1270, 1983.
155. Miliouskas, J. R. and Webber, B. L., Disseminated varicella at autopsy in children with cancer, *Cancer*, 53, 1518, 1984.
156. Feldhoff, C. M., Balfour, H. H., Simmons, R. L., Najarian, J. S., and Mauer, S. M., Varicella in children with renal transplants, *J. Pediatr.*, 98, 25, 1981.
157. Locksley, R. M., Flournoy, N., Sullivan, K. M., and Myers, J. D., Infection with varicella-zoster virus after marrow transplantation, *J. Infect. Dis.*, 152, 1172, 1985.
158. Gershon, A. A., Steinberg, S. P., and Gelb, L., Clinical reinfection with varicella-zoster virus, *J. Infect. Dis.*, 149, 137, 1984.
159. Weel-Sipman, M. H., Van der Meer, J. W. M., de Koning, J., and Versteeg, J., Severe atypical recurrent varicella in childhood leukemia, *Lancet*, 1, 147, 1981.
160. Fallers, C. J. and Ellis, E. F., corticosteroids and varicella, *Arch. Dis. Child.*, 40, 595, 1965.
161. Finkel, K. C., Mortality from varicella in children receiving adrenocorticosteroids and adrenocorticotropin, *Pediatrics*, 28, 436, 1961.
162. Gershon, A. A., Brunell, P. A., and Doyle, E. F., Steroid therapy and varicella, *J. Pediatr.*, 81, 1032, 1973.
163. Keene, J. K., Lowe, D. K., Grosfeld, J. L., Fitzgerald, J. F., and Gonzalez-Crussi, F., Disseminated varicella complicating ulcerative colitis, *JAMA*, 239, 45, 1978.
164. Amman, A. J. and Hong, R., disorders of the T-cell system in *Immunologic Disorders in Infants and Children*, Steim, E. R. and Fulginiti, V. A., Eds., W. B. Saunders, Philadelphia, 1980, 260.
165. McKusick, V. A., Elderidge, E., Hosteller, J. A., Ruangivt, S., and Egeland, J. A., Dwarfism in the Amish. II. Cartilage-hair hypoplasia, *Bull. Johns Hopkins Hosp.*, 116, 285, 1964.
166. Lux, S. E., Johnston, R. B., August, C. S., Say, B., Penchaszadeh, V. B., Rosen, F. S., and McKusick, V. A., Chronic neutropenia and abnormal cellular immunity in cartilage-hair hypoplasia, *N. Engl. J. Med.*, 282, 234, 1970.
167. Wade, N. A., Lepow, M. L., Veazey, J., and Meuwissen, H. J., Progressive varicella in three patients with Wiskott-Aldrich syndrome: treatment with adenine arabinoside, *Pediatrics*, 75, 672, 1985.
168. Hook, E. B., Orandi, M., ten Bensel, R. W., Schamber, W. F., and St. Geme, J. W., Familial fatal varicella, *JAMA*, 206, 305, 1968.
169. Sever, J. and White, L. R., Intrauterine viral infections, *Annu. Rev. Med.*, 19, 471, 1968.
170. fish, S., Maternal death due to disseminated varicella, *JAMA*, 173, 978, 1960.
171. Pickard, R. E., Varicella pneumonia in pregnancy, *Am. J. Obstet. Gynecol.*, 101, 504, 1963.
172. Pearson, H. E., Parturition varicella-zoster, *Obstet. Gynecol.*, 23, 21, 1964.
173. Harris, R. E. and Rhoades, E. R., Varicella pneumonia complicating pregnancy: report of a case and review of literature, *Obstet. Gynecol.*, 23, 734, 1965.
174. Purtilo, D. T., Bhawan, J., and Liao, S., Fatal varicella in a pregnant woman and a baby, *Am. J. Obstet. Gynecol.*, 94, 1144, 1966.
175. Brunell, P. A., Varicella-zoster infections in pregnancy, *JAMA*, 199, 315, 1967.
176. Mendelow, D. A. and Lewis, G. G., Varicella pneumonia during pregnancy, *Obstet. Gynecol.*, 33, 98, 1969.
177. Gershon, A. A., Varicella in mother and infant: Problems old and new, in *Infections of the Fetus and Newborn. Progress in Clinical and Biological Research*, Vol. 3, Krugman, S. and Gershon, A. A., Eds., Alan R. Liss, New York, 1975, 79.
178. Brunell, P. A., Fetal and neonatal varicella-zoster infections, in *Virus Infections in Pregnancy*, Amstey, M. S., Ed., Grune & Stratton, Orlando, Fla., 1984.
179. Siegel, M., Fuerst, H. T., and Peress, N. S., Comparative fetal mortality in maternal virus diseases: a prospective study on rubella, measles, mumps, chickenpox, and hepatitis, *N. Engl. J. Med.*, 274, 768, 1966.
180. Paryani, S. and Arvin, A. M., Clinical and immunologic evidence of intrauterine varicella-zoster virus infection, *N. Engl. J. Med.*, 314, 1542, 1986.
181. Enders, G., Varicella-zoster infections in pregnancy, *Prog. Med. Virol.*, 29, 166, 1984.

182. Laforet, E. G. and Lynch, C. L., Multiple congenital defects following maternal varicella, *N. Engl. J. Med.*, 236, 534, 1947.
183. Fox, M. J., Krumbiegel, E. R., and Teresi, J. L., Maternal measles, mumps, and chickenpox as a cause of congenital anomalies, *Lancet*, 1, 746, 1948.
184. Freud, P., Congenital varicella, *Am. J. Dis. Child.*, 96, 730, 1958.
185. Garcia, A. G. P., Fetal infection in chickenpox and alastrim with histopathologic study of the placenta, *Pediatrics*, 32, 895, 1963.
186. Rimvik, R., Congenital varicella encephalomyelitis in surviving newborn, *Am. J. Dis. Child.*, 117, 231, 1969.
187. McKendry, J. B., and Bailey, J. D., Congenital varicella associated with multiple defects, *Calif. Med. Assoc. J.*, 108, 66, 1973.
188. Dodion-Fransen, J., Dekegel, D., and Thiry, L., Congenital varicella-zoster infection related to maternal disease in early pregnancy, *Scand. J. Infect. Dis.*, 5, 149, 1973.
189. Savage, M. O., Moosa, A., and Gordon, R. R., Maternal varicella infecftion as a cause of fetal malformations, *Lancet*, 1, 352, 1973.
190. Strabstein, J. C., Morris, N., and Larke, R. P. B., et al., Is there a congenital varicella syndrome?, *J. Pediatr.*, 84, 239, 1974.
191. Brice, J. E. H., Congenital varicella resulting from infection during second trimester of pregnancy, *Arch. Dis. Child.*, 51, 474, 1976.
192. Fucillo, D. A., Congenital varicella, *Teratology*, 15, 329, 1977.
193. Charles, N. C., Bennett, T. W., and Margolis, S., Ocular pathology of the congenital varicella syndrome, *Arch. Ophthalmol.*, 95, 2034, 1977.
194. Frey, H. M., Bialkin, G., and Gershon, A. A., Congenital varicella: case report of a serologically proved long-term survivor, *Pediatrics*, 59, 110, 1977.
195. DeNicola, L. K. and Hanshaw, H. B., Congenital and neonatal varicella, *J. Pediatr.*, 94, 175, 1979.
196. Alexander, I., Congenital varicella, *Br. Med. J.*, 2, 1074, 1979.
197. Borzyskowski, M., Harris, R. F., and Jones, R. W. A., The congenital varicella syndrome, *Eur. J. Pediatr.*, 137, 335, 1981.
198. Essex-Cater, A. and Heggarty, H., Fatal congenital varicella syndrome, *J. Infect.*, 7, 77, 1983.
199. Hanshaw, J. B. and Dudgeon, J. A., Viral diseases of the fetus and newborn, in *Major Problems in Pediatrics*, W. B. Saunders, Philadelphia, 1978, 192.
200. Young, N. A. and Gershon, A. A., Chickenpox, measles, and mumps, in *Infections of the Fetus and Newborn Infant*, Remington, J. S. and Klein, J. O., Eds., W. B. Saunders, Philadelphia, 1983, 375.
201. Siegel, M., Congenital malformations following chickenpox, measles, mumps, and hepatitis, *JAMA*, 226, 1521, 1973.
202. Asha Bai, P. V. and John, T. J., Congenital skin ulcers following varicella in late pregnancy, *J. Pediatr.*, 98, 71, 1981.
203. Brunell, P. A., Zoster in infancy: failure to maintain virus latency following intrauterine infection, *J. Pediatr.*, 98, 71, 1981.
204. Hubbard, T. W., Varicella occurring in an infant 24 hours after birth, *Br. Med. J.*, 1, 822, 1878.
205. Pridham, F. C., Chickenpox during intrauterine life, *Br. Med. J.*, 1, 1054, 1913.
206. Lucchesi, P. F., LaBoccetta, A. C., and Peale, A. R., Varicella neonatorum, *Am. J. Dis. Child.*, 73, 44, 1947.
207. Middlekamp, J. N., Varicella in newborn twins, *J. Pediatr.*, 43, 573, 1953.
208. Abler, C., Neonatal varicella: occurrence in babies born of infected mothers, *Am. J. Dis. Child.*, 107, 492, 1964.
209. Hyatt, H. W., Neonatal varicella: report of a case in a 5-day old infant and review of the literature, *J. Natl. Med. Assoc.*, 59, 32, 1967.
210. Myers, J. D., Congenital varicella in term infants: risk reconsidered, *J. Infect. Dis.*, 129, 215, 1974.
211. Brunell, P. A., Placental transfer of VZ antibody, *Pediatrics*, 38, 1034, 1966.
212. Preblud, S. R., Age specific risks of varicella complications, *Pediatrics*, 68, 14, 1981.
213. Readett, M. D. and McGibbon, C., Neonatal varicella, *Lancet*, 1, 644, 1961.
214. Trumble, G. X., Spread of varicella in nurseries, *Am. J. Dis. Child.*, 105, 315, 1963.
215. Matseone, S. L. and Abler, C., Occurrence of neonatal varicella in a hospital nursery, *Am. J. Obstet. Gynecol.*, 92, 575, 1965.
216. Gustafson, T. L., Shehab, Z., and Brunell, P. A., Outbreak of varicella in a newborn intensive care nursery, *Am. J. Dis. Child.*, 138, 548, 1984.
217. Hope-Simpson, R. E., Studies on shingles: is the virus the ordinary chickenpox virus?, *Lancet*, 2, 1299, 1954.
218. Weller, T. H. and Coons, A. H., Fluorescent antibody studies with agents of varicella and herpes zoster propagated *in vitro*, *Proc. Soc. Exp. Biol.*, 86, 789, 1954.

219. Weller, T. H. and Witton, H. M., The etiologic agents of varicella and herpes zoster: serologic studies with the viruses as propagated *in vitro, J. Exp. Med.*, 108, 869, 1958.
220. Straus, S. E., Reinhold, W., Smith, H. A., Ruyechan, W. T., Henderson, D. K., Blaese, R. M., and Hay, J., Endonuclease analysis of viral DNA from varicella and subsequent zoster infections in the same patient, *N. Engl. J. Med.*, 311, 1362, 1984.
221. Pichini, B., Grose, C., Ecker, J. R., and Hyman, R. W., DNA mapping of paired varicella zoster virus isolates from patients with shingles, *Lancet*, 2, 1233, 1983.
222. Denny-Brown, D. and Adams, R., Pathologic features of herpes zoster: a note on "geniculate herpes", *Arch. Neurol. Psychiat.*, 51, 216, 1944.
223. Ghatak, N. R. and Zurenima, H. M., Spinal ganglion in herpes zoster, *Arch. Pathol.*, 95, 411, 1973.
224. Esiri, M. M. and Tomlinson, A. H., Herpes zoster. Demonstration of virus in trigeminal nerve and ganglion by immunofluorescence and electron microscopy, *J. Neurol. Sci.*, 15, 35, 1972.
225. Shibata, H., Ishikawa, T., and Hondo, R., Varicella virus isolation from spinal ganglion, *Arch. Gesamte. Virusforsch.*, 45, 382, 1974.
226. Bastain, F. O., Rabson, A. S., Yee, C. L., and Tralka, T. S., Herpes virus varicella: isolated from human dorsal root ganglia, *Arch. Pathol.*, 97, 331, 1974.
227. Hyman, R. W., Ecker, J. R., and Tenser, R. B., Varicella-zoster virus RNA in human trigeminal ganglia, *Lancet*, 2, 814, 1983.
228. McGregor, R. M., Herpes zoster, chickenpox and cancer in general practice, *Br. Med. J.*, 1, 84, 1957.
229. Hope-Simpson, R. E., The nature of herpes zoster: a long-term study and a new hypothesis, *Proc. R. Soc. Med.*, 58, 9, 1965.
230. Ragozzino, M. W., Melton, L. J., Kurland, L. T., Chu, C. P., and Perry, H. O., Population-based study of herpes zoster and its sequelae, *Medicine*, 61, 310, 1982.
231. Guess, H. A., Broughton, D. D., Melton, L. J., and Kurland, L. T., Epidemiology of herpes zoster in children and adolescents: a population-based study, *Pediatrics*, 76, 512, 1985.
232. Latif, R. and Shope, T. C., Herpes zoster in normal and immunocompromised children, *Am. J. Dis. Child.*, 137, 801, 1983.
233. Dworsky, M., Whitley, R. J., and Alford, C., Herpes zoster in early infancy, *Am. J. Dis. Child.*, 134, 618, 1980.
234. Burgoon, C. F., Burgoon, J. S., and Baldridge, G. D., The natural history of herpes zoster, *JAMA*, 164, 265, 1957.
235. Zaia, J. A., Clinical spectrum of varicella-zoster virus infection in *The Human Herpes Viruses: an Interdisciplinary Perspective*, Nahmias, A. J., Dowdle, W. R., and Schinazi, R. E., Eds., Elsevier, New York, 1981, 10.
236. Whitley, R. J., Varicella-zoster virus infections in *Antiviral Agents and Viral Diseases of Man*, Galasso, G. J., Merigan, T. C., and Buchanan, R. A., Eds., Raven Press, New York, 1984, 517.
237. Brunell, P. A., Miller, L. H., and Lovejoy, F., Zoster in children, *Am. J. Dis. Child.*, 115, 432, 1968.
238. Winkelmann, R. K., Perry, H. O., Herpes zoster in children, *JAMA*, 171, 876, 1959.
239. Rifkind, D., The activation of varicella-zoster virus infections by immunosuppressive therapy, *J. Labor. Clin. Med.*, 68, 463, 1966.
240. Schimpff, S., Serpick, A., Stolar, B., Varicella-zoster infection in patients with cancer, *Ann. Intern. Med.*, 76, 241, 1972.
241. Dolin, R., Reichman, R. C., Mazur, M. H., and Whitley, R. J., Herpes zoster-varicella infections in immunosuppressed patients, *Ann. Intern. Med.*, 89, 375, 1978.
242. Mazur, M. H. and Dolin, R., Herpes zoster at the NIH: a 20 year experience, *Am. J. Med.*, 65, 738, 1978.
243. Rosenberg, S. A., Diamond, H. D., and Jaslowitz, B., Lymphosarcoma: a review of 1269 cases, *Medicine*, 40, 31, 1961.
244. Sokal, J. E. and Firat, D., Varicella-zoster infections in Hodgkin's disease, *Am. J. Med.*, 39, 452, 1965.
245. Goffinet, D. R., Glatstein, E. J., and Merigan, T. C., Herpes zoster-varicella infections and lymphoma, *Ann. Intern. Med.*, 76, 235, 1972.
246. Wilson, J. F., Marsa, G. W., and Johnson, R. E., Herpes zoster in Hodgkin's disease: clinical, histologic, and immunologic correlations, *Cancer*, 29, 461, 1972.
247. Feld, R., Evans, W. K., and DeBoer, G., Herpes zoster in patients with small-cell carcinoma of the lung receiving combined modality therapy, *Ann. Intern. Med.*, 93, 282, 1980.
248. Pancoast, H. K. and Pendergrass, E. P., The occurrence of herpes zoster in Hodgkin's disease, *Am. J. Med. Sci.*, 168, 326, 1924.
249. Manning, D. M., Luparello, F. J., and Arena, V. C., Herpes zoster after splenectomy: a study of patients without malignancy, *N. Engl. J. Med.*, 243, 56, 1980.

250. Bacon, G. E., Oliver, W. J., and Shapiro, B. A., Factors contributing to severity of herpes zoster in children, *J. Pediatr.*, 67, 768, 1965.
251. Feldman, S., Hughes, W. T., and Kim, H. Y., Herpes zoster in children with cancer, *Am. J. Dis. Child.*, 126, 178, 1973.
252. Smith, K. L., Johnson, D., Hustu, O., Pratt, C., Fleming, I., and Holton, C., Concurrent chemotherapy and radiation therapy in the treatment of childhood and adolescent Hodgkin's disease, *Cancer*, 33, 38, 1974.
253. Reboul, F., Donaldson, S. S., and Kaplan, H. S., Herpes zoster and varicella infections in children with Hodgkin's disease: an analysis of contributing factors, *Cancer*, 41, 95, 1978.
254. Skovby, F. and Sullivan, M. P., Herpes zoster and varicella in children with Hodgkin's disease, *Acta Pediatr. Scand.*, 71, 269, 1982.
255. Ragozzino, M. W., Melton, L. J., Kurland, L. T., Chu, C. P., and Perry, H. O., Risk of cancer after herpes zoster: a population-based study, *N. Engl. J. Med.*, 307, 393, 1982.
256. Luby, J. P., Ramirez-Ronda, C., Rinner, S., Hull, A., and Verge-Marini, P., A longitudinal study of varicella-zoster virus infections in renal transplant recipients, *J. Infect. Dis.*, 135, 659, 1977.
257. Hurley, J. K., Greenslade, T., Lewy, P. R., Ahmadian, Y., and Firlit, C., Varicella-zoster infections in pediatric renal transplant recipients, *Arch. Surg.*, 115, 751, 1980.
258. Rand, K. H., Rasmussen, L. E., Pollard, R. B., Arvin, A. M., and Merigan, T. C., Cellular immunity and herpes virus infections in cardiac transplant patients, *N. Engl. J. Med.*, 296, 1372, 1977.
259. Montsopoulos, H. M., Gallagher, J. D., Decker, J. L., and Steinberg, A. D., Herpes zoster in patients with systemic lupus erythematosis, *Arthritis Rheum.*, 21, 798, 1978.
260. Payne, C. M. E. R., Farthing, C., Byron, N., and Staughton, R. C. D., Shingles in seven homosexuals, *Lancet*, 1, 103, 1984.
261. Gallagher, J. G. and Merigan, T. C., Prolonged herpes zoster infection associated with immunosuppressive therapy, *Ann. Intern. Med.*, 91, 842, 1979.
262. Feldman, S., Chaudary, S., Ossi, M., and Epp, E., A viremic phase for herpes zoster in children with cancer, *J. Pediatr.*, 91, 597, 1977.
263. Mersalis, J. G., Haye, D., and Hook, E. W., Disseminated herpes zoster: a report of 17 cases, *Arch. Intern. Med.*, 113, 679, 1964.
264. Rogers, R. S. and Tindall, J. P., Geriatric herpes zoster, *J. Am. Geriatr. Soc.*, 19, 491, 1971.
265. Gold, E. and Robbins, F. C., Isolation of herpes zoster virus from spinal fluid of a patient, *Virology*, 6, 293, 1958.
266. Evans, A. S. and Melnick, J. L., Electron microscopic studies of the vesicle and spinal fluids from a case of herpes zoster, *Proc. Soc. Exp. Biol. Med.*, 71, 283, 1949.
267. Andiman, W. A., White-Greenwald, M., and Trighitella, T., Zoster encephalitis. Isolation of virus and measurement of varicella-zoster specific antibodies in cerebrospinal fluid, *Am. J. Med.*, 73, 769, 1982.
268. Bieger, R. C., Van Scoy, R. E., and Smith, T. F., Antibodies to varicella-zoster in cerebrospinal fluid, *Arch. Neurol.*, 34, 489, 1977.
269. Gordon, I. R. S. and Tucker, S. F., Lesions of the central nervous system in herpes zoster, *J. Neurol. Neurosurg. Psychiat.*, 8, 40, 1945.
270. Appelbaum, E., Kreps, S. I., and Sunshine, A., Herpes zoster encephalitis, *Am. J. Med.*, 32, 25, 1962.
271. Norris, F. H., Leonards, R., Calanchini, P. R., and Calder, C. D., Herpes zoster meningoencephalitis, *J. Infect. Dis.*, 122, 335, 1970.
272. Reichman, R. C., Neurologic complications of varicella-zoster infections, *Ann. Intern. Med.*, 89, 375, 1978.
273. Thomas, J. E. and Howard, F. M., Segmental zoster paresis — a disease profile, *Neurology*, 22, 459, 1972.
274. Jamsek, J., Greenberg, S., and Taber, L., Herpes zoster associated encephalitis: clinicopathologic report of 12 cases and review of the literature, *Medicine*, 62, 81, 1983.
275. Boughton, C. R., Varicella-zoster in Sydney: II. Herpes zoster and complications, *Med. J. Aust.*, 2, 502, 1966.
276. O'Donnell, P. P., Pula, T. P., and Sellman, M., Recurrent herpes zoster encephalitis, *Arch. Neurol.*, 38, 49, 1981.
277. Dix, R. D., Bredesen, D. E., Erlich, K. S., and Mills, J., Recovery of herpes viruses from cerebrospinal fluid of immunodeficient homosexual men, *Ann. Neurol.*, 18, 611, 1985.
278. Laws, H. W., Herpes zoster ophthalmicus complicated by contralateral hemiplegia, *Arch. Ophthalmol.*, 63, 273, 1960.

279. Acers, T. E., Herpes zoster ophthalmicus with contralateral hemiplegia, *Arch. Ophthalmol.*, 71, 371, 1964.
280. Walker, R. J., Gammal, T. E., and Allen, M. B., Cranial arteritis associated with herpes zoster, *Radiology*, 107, 109, 1973.
281. Gilbert, G. J., Herpes zoster ophthalmicus and delayed contralateral hemiparesis: relationship of the syndrome to central nervous system granulomatous angiitis, *JAMA*, 229, 302, 1974.
282. Reshef, E., Greenberg, S. B., and Jankovich, J., Herpes zoster ophthalmicus followed by contralateral hemiparesis: report of two cases and review of literature, *J. Neurol. Neurosurg. Psych.*, 48, 122, 1985.
283. MacKenzie, R. A., Forbes, G. S., and Karnes, W. E., Angiographic findings in herpes zoster arteritis, *Ann. Neurol.*, 10, 458, 1981.
284. Hilt, D. C., Buchholz, D., Krumholz, A., Weiss, H., and Wolinsky, J. S., Herpes zoster ophthalmicus and delayed contralateral hemiparesis caused by cerebral angiitis: diagnosis and management approaches, *Ann. Neurol.*, 14, 543, 1983.
285. Rosenblum, W. I. and Hadfield, M. G., Granulomatous angiitis of the nervous system in cases of herpes zoster and lymphosarcoma, *Neurology*, 22, 348, 1972.
286. Linneman, C. C. and Alvira, M. M., Pathogenesis of varicella-zoster angiitis in the CNS, *Arch. Neurol.*, 37, 239, 1980.
287. Edgerton, A. E., Herpes zoster ophthalmicus: report of cases and a review of the literature, *Trans. Am. Ophthalmol. Soc.*, 40, 390, 1942.
288. Pemberton, J. W., Optic atrophy in herpes zoster ophthalmicus, *Am. J. Ophthalmol.*, 58, 852, 1964.
289. Harrison, E. Q., Complications of herpes zoster ophthalmicus: a case report, *Am. J. Ophthalmol.*, 60, 1111, 1965.
290. Womack, L. W. and Liesegang, T. J., Complications of herpes zoster ophthalmicus, *Arch. Ophthalmol.*, 101, 42, 1983.
291. Aitken, R. S. and Brain, R. T., Facial palsy and infection with zoster virus, *Lancet*, 1, 19, 1933.
292. Mair, I. W. S. and Flugsrud, L. B., Peripheral facial palsy and herpes zoster infection, *J. Laryngol. Otol.*, 90, 373, 1976.
293. Peiterson, E. and Caunt, A. E., The incidence of herpes zoster antibodies in patients with peripheral facial palsy, *J. Laryngol. Otol.*, 84, 65, 1970.
294. Payten, R. J. and Dawes, J. D. K., Herpes zoster of the head and neck, *J. Laryngol. Otol.*, 86, 1031, 1972.
295. Crabtree, J. A., Herpes zoster oticus, *Laryngoscope*, 78, 1853, 1968.
296. Hunt, J. R., On herpetic inflammation of the geniculate ganglion: a new syndrome and its complications, *J. Nerv. Ment. Dis.*, 34, 73, 1907.
297. Tomita, H., Idiopathic facial palsy and varicella-zoster virus, *Jpn. Med. J.*, 2459, 31, 1971.
298. Ohsaski, M., Chiba, S., and Nakao, T., Bell's palsy in infants associated with varicella-zoster virus infection, *J. Pediatr.*, 84, 103, 1974.
299. Lewis, G. W., Zoster sine herpete, *Br. Med. J.*, 2, 418, 1958.
300. Broadbent, W. H., Case of herpetic eruption in the course of branches of the brachial plexus followed by partial paralysis in corresponding motor nerves, *Br. Med. J.*, 2, 460, 1860.
301. Gupta, S. K., Helal, B. H., and Kiely, P., The prognosis in zoster paralysis, *J. Bone J. Surg.*, 51, 593, 1969.
302. Anderson, J. P., Paralysis in herpes zoster, *Br. Med. J.*, 3, 587, 1970.
303. Gibbon, N. K. O., A case of herpes zoster with involvement of the urinary bladder, *Br. J. Urol.*, 2, 417, 1956.
304. Izumi, A. I. and Edwards, J., Herpes zoster and neurogenic bladder dysfunction, *JAMA*, 224, 1748, 1973.
305. Richmond, W., The genito-urinary manifestations of herpes zoster: three case reports and a review of the literature, *Br. J. Urol.*, 46, 193, 1974.
306. Syburn-Mason, R., Visceral lesions in herpes zoster, *Br. Med. J.*, 1, 618, 1957.
307. Figiel, S. J. and Figiel, L. S., Herpes zoster with ileus stimulating intestinal obstruction, *Am. J. Med.*, 23, 999, 1957.
308. Anderson, J. P. and Keal, E. E., Cervical herpes zoster and diaphragmatic paralysis, *Br. J. Dis. Chest.*, 63, 222, 1969.
309. Nachman, A. R., Neurologic complications of herpes zoster, *Pediatrics*, 7, 200, 1951.
310. Rose, F. C., Brett, E. M., and Burston, J., Zoster encephalomyelitis, *Arch. Neurol.*, 11, 155, 1964.
311. Grant, D. B. and Rowe, C. R., Motor paralysis of the extremities in herpes zoster, *J. Bone J. Surg.*, 43, 885, 1961.
312. Kendall, D., Motor complications of herpes zoster, *Br. Med. J.*, 2, 616, 1957.
313. Hogan, E. L. and Krigman, M. R., Herpes zoster myelitis, *Arch. Neurol.*, 29, 309, 1973.

314. Goodman, C. E. and Kendrick, M. M., Herpes zoster with motor paresis, *South. Med. J.*, 67, 1171, 1974.
315. Nyberg-Hansen, R., Relapsing herpes zoster myelitis, *Acta Neurol. Scand.*, 48, 397, 1972.
316. Knox, J. D. E., Levy, R., and Simpson, J. A., Herpes zoster and the Landry-Guillain-Barre syndrome, *J. Neurol. Neurosurg. Psychiat.*, 24, 167, 1961.
317. Singh, S., Malhotra, V., and Malhotra, R. P., Guillain-Barre syndrome in herpes zoster, *N.Y. State J. Med.*, 72, 2094, 1972.
318. Eyal, A., Friedman, M., Peretz, B. A., and Paldi, E., Pregnancy complicated by herpes zoster. A report of two cases and literature review, *J. Reprod. Med.*, 28, 600, 1983.
319. Duehr, P., Herpes zoster as a cause of congenital cataract, *Am. J. Ophthalmol.*, 30, 157, 1955.
320. Webster, M. H. and Smith, C. S., Congenital abnormalities and maternal herpes zoster, *Br. Med. J.*, 2, 1193, 1977.
321. Asano, Y., Itakura, N., Hiroishi, Y., Hirose, S., Ozaki, T., Kuno, K., Nagai, T., Yazaki, T., Yamanishi, K., and Takahasi, M., Viral replication and immunologic responses in children naturally infected with varicella-zoster virus and in varicella vaccine recipients, *J. Infect. Dis.*, 152, 863, 1985.
322. Arvin, A. M., Koropchak, C. M., Williams, B. R. G., Grumet, F. C., and Foung, S. K. H., The early immune response in healthy and immunocompromised subjects with primary varicella-zoster virus infection, *J. Infect. Dis.*, 154, 422, 1986.
323. Gold, E., Serologic and virus isolation studies of patients with varicella or herpes-zoster infection, *N. Engl. J. Med.*, 274, 181, 1966.
324. Brunell, P. A., Gershon, A. A., Uduman, S. A., and Steinberg, S., Varicella-zoster immunoglobulins during varicella, latency and zoster, *J. Infect. Dis.*, 132, 49, 1975.
325. Friedman, M. G., Leventon-Kriss, S., and Sarov, I., Sensitive solid phase radioimmunoassay for detection of human immunoglobulin G antibodies to varicella-zoster virus, *J. Clin. Microbiol.*, 9, 1, 1979.
326. Arvin, A. M. and Koropchak, C. M., Immunoglobulins M and G to varicella and herpes zoster infections, *J. Clin. Microbiol.*, 12, 367, 1980.
327. Bogger-Goren, S., Baba, K., Hurley, P., Yabuchi, H., Takahashi, M., and Ogra, P. L., Antibody response to varicella-zoster virus after natural or vaccine-induced infection, *J. Infect. Dis.*, 146, 260, 1982.
328. Williams, V., Gershon, A. A., and Brunell, P., Serologic response to varicella-zoster membrane antigens measured by indirect immunofluorescence, *J. Infect. Dis.*, 130, 669, 1974.
329. Zaia, J. A. and Oxman, M. N., Antibody to varicella-zoster virus-induced membrane antigen: immunofluorescence assay using monodisperse gluteraldehyde-fixed target cells, *J. Infect. Dis.*, 136, 519, 1977.
330. Sundquist, V. A., Linde, A., and Wahren, B., Virus-specific immunoglobulin G subclass in herpes simplex and varicella-zoster virus infections, *J. Clin. Microbiol.*, 20, 94, 1984.
331. Shore, S. L. and Cross, G. D., Lysis of varicella-zoster virus infected cells by human antiviral antibody and complement, *Am. Soc. Microbiol.*, (Abstr. E106), 1981.
332. Leonard, L. L., Schmidt, N. J., and Lennette, E. H., Demonstration of viral antibody activity in two immunoglobulin subclasses in patients with varicella-zoster infection, *J. Immunol.*, 104, 23, 1970.
333. Schmidt, N. J. and Lennette, E. H., Neutralizing antibody responses to varicella-zoster virus, *Infect. Immun.*, 12, 606, 1975.
334. Grose, C., Edmond, B. J., and Brunell, P. A., Complement enhanced neutralizing antibody response to varicella-zoster virus, *J. Infect. Dis.*, 139, 432, 1979.
335. Gershon, A. A. and Steinberg, S. P., Inactivation of varicella-zoster virus in vitro: effect of leukocytes and specific antibody, *Infect. Immun.*, 33, 507, 1981.
336. Kamiya, H., Starr, S. E., Arbeter, A. M., and Plotkin, S. A., Antibody-dependent cell-mediated cytotoxicity against varicella-zoster virus-infected targets, *Infect. Immun.*, 38, 554, 1982.
337. Babbage, J., Sigfusson, A., and Souhami, R. L., Antibody-dependent cell-mediated cytotoxicity to varicella-zoster, *Clin. Exp. Immunol.*, 58, 217, 1984.
338. Ihara, T., Starr, S. E., Ito, M., Douglas, S. D., and Arbeter, A. M., Human polymorphonuclear leukocyte-mediated cytotoxicity against varicella-zoster virus-infected fibroblasts, *J. Virol.*, 51, 110, 1984.
339. Szanton, E. and Sarov, I., Interaction between polymorphonuclear leukocytes and varicella-zoster virus-infected cells, *Intervirology*, 24, 119, 1985.
340. Ihara, T., Ito, M., and Starr, S. E., Human lymphocyte, monocyte, and polymorphonuclear leukocyte mediated antibody-dependent cellular cytotoxicity against varicella-zoster virus-infected targets, *Clin. Exp. Immunol.*, 63, 179, 1986.
341. Weller, T. and Witton, H., The etiologic agents of varicella and zoster. Serologic studies with the virus as propagated in vitro, *J. Exp. Med.*, 108, 869, 1958.

342. Stevens, D. A., Ferrington, R. A., Jordan, G. W., and Merigan, T. C., Cellular events in zoster vesicles: relation to clinical course and immune parameters, *J. Infect. Dis.*, 131, 509, 1975.
343. Schmidt, N. J., Dennis, J., and Lennette, E. H., Complement-fixing reactivity of varicella-zoster virus subunit antigens with sera from monotypic infections and heterotypic herpes simplex virus infections, *Infect. Immun.*, 15, 850, 1977.
344. Gerna, G., Cereda, P. M., Cattaneo, S., Achilli, G., and Gerna, M. T., Antibody to early antigens of varicella-zoster virus during varicella and zoster, *J. Infect. Dis.*, 140, 33, 1979.
345. Schmidt, N. J. and Gallo, D., Class-specific antibody responses to early and late antigens of varicella and herpes simplex viruses, *J. Med. Virol.*, 13, 1, 1984.
346. Zweerink, H. J. and Neff, B. J., Immune response after exposure to varicella-zoster virus: characterization of virus-specific antibodies and their corresponding antigens, *Infect. Immun.*, 31, 436, 1981.
347. Weigle, K. A. and Grose, C., Molecular dissection of the humoral immune response to individual varicella-zoster viral proteins during chickenpox, quiescence, reinfection, and reactivation, *J. Infect. Dis.*, 149, 741, 1984.
348. Palumbo, P. G., Arvin, A. M., Koropchak, C. M., and Wittek, A. E., Investigation of varicella-zoster virus-infected cell proteins that elicit antibody production during primary varicella using the immune transfer method, *J. Gen. Virol.*, 65, 2141, 1984.
349. Killander, C. F. R., Gronowitz, J. S., and Torfason, E. G., Human serum antibodies to varicella-zoster virus thymidine kinase, *Infect. Immun.*, 36, 30, 1982.
350. Arvin, A. M., Kinney-Thomas, E., Shriver, K., Grose, C., Koropchak, C. M., Scranton, E., Wittek, A. E., and Diaz, P. S., Immunity to varicella-zoster viral glycoproteins, gpI (gp90/58), gpIII (gp 118), and to a nonglycosylated protein, *J. Immunol.*, 137, 1346, 1986.
351. Foung, S. K. H., Perkins, S., Koropchak, C. M., Fishwild, D. M., Wittek, A. E., Engleman, E. G., Grumet, F. C., and Arvin, A. M., Human monoclonal antibodies neutralizing varicella-zoster virus, *J. Infect. Dis.*, 152, 280, 1985.
352. Ito, M., Ihara, T., Grose, C., and Starr, S. E., Human leukocytes kill varicella-zoster virus-infected fibroblasts in the presence of murine monoclonal antibodies to virus-specific glycoproteins, *J. Virol.*, 54, 98, 1985.
353. Levy, E. and Sarov, I., Detection of specific IgA antibodies in serum of patients with varicella and zoster infections, *Intervirology*, 15, 103, 1981.
354. Haikiu, H. and Sarov, I., Determination of specific IgA antibodies to varicella-zoster virus by immunoperoxidase assay, *J. Clin. Pathol.*, 35, 645, 1982.
355. Levy, E., Mosovitz, B., Friedman, M., and Sarov, I., Detection of varicella-zoster virus-specific IgA antibodies in varicella and zoster patients and in healthy adults of various ages by solid phase radioimmunoassay, *Intervirology*, 20, 123, 1983.
356. Wittek, A. E., Arvin, A. M., and Koropchak, C. M., Serum immunoglobulin A antibody to varicella-zoster virus in subjects with primary varicella and herpes zoster infections and in immune subjects, *J. Clin. Microbiol.*, 18, 1146, 1983.
357. Gerna, G., Achilli, G., and Chambers, R. W., Determination of neutralizing antibody and IgG antibody to varicella-zoster virus and IgG antibody to membrane antigens by the immunoperoxidase technique, *J. Infect. Dis.*, 135, 975, 1977.
358. Forghani, B., Schmidt, N. J., and Dennis, J., Antibody assays for varicella-zoster virus: comparison of enzyme immunoassay with neutralization, immune adherence hemagglutination and complement fixation, *J. Clin. Micro.*, 8, 545, 1978.
359. Gershon, A. A., Frey, H. M., and Steinberg, S. P., Determination of immunity to varicella using an enzyme-linked immunosorbent assay, *Arch. Virol.*, 70, 169, 1981.
360. Shehab, Z. and Brunell, P. A., Enzyme-linked immunosorbent assay for susceptibility to varicella, *J. Infect. Dis.*, 148, 472, 1983.
361. Souhani, R. L., Babbage, J., and Callard, R. E., Specific *in vitro* antibody response to varicella-zoster, *Clin. Exp. Immunol.*, 46, 98, 1981.
362. Bardy, B., Seuna, L., and Hovi, T., Age-specific prevalence of antibodies to pyrimidine kinase enzymes of herpes simplex virus types 1 and 2 and of varicella-zoster virus, *J. Med. Virol.*, 18, 61, 1986.
363. Arvin, A. M., Koropchak, C. M., and Wittek, A. E., Immunologic evidence for re-infection with varicella-zoster virus, *J. Infect. Dis.*, 148, 200, 1983.
364. Baba, K., Yabuuchi, H., Takahashi, M., and Ogra, P. L., Increased incidence of herpes zoster in normal children infected with varicella-zoster virus during infancy: community-based follow-up study, *J. Pediatr.*, 108, 373, 1986.
365. Gershon, A. A., Steinberg, S. P., Borkowsky, W., Lennette, D., and Lennette, E., IgM to varicella-zoster virus: demonstration in patients with and without clinical zoster, *Ped. Infect. Dis.*, 1, 164, 1982.
366. Caunt, A. E. and Shaw, D. G., Neutralization tests with varicella-zoster virus, *J. Hyg.* (Cambridge), 67, 343, 1969.

367. Grose, C., Zoster in children with cancer: radioimmune precipitation profiles of sera before and after illness, *J. Infect. Dis.*, 147, 47, 1983.
368. Miller, L. H. and Brunell, P. A., Zoster reinfection or activation of latent virus: observations on the antibody response, *Am. J. Med.*, 49, 480, 1970.
369. Ross, C. A. C. and McDaid, R., Specific IgM antibody in serum of patients with herpes zoster infections, *Br. Med. J.*, 2, 522, 1972.
370. Tedder, R. S., Mortimer, P. P., and Lord, R. B., Detection of antibody to varicella-zoster virus by competitive and IgM antibody capture immunoassay, *J. Med. Virol.*, 8, 89, 1981.
371. Cradock-Watson, J. E., Ridehalgh, M. K. S., and Bourne, M. S., Specific immunoglobulin responses after varicella and herpes zoster, *J. Hyg.* (Cambridge), 82, 319, 1979.
372. Sundquist, V. A., Frequency and specificity of varicella-zoster virus IgM response, *J. Virol. Meth.*, 5, 219, 1982.
373. Forghani, B., Myoraku, C. K., Dupuis, K. W., and Schmidt, N. J., Antibody class capture assays for varicella-zoster virus, *J. Clin. Microbiol.*, 19, 606, 1984.
374. Schmidt, N. J. and Arvin, A. M., Immunoglobulin M responses to varicella-zoster virus in reactivated infections (zoster): sensitivity of different assay systems, *J. Clin. Microbiol.*, 23, 978, 1986.
375. Orenstein, W. A., Heymann, D. L., Ellis, R. J., Rosenberg, R. L., Nakaro, J., Halsey, N. A., Overturf, G. D., Hayden, G. F., and Witte, J. J., Prophylaxis of varicella in high-risk children: dose response effect of zoster immunoglobulin, *J. Pediatr.*, 98, 368, 1981.
376. Baba, K., Yabunchi, H., Takahashi, M., and Ogra, P. L., Immunologic and epidemiologic aspects of varicella infection acquired during infancy and early childhood, *J. Pediatr.*, 100, 881, 1982.
377. Hanngren, K., Grandien, M., and Granstrom, G., Effect of zoster immunoglobulin for varicella prophylaxis in the newborn, *Scand. J. Infect. Dis.*, 17, 343, 1985.
378. Gershon, A. A. and Steinberg, S. P., Cellular and humoral immune responses to varicella-zoster virus in immunocompromised patients during and after varicella-zoster infection, *Infect. Immun.*, 25, 170, 1979.
379. Gershon, A. A. and Steinberg, S. P., Antibody responses to varicella-zoster virus and the role of antibody in host defense, *Am. J. Med. Sci.*, 282, 12, 1981.
380. Arvin, A. M., Pollard, R. B., Rasmussen, L. E., and Merigan, T. C., Cellular and humoral immunity in the pathogenesis of recurrent herpes viral infections in patients with lymphoma, *J. Clin. Invest.*, 65, 869, 1980.
381. Pollard, R. B., Arvin, A. M., Ganberg, P., Rand, K. H., Gallagher, J. G., and Merigan, T. C., Specific cell-mediated immunity and infections with herpes viruses in cardiac transplant recipients, *Am. J. Med.*, 73, 679, 1982.
382. Mazur, M. H., Whitley, R. J., and Dolin, R., Serum antibody levels as risk factors in the dissemination of herpes zoster, *Arch. Intern. Med.*, 139, 1341, 1979.
383. Larkin, M., Heckels, J. E., and Ogilvie, M. M., Antibody response to varicella-zoster virus surface glycoproteins in chickenpox and shingles, *J. Gen. Virol.*, 66, 1785, 1985.
384. Stevens, D. A. and Merigan, T. C., Zoster immunoglobulin prophylaxis of disseminated zoster in compromised hosts: a randomized trial, *Arch. Intern. Med.*, 140, 28, 1980.
385. Ihara, T., Starr, S. E., Arbeter, A. M., and Plotkin, S. A., Effects of interferon on natural killing and antibody dependent cellular cytotoxicity against varicella-zoster virus infected and uninfected cells, *J. Interferon Res.*, 3, 263, 1983.
386. Bowden, R. A., Levin, M. J., Giller, R. H., Tubergen, D. G., and Hayward, A. R., Lysis of varicella-zoster virus infected cells by lymphocytes from normal humans and immunosuppressed pediatric leukemic patients, *Clin. Exp. Immunol.*, 60, 387, 1985.
387. Ito, M., Bandyopadhyay, S., Matsumoto-Kobayashi, M., Clark, S. C., Miller, D., and Starr, S. E., Interleukin 2 enhances natural killing of varicella-zoster virus infected targets, *Clin. Exp. Immunol.*, 65, 182, 1986.
388. Hayward, A. R., Herberger, M., Lazslo, M., Cellular interactions in the lysis of varicella-zoster virus infected human fibroblasts, *Clin. Exp. Immunol.*, 63, 141, 1986.
389. Rasmussen, L. E., Holmes, A., Hofmeister, B., and Merigan, T. C., Multiplicity dependent replication in interferon-treated cells of varicella-zoster virus, *J. Gen. Virol.*, 35, 361, 1977.
390. Arvin, A. M., Kushner, J. H., Feldman, S., Baehner, R. L., Hammand, D., and Merigan, T. C., Human leukocyte interferon for the treatment of varicella in children with cancer, *N. Engl. J. Med.*, 306, 761, 1982.
391. Arvin, A. M., Martin, D. P., Gard, E. A., and Merigan, T. C., Interferon prophylaxis for simian varicella infection in *Erythrocebus patas* monkeys, *J. Infect. Dis.*, 147, 149, 1983.
392. Armstrong, R. W., Gurwith, M. J., Waddell, D., and Merigan, T. C., Cutaneous interferon production in patients with Hodgkin's disease and other cancers infected with varicella or vaccinia, *N. Engl. J. Med.*, 283, 1182, 1970.

393. Patel, P. A., Yoonessi, S., O'Malley, J., Freeman, A., Gershon, A. A., and Ogra, P. L., Cell-mediated immunity to varicella-zoster virus infection in subjects with lymphoma and leukemia, *J. Pediatr.*, 94, 223, 1979.
394. Williams, B. R. G., Read, S. E., Freedman, M. H., Carver, D. H., and Gelfand, E. W., The assay of 2-5A synthetase as an indicator of interferon activity and virus infection in vivo, in *Chemistry and Biology of Interferons*, Vol. 25, UCLA Symp. Mol. Cell. Biol., Merigan, T., Friedman, R., and Fox, C. F., Eds., Academic Press, New York, 1982, 253.
395. Bertotto, G., Genitli, F., and Vaccaro, R., Lymphocyte surface phenotypes in children with varicella, *N. Engl. J. Med.*, 307, 1271, 1982.
396. Arneborn, P. and Biberfield, G., T-lymphocyte subpopulations in relation to immunosuppression in measles and varicella, *Infect. Immun.*, 39, 29, 1983.
397. Mitchell, A. G., Nelson, W. E., and Leblanc, T. J., Studies in immunity. V. Effect of acute diseases on the reaction of the skin to tuberculin, *Am. J. Dis. Child.*, 49, 695, 1935.
398. Starr, S., Berkovich, S., The depression of tuberculin reactivity during chicken pox, *Pediatrics*, 33, 769, 1964.
399. Belsey, M. A., Tuberculosis and varicella infections in children, *Am. J. Dis. Child.*, 113, 443, 1967.
400. Hayes, F. A. and Feldman, S., Cell-mediated immunity to varicella-zoster virus in children being treated for cancer, *Cancer*, 42, 159, 1978.
401. Kumagai, T., Chiba, Y., Wataya, Y., Hanazono, H., Chiba, S., and Nakao, T., Development and characteristics of the cellular immune response to infection with varicella-zoster virus, *J. Infect. Dis.*, 141, 7, 1980.
402. Steele, R. W., Transfer factor and cellular reactivity to varicella-zoster antigens in childhood leukemia, *Cell. Immunol.*, 50, 282, 1980.
403. Gershon, A. A., Steinberg, S. P., and Smith, M., Cell-mediated immunity to varicella-zoster virus demonstrated by viral inactivation with human leukocytes, *Infect. Immun.*, 13, 1549, 1976.
404. Gershon, A. A. and Steinberg, S. P., Cell-mediated immunity to varicella zoster virus measured by virus inactivation: mechanism and blocking of the reaction by specific antibody, *Infect. Immun.*, 25, 164, 1979.
405. Hayward, A. R., Pontesilli, O., Herberger, M., Laszlo, M., and Levin, M., Specific lysis of varicella-zoster virus infected B-lymphocytes by human T cells, *J. Virol.*, 58, 179, 1986.
406. Arbeit, R. D., Zaia, J. A., Valerio, M. A., Levin, M. J., Infection of human peripheral blood mononuclear cells by varicella-zoster virus, *Intervirol.*, 18, 56, 1982.
407. Baba, M. and Shigeta, S., Incomplete growth of varicella-zoster virus in human monocytes, *Microbiol. Immunol.*, 27, 767, 1983.
408. Gresser, I., Vignaux, F., Belardelli, F., Tovey, M., Maunoury, M., Injection of mice with antibody to mouse interferon a/b decreases the level of 2'5'oligoadenylate synthetase in peritoneal macrophages, *J. Virol.*, 53, 221, 1985.
409. Jordan, G. W. and Merigan, T. C., Cell mediated immunity to varicella-zoster virus: in vitro lymphocyte responses, *J. Infect. Dis.*, 130, 495, 1974.
410. Ruckdeschel, J. C. and Mardiney, M. R., The demonstration of cell-associated immunity to viruses. In vitro lymphocyte responsiveness to varicella-zoster antigen, *J. Immunol. Meth.*, 13, 131, 1976.
411. Zaia, J. A., Leary, P. L., and Levin, M. J., Specificity of the blastogenic response of human mononuclear cells to herpes virus antigens, *Infect. Immun.*, 20, 646, 1978.
412. Bogger-Goren, S., Bernstein, J. M., Gershon, A. A., and Ogra, P. L., Mucosal cell mediated immunity to varicella-zoster virus: role in protection against disease, *J. Pediatr.*, 105, 195, 1984.
413. Kamiya, H., Ihara, T., Hattori, A., Iwasa, T., Sakurai, M., Izawa, T., Yamada, A., and Takahashi, M., Diagnostic skin test reactions with varicella-zoster virus antigen and clinical application of the test, *J. Infect. Dis.*, 136, 784, 1977.
414. LaRussa, P., Steinberg, S. P., Seeman, M. D., and Gershon, A. A., Determination of immunity to varicella-zoster virus by means of an intradermal skin test, *J. Infect. Dis.*, 152, 869, 1985.
415. Shiraki, K., Yamanishi, K., and Takahashi, M., Biologic and immunologic characterization of the soluble skin-test antigen of varicella-zoster virus, *J. Infect. Dis.*, 149, 501, 1984.
416. Miller, A. E., Selective decline in cellular immune responses to varicella-zoster in the elderly, *Neurology*, 30, 582, 1980.
417. Berger, R., Florent, G., and Just, M., Decrease of the lymphoproliferative response to varicella-zoster virus antigen in the aged, *Infect. Immun.*, 32, 24, 1981.
418. Burke, B. L., Steele, R. W., Beard, O. W., Wood, J. S., Cain, T. D., and Marmor, D. J., Immune response to varicella-zoster in the aged, *Arch. Intern. Med.*, 142, 291, 1982.
419. Arvin, A. M., Pollard, R. B., Rasmussen, L. R., and Merigan, T. C., Selective impairment of lymphocyte reactivity to varicella-zoster virus antigen among untreated patients with lymphoma, *J. Infect. Dis.*, 137, 531, 1978.

420. Ruckdeschel, J. C., Schmipff, S. C., and Smith, A. C., Herpes zoster and impaired cell associated immunity to varicella-zoster virus in patients with Hodgkin's disease, *Am. J. Med.*, 62, 77, 1977.
421. Myers, J. D., Fluornoy, N., and Thomas, E. D., Cell mediated immunity to varicella-zoster virus after allogeneic bone marrow transplant, *J. Infect. Dis.*, 141, 479, 1980.
422. Russell, A. S., Maini, R. A., Bailey, M., and DuMonde, D. C., Cell mediated immunity to varicella-zoster antigen in acute herpes zoster, *Clin. Exp. Immunol.*, 14, 181, 1972.
423. Sorensen, O. S., Haahr, S., Moller-Larsen, A., and Wildenhoff, K., Cell-mediated and humoral immunity to herpesviruses during and after herpes zoster infections, *Infect. Immunol.*, 29, 369, 1980.
424. Cooper, E., Vujcic, L., Eron, L., Quinnan, G., Varicella-zoster specific HLA-restricted cytotoxic responses in vitro of lymphocytes from patients with acute herpes zoster infection, 23rd Interscience Conf. Antimicrob. Agents and Chemother., Abstr. 394, 1983.
425. Stevens, D. A. and Merigan, T. C., Interferon, antibody, and other host factors in herpes zoster, *J. Clin. Investig.*, 51, 1170, 1972.
426, Arvin, A. and Diaz, P., unpublished data.

Chapter 3

THE MOLECULAR GENETICS OF VARICELLA-ZOSTER VIRUS

Richard W. Hyman

TABLE OF CONTENTS

I. INTRODUCTION

Because of the highly cell-associated nature and poor growth characteristics of varicella-zoster virus (VZV) in cell culture (discussed in Chapter 1), traditional virological approaches to study VZV have yielded only limited success. The advances in our understanding of VZV on the molecular level have been derived (almost exclusively) by using the powerful new tools of contemporary biology: monoclonal antibodies, immunoprecipitation, recombinant DNA, hybrid selection coupled with in vitro translation and immunoprecipitation, prokaryotic open reading frame expression vectors, blot and *in situ* hybridization, and direct DNA sequencing. In addition, knowledge concerning VZV has been gained by drawing parallels to herpes simplex virus (HSV), another neurotropic human herpesvirus. VZV and HSV cross-react immunologically,[1] and the DNAs of these two herpesviruses hybridize to each other under relaxed hybridization conditions.[2] Thus, there is intellectual support for drawing parallels. (This literature review was completed on December 1, 1985.)

II. VZV DNA

VZV DNA is a linear duplex molecule. Ludwig et al.[3] measured the buoyant density in CsCl of VZV DNA to be 1.705 g/cm^3. This value was later confirmed by Iltis et al.[4] and Richards et al.[5] Assuming there are no unusual or modified bases in VZV DNA, the equation of Schildkraut et al.[6] could be used to relate buoyant density to guanine-plus-cytosine (G + C) content. (The equation is invalid, for example, if there are methylated bases.) VZV DNA was calculated to have a G + C content of 46 to 47%, which is in excellent agreement with the recent sequence data (46% G + C) of Davison.[7]

HSV type 1 (HSV-1) and HSV type 2 (HSV-2) can be distinguished from each other on the basis of a small but reproducible difference in the buoyant densities of HSV-1 and HSV-2 DNAs.[8] By analogy, could VZV isolated from varicella be distinguished from VZV isolated from herpes zoster on the basis of the buoyant densities of the VZV DNAs? Richards et al.[5] collected VZV isolates from the two clinical conditions and from different geographic locations (California, Japan, Pennsylvania). When looking for small differences in buoyant density, one must be aware that ^{14}C is inherently heavier than ^{12}C. Therefore, Richards et al.[5] radiolabeled each VZV DNA separately with [^3H]TdR and [^{14}C]TdR before the mixing experiments and also included in each gradient a mixture of ^3H-labeled HSV-1 DNA and ^{14}C-labeled HSV-1 DNA as controls for the isotope effect. The results are shown in Figure 1. No reproducible difference in buoyant density of VZV DNA could be observed ragardless of the medical condition or geographic location of the VZV isolate. Therefore, VZV isolated from varicella could not be distinguished from VZV isolated from herpes zoster on the basis of virus DNA buoyant density, unlike the HSV analogy.

The molecular weight of the linear duplex VZV DNA was measured by electron microscopy by Dumas et al.,[9] and found to be $80 \pm 4 \times 10^6$. (Note that molecular weight has no units. Molecular size or mass has units and is reported as either megadaltons (MDa), or kilobase pairs (kbp). The conversion between the two size units is 1 kbp = 0.66 MDa or 1 MDa = 1.5 kbp). From cloning and mapping data, Ecker and Hyman[10] reported a molecular weight for VZV DNA of $80 \pm 3 \times 10^6$, in good agreement with the value of Dumas et al.[9] A molecular weight of 80×10^6 converts to a molecular size of 122 kbp, which is well within experimental error of the recent size determination by direct DNA sequencing (the rounded-off molecular size is 125 kbp).[7] Thus, VZV DNA is the smallest human herpesvirus DNA.

One of the unusual properties of herpesvirus DNAs is their heterogeneous sedimen-

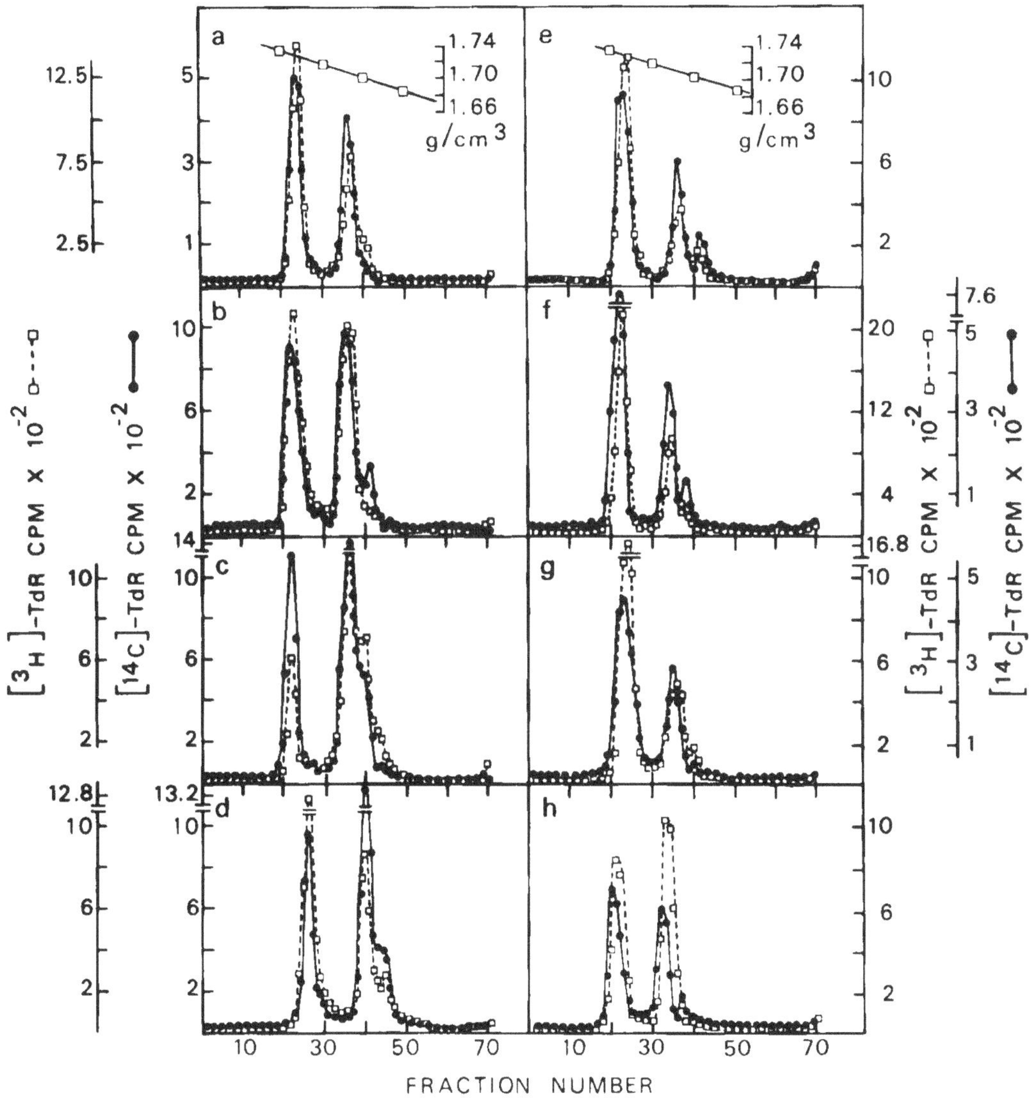

FIGURE 1. The buoyant density of VZV DNA. The varicella isolates were VZV (Ma), (Rc), (Kawaguchi), and (Oka). The herpes zoster isolates were VZV (Caufield), (Ludwig), and (Jab). ^3H- and ^{14}C-labeled HSV-1 (KOS) DNAs were included in all gradients (left-most bands, density 1.727 g/cm³). The refractive index of selected fractions was measured and converted to density using standard tables. To avoid cluttering the figure, the density values are plotted only for the gradient at the top of each column. A small amount of contaminating cell DNA is seen as the lightest (shoulder) DNA in some gradients. (a) VZV (Jab) [^3H] DNA and VZV (Oka) [^{14}C] DNA; (b) VZV (Oka) [^3H] DNA and VZV (Jab) [^{14}C] DNA; (c) VZV (Ludwig) [^3H] DNA and VZV (Oka) [^{14}C] DNA; (d) VZV (Oka) [^3H] DNA and VZV (Ludwig) [^{14}C] DNA; (e) VZV (Caufield) [^3H] DNA and VZV (Ma) [^{14}C] DNA; (f) VZV (Ma) [^3H] DNA and VZV (Caufield) [^{14}C] DNA; (g) VZV (Caufield) [^3H] DNA and VZV (Kawaguchi) [^{14}C] DNA; and (h) VZV (Kawaguchi) [^3H] DNA and VZV (Caufield) [^{14}C] DNA. Symbols: open squares, ^3H cpm; closed circles, ^{14}C cpm. (From Richards, J. C., Hyman, R. W., and Rapp, F., *J. Virol.*, 32, 812, 1979. With permission.)

tation pattern in alkaline sucrose gradients. This observation was first made by Kieff et al.[8] for HSV DNA and is now known to result from preexisting interruptions in the phosphodiester backbone of the herpesvirus DNA.[11] VZV DNA, like all studied herpesvirus virion DNAs, also shares this property.[4,12] Neither the molecular mechanism nor the philosophical reason for this unusual DNA property has been elucidated.

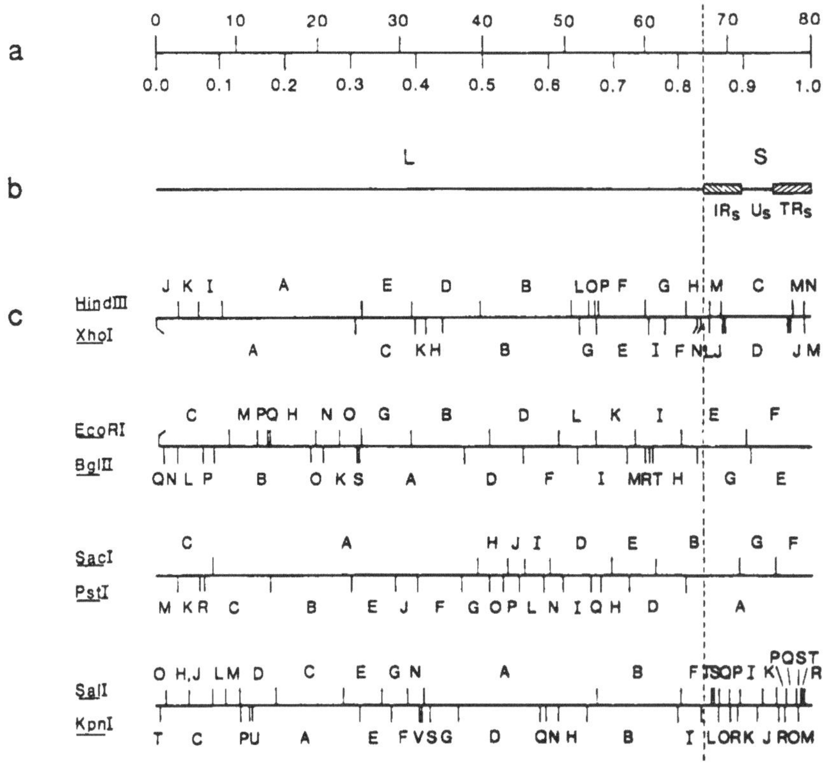

FIGURE 2. Genome organization and restriction enzyme cleavage maps of VZV DNA. All maps are shown in the prototype orientation; that means that the *Eco*RI-A fragment, for example, is not shown. (a) Map scales, giving genome size in MDa (top) and fractional distance (bottom). (b) Structural organization of the VZV genome, showing the division into L and S and the joint region (vertical broken line), as well as IR_S, U_S, TR_S. The repeated sequences are represented as cross-hatched boxes. (c) Cleavage maps of VZV DNA. *Hin*dIII,[10,19] *Eco*RI,[10,14,19] *Bgl*II,[13] *Pst*I[13], and *Xho*I, *Sal*I, and *Kpn*I[20,21] maps are shown. Strain-dependent cleavage polymorphisms and size heterogeneities are not shown, except as stated by Vlazny and Hyman.[22] The short vertical lines indicate cleavage sites. A bend in a line signifies that the site is not the physical end or joint region.

III. VZV GENOME ORGANIZATION

As determined by blot hybridization,[13] cloning, and mapping,[10,14] VZV DNA exists predominantly as two isomers. In the conventional (prototype) arrangement, following the nomenclature established earlier for HSV DNA,[15] VZV DNA can be divided into two segments, long (L) and short (S), separated by a "joint region". The L segment is composed almost exclusively of a unique sequence (U_L) 67 MDa in size,[7,10] The S segment is composed of a central unique sequence (U_S) 3 MDa in size[10] or 5232 base pairs (bp)[16] flanked by inverted, repeated sequences (IR_S/TR_S) 5 MDa in size[10] or 7319 bp.[17] Other size measurements are apparently in error.[14,18]

Because of the presence of the repeated sequences, U_S can be found in either of two equimolar orientations relative to U_L.[10,13,14] For restriction enzymes that cleave within U_S (e.g., *Eco*RI), there are two right-end fragments, each 0.5 *M*.[13,14] Figure 2 shows a schematic diagram of the VZV genome and includes several restriction enzyme cleavage maps. Table 1 gives the sizes and molarities of the VZV DNA *Eco*RI and *Hin*dIII fragments. Blot hybridization can determine whether two fragments share base sequence homology (e.g., VZV DNA *Hin*dIII-H and -N, Figure 2c), but cannot easily determine the orientation of the shared sequences. The latter is accomplished by elec-

Table 1
MOLECULAR SIZES AND MOLARITIES OF VZV DNA *Eco*RI AND *Hin*dIII FRAGMENTS

Fragment	*Eco*RI	*Hin*dIII
A	10.8	17.0
B	9.4	11.0
C	9.0	8.6
D	8.3	8.0
E	8.25	6.2
F	7.7	5.5
G	5.8	4.6
H	5.7	3.0
I	5.3	2.95
J	5.25	2.8
K	4.9	2.5
L	4.6	2.2
M	3.4	1.25
N	3.0	0.9
O	2.6	0.7
P	1.2	0.65
Q	0.6	

Note: Data for VZV (80-2) DNA, adapted from Ecker and Hyman.[10] All fragments are 1.0 *M* except for *Eco*RI-A, -E, -F, and -J, which are 0.5 *M*, and *Hin*dIII-M, which is 2.0 *M*. Fragment sizes can vary slightly from strain to strain; molarities usually do not. Molecular mass in MDa.

tron microscopy. For example, when the cloned VZV DNA *Hin*dIII-C fragment (Figure 2c) was denatured, briefly renatured, and mounted for electron microscopic examination, a stem-plus-loop structure characteristic of inverted repeated sequences was observed (Figure 3).

From blot hybridization experiments and electron microscopy, Ecker et al.[21] concluded that VZV DNA did not possess a detectable terminal repetition, with a minimum detection limit of approximately 100 bp. Davison[24] cloned the genomic termini of VZV DNA and by direct sequence analysis determined that VZV DNA does not possess a terminal repetition, with a minimum detection limit of 1 bp. Therefore, VZV DNA is alone among the human herpesvirus DNAs in not possessing a terminal repetition. Davison[24] discovered that the termini of VZV DNA possess a 3′ single base extension, a cytosine on one strand and a guanosine on the other, such that a correct base pair could be formed upon circularization.

Davison[24] achieved an extremely interesting result by comparing directly the base sequences of the cloned VZV genome termini and joint regions. In addition to the expected left terminus and joint region, an additional left end and a novel joint region were found at low frequency. Comparison of these sequences defined an 88-bp terminal and inverted repeat bounding the VZV DNA L region, so that VZV DNA L comprises TR_L-U_L-IR_L. Davison[24] proposed that the novel left end and joint region arose from a low frequency inversion of L, comprising no more than 5% of the population

FIGURE 3. Electron micrograph of the stem-plus-loop structure of VZV DNA *Hin*dIII-C. The cloned VZV DNA *Hin*dIII-C fragment was cleaved by *Hin*dIII, denatured, briefly renatured, and mounted for examination in the electron microscope. The arrow points to the transition from single-stranded to duplex DNA. The single-stranded DNA represents U_s. The duplex DNA represents part of IR_s/TR_s (Figure 2). (From Ecker and Hyman.[23])

of VZV DNA molecules. Earlier, Ecker et al.[21] had identified by size and hybridization two low molarity fragments in *Hin*dIII digests of VZV DNA. These had been termed *Hin*dIII-"J/N" and -"N/N", which was a misinterpretation. These fragments are now recognized to represent the low frequency inversion of L and the novel joint region, J_{IL}, and left end, T_{IL}.[22] These observations have been confirmed by Kinchington et al.[25] That the inversion of L is not an artifact of the cell culture passage of VZV was shown by Ecker et al.,[21] who identified J_{IL} and T_{IL} in VZV DNA isolated directly from vesicle fluid and never passed in cell culture.

The presence of an 88-bp TR_L/IR_L bounding the L region of VZV DNA raises a fascinating question: why are there not four equimolar isomers of VZV DNA, rather than predominantly two isomers?[10,13,14] HSV-1 and HSV-2 DNAs possess a terminal repetition, termed *a*, which is present internally in inverted form.[15] The HSV *a* domain contains the signal sequences "necessary and sufficient" to produce the observed four equimolar isomers of HSV DNA.[26-28] The size of *a* is variable from strain to strain: for HSV-1 (F), *a* is 501 bp[27] and for HSV-1 (Justin), *a* is 264 bp.[29] Chou and Roizman[30] have delineated a combined sequence of less than 50 bp within HSV-1 DNA *a* that defines a *cis*-acting inversion element. These sequences are present within *a* for all studied HSV-1 DNAs.[27,30,31] Human cytomegalovirus (HCMV) DNA possesses a terminal repetition, also termed *a*, and is present as four equimolar isomers.[32,33] The HSV-1 DNA *a* sequence[27] shares almost no homology with the HSV-2 DNA *a* sequence[31] nor with the HCMV DNA *a* sequence.[34] The *cis*-acting inversion sequence defined by Chou and Roizman[30] for HSV-1 DNA *a* is not found within the HSV-2 or HCMV DNA *a* domains. The only conserved sequence, G_{6-9}, T-rich$_{8-10}$, G_{7-9}, is apparently (at least part of) the signal for cleavage/packaging,[34] as it is also found at the appropriate position on Epstein-Barr virus DNA[35] and tupaia herpesvirus DNA.[36] This conserved sequence is found close to the right end (Figure 2) of VZV DNA,[17] where it presumably also functions as (part of) the cleavage/packaging signal.[22] As VZV DNA does not possess a terminal repetition,[21,24] this sequence does not appear at the left end of VZV DNA.[24] Thus, for HSV-1, HSV-2, and HCMV DNAs, no conserved isomerization sequence element has been identified, although the domain required for isomerization has been identified as *a*. The VZV DNA 88-bp TR_L/IR_L cannot be said, at present, either to contain or not to contain an element required to produce four equimolar isomers. One hypothesis would be that the VZV DNA TR_L/IR_L contains an inexact, inefficient, or partial isomerization element, such that the inversion of L occurs only at low frequency.

Pseudorabies virus (PrV) is a herpesvirus of swine. Standard PrV DNA has a structure analogous to VZV DNA and exists as two isomers, with U_S inverting with respect to U_L.[37] Lomniczi et al.[38] discovered that PrV vaccine (Norden) DNA contained two novel features: (1) the left end of L had been duplicated and inverted at the right end of L forming a TR_L-U_L-IR_L structure for PrV (Norden) DNA; and (2), the PrV (Norden) DNA, unlike wild type, existed as four equimolar isomers. The PrV (Norden) DNA TR_L/IR_L was estimated to be 440 bp in size.[38] The sizes of the PrV, HSV-1, HSV-2, and HCMV DNA *a* sequences are larger than the 88 bp for VZV DNA TR_L/IR_L. Therefore, a second hypothesis to explain why VZV DNA does not exist as four equimolar isomers, despite containing a TR_L/IR_L, is that 88 bp is too small to allow efficient isomerization. This hypothesis is not very satisfactory, given the variations in the size of the HSV-1, HSV-2, and HCMV DNA *a* sequence.

A third hypothesis to explain why VZV DNA exists predominantly as two isomers despite possessing a TR_L/IR_L is a mechanism of negative control. In this hypothesis, the VZV DNA 88-bp TR_L/IR_L contains the appropriate signal element for efficient isomerization and is of an appropriate size. However, a host-coded or VZV-coded (or both) protein(s) acts to prevent efficient isomerization, but does not prevent isomerization completely, so that a low level (5%) of inversion of VZV DNA U_L occurs. These three hypotheses (inexact signal sequence, insufficient TR_L/IR_L size, and negative control) to explain the presence of predominantly two VZV DNA isomers are not mutually exclusive and remain open to experimental testing.

One group of investigators has reported an additional, low frequency form of VZV DNA: encapsidated circular VZV DNA.[25,39] Circular VZV DNA would be expected to be found in VZV-infected cells as a normal intermediate in VZV DNA replication, analogous to HSV DNA replication[40] and PrV DNA replication.[41] However, there is insufficient evidence that full-length, genetically complete, circular VZV DNA is encapsidated.[25,39] Circular DNA, far less than full length, has been observed in the electron microscope.[21]

IV. MOLECULAR EPIDEMIOLOGY

HSV-1 and HSV-2 DNAs can be distinguished from each other easily by a single restriction enzyme cleavage pattern.[42,43] Even within HSV-1 and HSV-2 strains, there are sufficient restriction enzyme cleavage site polymorphisms to distinguish the DNAs of individual isolates and follow the spread of the virus.[44] As implied by the data in Figure 2, the same is not true for VZV DNA. Restriction enzyme cleavage patterns of the DNAs from varicella isolates cannot be distinguished systematically from the patterns of the DNAs from herpes zoster isolates.[5,45] For example, Figure 4 shows the *Hin*dIII digest patterns of the VZV DNAs from three varicella and two herpes zoster isolates. There are no clear or systematic differences in the pattern of bands. (The apparent diffuse or minor differences in *Hin*dIII-E are discussed below.) In some cases, the novel joint region fragment, J_{IL}, representing the low frequency inversion of the VZV DNA L region is observed (Figure 4, lane 5). The same VZV DNAs were cleaved by *Eco*RI; those restriction enzyme cleavage patterns are shown in Figure 5. The only clear difference is in the pattern of VZV (Rc) DNA, a varicella isolate (Figure 5, lane 2). For *Eco*RI-digested VZV (Rc) DNA, a novel low mobility band is observed (Figure 5, lane 2, solid circle) as well as the disappearance of two standard bands (Figure 5, lane 2, open circles). The sum of the sizes of the two missing bands is equal, within experimental error, to the size of the novel band; the novel band is presumably the fusion of *Eco*RI-B and -D (Figure 2c). Beyond a sometimes diffuse *Eco*RI-G band (discussed below), there are no other clear differences in the *Eco*RI cleavage patterns of the VZV DNAs from four varicella and two herpes zoster isolates (Figure 5).

The *Eco*RI cleavage patterns (Figure 5), but not the *Hin*dIII cleavage patterns (Figure 4), showed one clear difference in the independent VZV DNAs. It is from restriction enzyme cleavage site polymorphisms as infrequent as these that VZV molecular epidemiological studies have been, and will be made. As examples, Ecker and Hyman[48] demonstrated significant differences in the endonuclease cleavage patterns of VZV (parental Oka) and VZV (vaccine Oka) DNAs. These differences have been defined further by the use of recombinant VZV DNA probes and blot hybridization.[22] Pichini et al.[49] found the restriction enzyme cleavage patterns of the DNAs from paired (two separate locations on each patient) VZV isolates from patients with herpes zoster to be indistinguishable. Gershon et al.[50] demonstrated the clinical reinfection by a wild-type VZV in a child previously given the live, attenuated VZV vaccine (see Chapter 5). Straus et al.[51] demonstrated the reemergence as herpes zoster of the VZV strain that had 3 months earlier caused varicella in a child with Wiskott-Aldrich syndrome. Williams et al.[52] showed that the VZV vaccine reemerged as herpes zoster 21 months after inoculation. All of these studies relied on the small number of polymorphisms in VZV DNA. For safety, it is wise to examine the VZV DNAs under code, along with standard VZV DNAs (also under code), as has been used successfully in HSV molecular epidemiology.[53]

Instabilities in VZV DNA should not be confused with polymorphisms. Straus et al.[54] identified four regions of VZV DNA that they considered to be hypervariable among VZV isolates, but that is an understatement. The VZV DNA IR_S/TR_S (two of the four regions) varies in size, but includes tandem 27-bp repeats.[17] The number of 27-bp repeats found within the VZV DNA *Eco*RI-E fragment (Figure 2c) would be

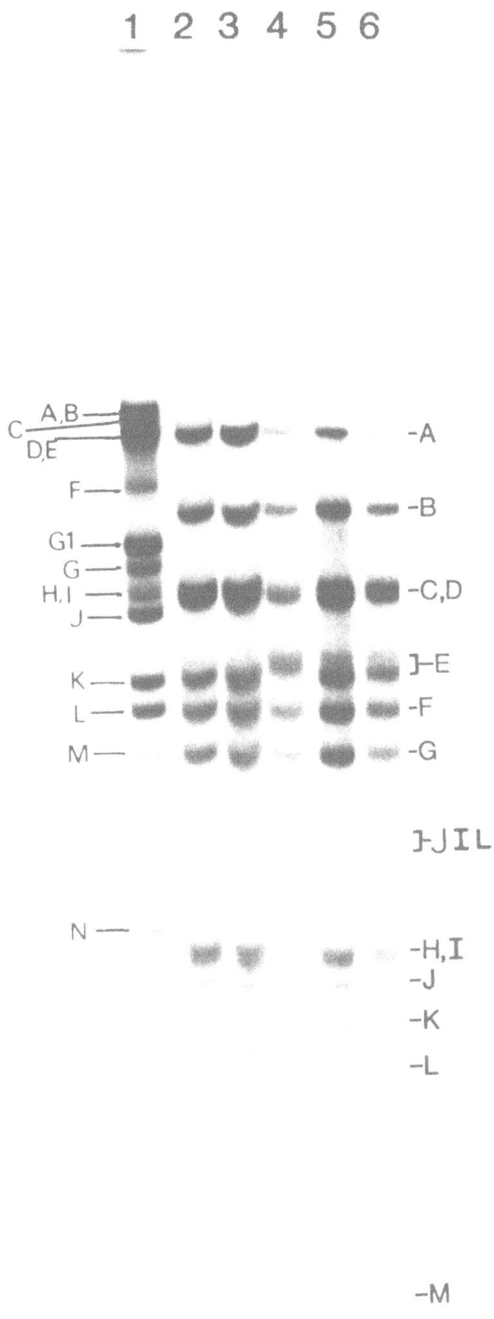

FIGURE 4. Autoradiograph showing the *Hin*dIII cleavage patterns of VZV DNAs. VZV (Ma), (Rc), and (Oka) are varicella isolates. VZV (Ludwig) and (Jab) are herpes zoster isolates. The DNAs were radiolabeled in virus-infected cells by the addition of [^{32}P]orthophosphate, in an attempt to achieve radiolabeling independent of base composition. However, the combined damage caused by radiation and shearing reduced the recovery of the larger bands and made calculation of the relative molarities unsatisfactory. Lane 1 is the control HSV-1 (KOS) DNA cleaved by *Hin*dIII.[46,47] The letters designating *Hin*dIII-cleaved HSV-1 (KOS) DNA bands are to the left. The VZV isolates are: lane 2, VZV (Ma) DNA; lane 3, VZV (Rc) DNA; lane 4, VZV (Oka) DNA; lane 5, VZV (Ludwig) DNA; lane 6, VZV (Jab) DNA. The letters designating the VZV DNA bands are to the right and reflect the *Hin*dIII map (Figure 2c). J_{IL} is the novel joint region fragment caused by the inversion of L. (Adapted from Richards, J. C., Hyman, R. W., and Rapp, F., *J. Virol.*, 32, 812, 1979. With permission.)

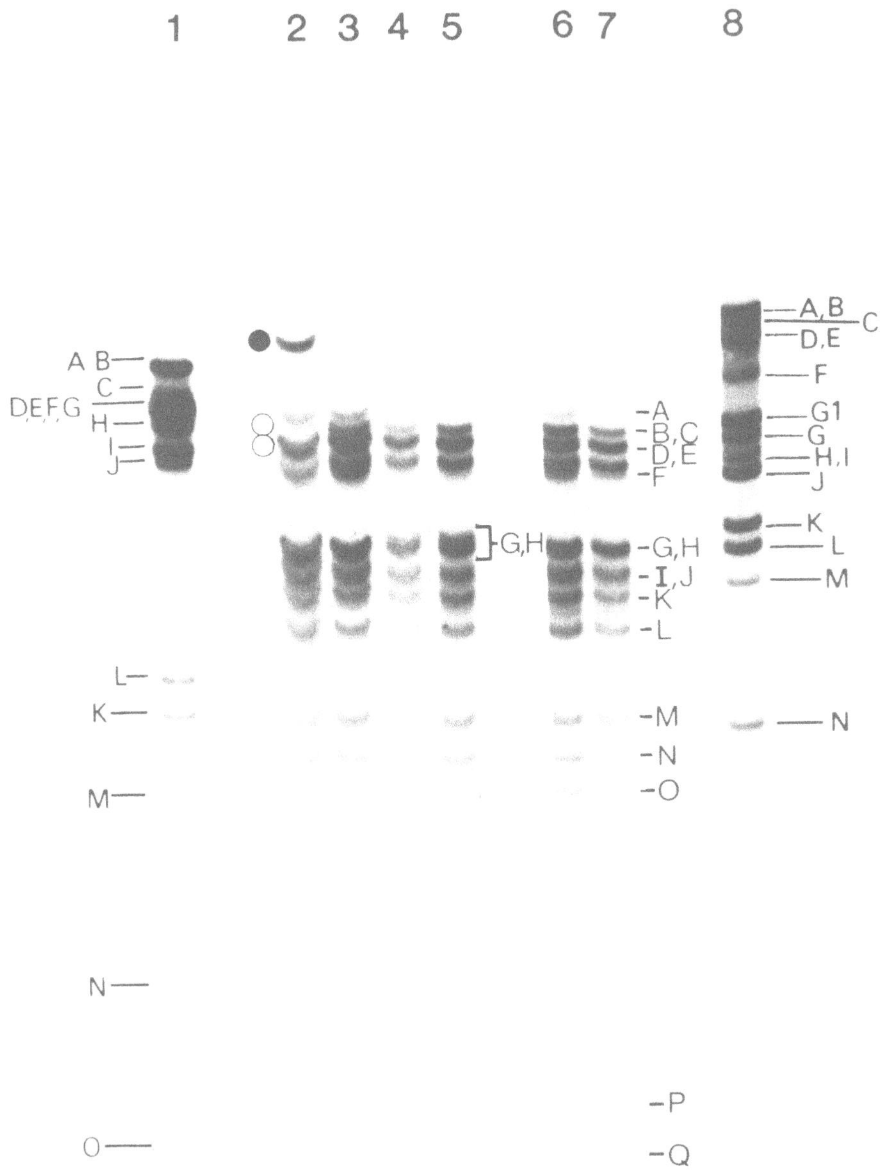

FIGURE 5. Autoradiograph showing the *Eco*RI cleavage patterns of VZV DNAs. VZV (Rc), (Ma), (Kawaguchi), and (Oka) are varicella isolates. VZV (Ludwig) and (Jab) are herpes zoster isolates. The DNAs were radiolabeled in vivo by inclusion of [^{32}P]orthophosphate in the culture medium. Lanes 1 and 8 are the controls: HSV-1 (KOS) DNA digested by *Eco*RI and *Hin*dIII, respectively.[46,47] The letters designating the *Eco*RI and *Hin*dIII fragments of HSV-1 (KOS) DNA are to the left of lane 1 and the right of lane 8, respectively. The VZV isolates are: lane 2, VZV (Rc) DNA; lane 3, VZV (Ma) DNA; lane 4, VZV (Kawaguchi) DNA; lane 5, VZV (Oka) DNA; lane 6, VZV (Ludwig) DNA; lane 7, VZV (Jab) DNA. The letters designating VZV DNA bands are to the right of lane 7 and reflect the *Eco*RI map (Figure 2c). To the left of lane 2, the closed circle represents the fusion band of *Eco*RI-B and -D, and the open circles represent the absence of those bands for VZV (Rc) DNA. (Adapted from Richards, J. C., Hyman, R. W., and Rapp, F., *J. Virol.*, 32, 812, 1979. With permission.)

expected to vary not only from strain to strain but also upon VZV passage. Analogously, it has been shown that the HSV DNA joint region (IR_L/IR_S) is heterogeneous in size within an HSV strain, varies in size on passage, and is even unstable as recombinant HSV DNA in an *E. coli* host.[55-57] The VZV DNA *Eco*RI-P fragment (third of the four regions) has been found to contain tandem repeats of a 42-bp sequence.[58] The number of 42-bp repeats would be expected to vary among VZV isolates and to be unstable on passage for any given isolate. The VZV DNA *Eco*RI-P fragment does show variation in size in Figure 5, although the bands are very faint for the particular exposure shown (chosen so as not to overexpose the higher molecular weight bands). The VZV DNA *Eco*RI-P fragment is contained entirely within the *Hin*dIII-A fragment (Figure 2c), which is so large that differing numbers of 42-bp tandem repeats would not be noticeable (Figure 4). The fourth hypervariable region is defined by the colinear VZV DNA *Hin*dIII-E and *Eco*RI-G fragments (Figure 2c). This region of the VZV genome possesses several peculiarities, including extensive single-strandedness[21] and low frequency construction of recombinant VZV DNA.[19] On VZV passage in cell culture, the VZV DNA *Eco*RI-G band becomes broader and fuzzy, as seen by comparing lanes 4—5 (Figure 5, bracket to right) to lanes 6—7 (Figure 5). The colinear VZV DNA *Hin*dIII-E fragment often appears as a doublet (Figure 4, bracket to the right of lane 6). In Figure 4, lane 2, *Hin*dIII-E appears as a reasonably sharp band; in lane 3, a fuzzy band of increased mobility; in lane 4, fuzzy bands of decreased mobility; in lane 5, a clear doublet; and in lane 6, a less distinct doublet. These internal instabilities are of limited use for epidemiological studies.

V. VZV TRANSCRIPTS

There are two classical approaches to define the temporal transcription of a virus genome in virus-infected cells. In the first approach, following synchronous infection of the cells, inhibitors of protein synthesis, and RNA and DNA syntheses, are used to define immediate early (virus protein synthesis independent), early (virus protein synthesis dependent, and virus DNA replication independent), and late (virus DNA replication dependent) transcripts. In the second approach, conditional-lethal mutants with blocks in transcription are derived and studied to examine the regulation of virus gene expression. Unfortunately for VZV, neither of these approaches is readily available. As discussed in Chapter 1, stable high-titer VZV stocks are almost unobtainable. Therefore, synchronous VZV infection at a known multiplicity of infection is technically out of reach. Temperature-sensitive mutants of VZV, if produced, would be difficult to plaque-purify and assign to complementation groups. Therefore, alternative, somewhat less satisfying approaches have been used to define VZV transcripts, but these are the only approaches currently available.

From direct sequence analysis, open reading frames (ORFs) and potential transcription signals have been identified.[16,17] For example, for the VZV DNA U_s region, the sequence was examined for the presence of ORFs starting with an ARG codon; four were found.[16] Transcripts were predicted on the basis of the consensus promoter sequence TATA (A/T)A(T/A) upstream from the start codon and the consensus polyadenylation signal sequence AATAAA downstream from the stop codon.[16] The limitation of this type of analysis was immediately apparent, as there were ambiguities in the promoter for three of the four potential transcripts.[16] The prediction of VZV transcripts from the base sequence is aided by the observation for other herpesviruses that RNA splicing is rare (though not absent) for viral transcripts. On the other hand, HSV-1 transcripts often form a nested set, sharing a 3′ end but differing in 5′ end, as is true of the transcripts from the HSV-1 DNA U_s region.[59] If the same pattern is true for VZV transcripts, it would be very difficult to predict a nested set from base sequence alone.

Ostrove et al.[60] isolated total RNA from VZV-infected cells and subjected the RNA to blot hybridization analysis. Total RNA provides a useful first approximation of VZV transcripts but suffers from two potential problems. One is that most of the RNA is ribosomal RNA (rRNA). The 28S- and 18S-rRNA will form concentrated bands in two regions of the gel, possibly giving rise to artifacts. In addition, the minimum detection limit for mRNA will be reduced significantly compared with isolating and analyzing poly(A$^+$) RNA. The second problem is that the total RNA will contain the precursors and products of splicing and/or processing reactions. These relationships will be difficult to discern. Nevertheless, Ostrove et al.[60] detected and mapped 58 VZV transcripts. The VZV transcripts ranged in size from 6.5 kilobases (kb) to 0.8 kb. Several potential sets of nested transcripts were observed, particularly between map positions 0.3 and 0.5 (Figure 2). Ostrove et al.[60] carried out a detailed mapping of the transcripts from the right-most (Figure 2) 20% of the VZV genome. Twelve VZV transcripts were mapped to this region, which includes the VZV DNA IR_S/TR_S. Three VZV transcripts mapped entirely within the repeated sequences: 1.65, 2.05, and 3.7 kb. With the important caveat that Ostrove et al.[60] did not determine the polarity of the VZV transcripts, there appears to be a similarity of transcript organization within the VZV and HSV-1 DNAs IR_S/TR_S.[59-62] Curiously, the VZV DNA around map position 0.3 (Figure 2) appeared not to give rise to transcripts.[60] Whether this region is transcriptionally inactive or the transcripts were present but below the minimum detection limit remains to be determined.

Maguire and Hyman[63] examined cytoplasmic, polyadenylated RNA from VZV-infected cells. The VZV-specific transcripts are likely to be VZV mRNAs. Figure 6 presents an example of an autoradiograph of an RNA blot hybridization experiment. The probe was pVHF, the VZV DNA *Hin*dIII-F fragment (Figure 2c), map position 0.68-0.75. The control lanes contain the total cytoplasmic RNA (Figure 6, lane 1), poly(A$^-$) RNA (Figure 6, lane 2), and poly(A$^+$) RNA (Figure 6, lane 3) from uninfected human cells. No bands were seen. Therefore, neither the insert nor the vector hybridized detectably to human cellular RNA. The VZV-specific transcripts can be seen in the lane containing total cytoplasmic RNA from VZV-infected human cells (Figure 6, lane 4). The VZV mRNAs can be seen in the lane containing poly(A$^+$) RNA from VZV-infected cells (Figure 6, lane 6). The lane containing poly(A$^-$) RNA is negative, demonstrating an absence of artifacts caused by nonspecific sticking of the probe to rRNAs. Five VZV mRNAs, with sizes of 3.9, 2.0, 1.6, 1.2, and 1.0 kb, were detected. The 2.0-kb band was somewhat broader than the others and may have represented more than one mRNA. Because of the uncoordinated nature of the VZV infection, the relative intensities of the bands cannot be interpreted unambiguously. Using different probes and RNA preparations, Ostrove et al.[60] mapped eight VZV transcripts to VZV DNA map position 0.68 to 0.75: (1) 4.1- and 1.9-kb RNAs mapped entirely within pVHF and probably correlate with the 3.9- and 2.0-kb RNAs, respectively, mapped by Maguire and Hyman[63]; (2) the 2.4- and 2.2-kb RNAs are only partially within pVHF and may have contributed to the broadening of the 2.0-kb RNA band; (3) the 1.5-kb RNA may correspond to the 1.6-kb RNA; (4) the 3.6- and 2.8-kb RNAs, which partially overlap pVHF, were not detected by Maguire and Hyman[63]; and (5) the 1.1-kb RNA may correspond to one of the 1.2- or 1.0-kb RNAs. These correlations seem reasonable, but remain to be proven.

VI. VZV PROTEINS

Early studies of VZV proteins were marred seriously by the inability to separate VZV from the host cell, despite the hard work of the investigators.[64-66] The combined use of monospecific, polyclonal sera and, particularly, monoclonal antibodies (MAbs), coupled with immunoprecipitation, has proved to be invaluable in the analysis of VZV proteins.

FIGURE 6. RNA blot hybridization analysis of VZV mRNA. The RNA (and lambda DNA) was denatured by reaction with heat and formaldehyde, and electrophoresis through the agarose gel was conducted in the presence of formaldehyde. The gel was blotted to nitrocellulose. The negative control RNAs were from uninfected human cells: lane, 1, total cytoplasmic RNA; lane 2, poly(A^-); and lane 3, poly(A^+) RNA. The poly(A^+) RNA was prepared by two cycles of chromatography on oligo(dT)-cellulose. The VZV-infected human cell RNAs were: lane 4, total cytoplasmic RNA; lane 5, poly(A^-) RNA; and lane 6, poly(A^+) RNA. The probe was pVHF DNA (Figure 2c) nick-translated in vitro using [^{32}P] TTP and [^{32}P]dCTP as substrates. Following hybridization, the blot was washed extensively at high temperature and low salt. Autoradiography was performed with the assistance of enhancer screens. The arrowheads to the left of lane 6 mark the VZV mRNAs. The horizontal lines to the right of lane 6 mark the migration positions of the size markers, lambda DNA cleaved by *Hin*dIII and *Eco*RI and then denatured. From top to bottom, the sizes of the markers are 21.2, a doublet of 5.15 and 4.97, 4.28, 3.53, a doublet of 2.03 and 1.90, 1.58, 1.33, and 0.98 kb. (From Maguire and Hyman.[63])

Table 2
VZV GLYCOPROTEIN NOMENCLATURE

New[a] designations	Previous Designations				
	Grose[b]	Keller[c]	Forghani[d]	Yamanishi[e]	Edson[f]
gpI	gp98/62	gC	group 3	gp2	gp98/59/47
gpII	gp140	gB	group 2	gp3	gp63
gpIII	gp118	gA	group 1	gp1	gp118

[a] Nomenclature agreed upon at the 10th Herpesvirus Workshop and published by Davison et al.[74]
[b] Data from Figure 7 and Grose et al.[67,68]
[c] Data from Figure 7 and Keller et al.[72]
[d] Data from Figure 7 and Forghani et al.[70]
[e] Data from Okuno et al.[71] and Namazue et al.[75] This column was generously filled in by Dr. Koichi Yamanishi.
[f] Data from Edson et al.[73] VZV gps placed in "best-fit" positions.

Several investigators have produced MAbs against the highly antigenic VZV glycoproteins.[67-73] Glycoproteins are not globular, and, therefore, migrate anomalously in polyacrylamide gels, yielding different apparent molecular sizes, depending on gel conditions. It has been very difficult to correlate the data published by different investigators. Therefore, a side-by-side comparison was undertaken to compare the VZV glycoproteins (gps) immunoprecipitated by MAbs and subjected to polyacrylamide gel electrophoresis (PAGE) on the same gel. Figure 7 presents a fluorograph showing the comparison of three sets of independently derived MAbs. Predictions of correlations based on published literature proved to be wrong in one case; only a side-by-side comparison demonstrated the correct correlation (Figure 7). In an attempt to standardize this field of investigation, a nomenclature for VZV gps was adopted at the 10th Herpesvirus Workshop, Ann Arbor, Michigan, U.S., (1985) and published.[74] This nomenclature employs roman numerals and supersedes designations by letters or apparent molecular size. Table 2 presents a compilation of the old and new nomenclatures for VZV gps.

Once the correlations have been made, it can be seen that there are three prominent groups of VZV gps. The first group, gpI, consists of at least three and perhaps five related gps in VZV-infected cells. The VZV gene for this protein family has been mapped to the 3′ end of the U_S region.[76] For mapping, two complementary but independent approaches were used. (1) An *E. coli* plasmid ORF expression vector was employed in conjunction with an anti-gpI MAb. The VZV DNA insert corresponding to the MAb^+ fusion protein was mapped to the 3′ end of U_S by DNA blot hybridization. (2) Recombinant VZV DNA representing U_S (the *Hin*dIII-C fragment; Figure 2c) was employed to select by hybridization RNA from VZV-infected cells. The selected RNA was translated in vitro, and the protein products were immunoprecipitated and subjected to PAGE and fluorography. MAb anti-gpI immunoprecipitated a protein whose RNA mapped to U_S.[76] The in vitro translation product is the primary protein and has not undergone the extensive processing of VZV gps described in Chapter 1. Ostrove et al.[60] identified and mapped a 3.7-kb VZV transcript from VZV-infected cells that is consistent with, but not yet proved to be, the VZV mRNA for VZV gpI.

The second prominent VZV gp, gpII, is unusual in that the mature gp is composed of two disulfide-linked protein chains.[68] The 140-K dalton disulfide-linked gp can be reduced to two 66-K dalton gps.[68] The processing of gpII is described in Chapter 1. The VZV gene for gpII has been mapped to the center of VZV DNA, within U_L.[7,77]

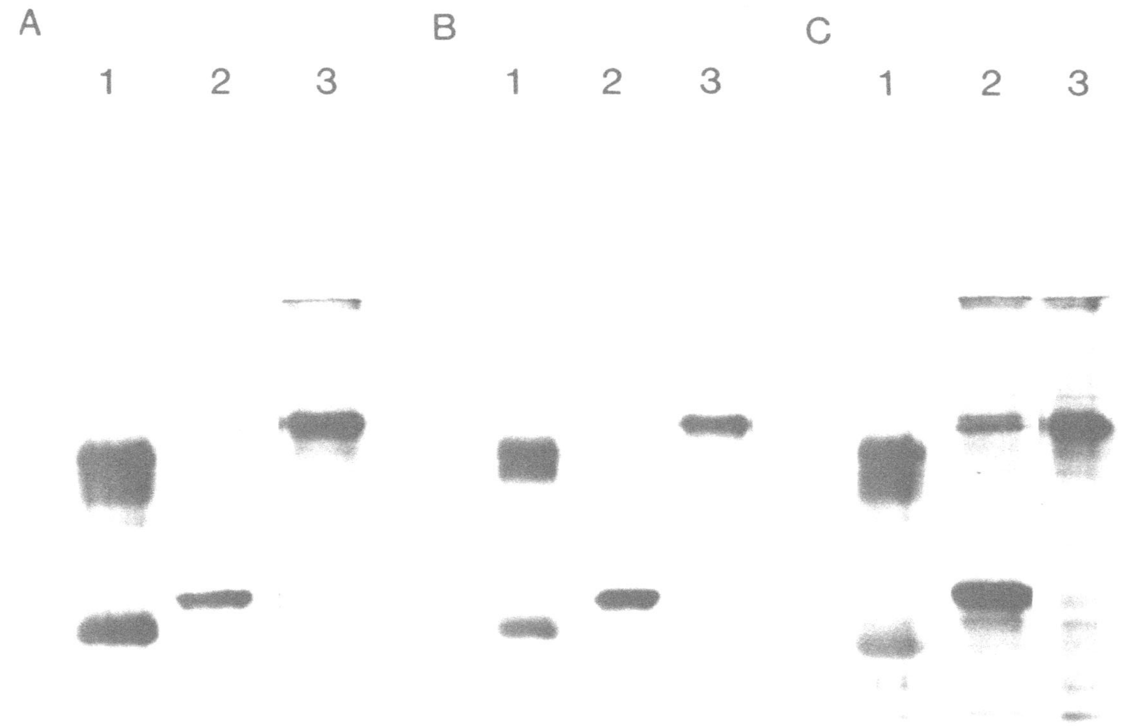

FIGURE 7. Fluorograph comparing VZV glycoproteins immunoprecipitated by monoclonal antibodies and subjected to PAGE. VZV-infected cells were radiolabeled with [^{35}S]methionine. The soluble proteins were immunoprecipitated with the following MAbs. Panel A: lane 1, MAb 3B3; lane 2, MAb 158; lane 3, MAb 206B9, all generously supplied by Dr. Charles Grose. Panel B: lane 1, MAb 7C33; lane 2, MAb 1E227; lane 3, MAb 1A214, all generously supplied by Drs. Paul Keller and Ronald Ellis. Panel C: lane 1, MAb 3G8; lane 2, MAb 3C7; lane 3, MAb 4B7, all generously supplied by Dr. Bagher Forghani. The figure is a composite of multiple exposures of a single gel. Under the recently adopted nomenclature, lanes 1 are VZV gpI; lanes 2, VZV gpII; and lanes 3, VZV gpIII. (From Maguire and Hyman.[63])

VZV gpII has been shown to cross-react immunologically with HSV gB.[78,79] This cross-reactivity substantiates the earlier finding of Schmidt.[1] It is not clear, at present, how far to push the implications of the relatedness of VZV gpII and HSV-1 gB. HSV gB is the most abundant HSV gp.[80,81] VZV gpI, and not gpII, is the most abundant VZV gp,[66,67,71] although both VZV gpI and gpII are highly antigenic.[82]

The third major VZV gp, gpIII, appears as a single band,[67,70-72] has neutralizing epitopes, and has been purified by affinity chromatography.[83] In addition to these three prominent VZV gps, the Davison[7] VZV DNA sequence predicts two possible additional VZV gps. Working from the sequence[16] of the VZV DNA U_S region, Davison et al.[84] synthesized a peptide composed of eight amino acids and coupled the peptide to bovine serum albumin. Seven of the eight amino acids were, as predicted by the sequence, encoded within the VZV DNA U_S. The peptide conjugate was used to immunize rabbits, and the rabbit immune sera were employed to immunoprecipitate [^{3}H]mannose- or [^{35}S]methionine-labeled proteins from VZV-infected cells. The proteins were subjected to PAGE and fluorography. The results for the [^{35}S]methionine-labeled proteins were not easily interpretable. The [^{3}H]mannose-labeled protein(s) appeared as one or two (gp55, gp45) very faint bands after long exposure times. Davison et al.[84] concluded that they had identified VZV gpIV and the VZV gp IV gene, as predicted from the VZV DNA sequence.[16]

Vafai et al.[85] produced a MAb directed against gpII and confirmed the previous work of Okuno et al.[71] and Grose et al.[68] They also produced a MAb that immunoprecipitated a nonglycosylated, 125-K dalton protein found in the nuclei of VZV-infected cells late after infection.[85] The MAb directed against VZV gpII was employed by Wroblewska et al.[86] to purify by affinity VZV gpII. When this preparation of VZV gpII was used to immunize rabbits, the animals developed specific anti-VZV neutralizing antibody.[86]

Lopetegui et al.[87] obtained one reasonably stable, cell-free VZV preparation of titer 1.6×10^6 plaque-forming units (PFU)/m*l*. This VZV preparation was unusual because of its stability and relatively high titer. Lopetegui et al.[87] employed this VZV stock for the classic inhibitor studies in an attempt to define immediate early (IE), early (E), and late (L) (all defined above) VZV proteins in VZV-infected cells. Even with this virus stock, Lopetegui et al.[87] were forced to use a multiplicity of infection (m.o.i.) of only 0.3 PFU/cell, which is significantly lower than the usual 10 PFU/cell employed for the equivalent HSV-1 studies; however, this was the best they could do under difficult experimental circumstances. At that low m.o.i., many cells were not infected, so that immunoprecipitation with a polyclonal anti-VZV monkey serum was employed before PAGE and fluorography. Seven presumptive IE proteins were identified (Table 3). The 33-K dalton protein was the most prominent. Six presumptive E-proteins were detected (Table 3). The 68- and 35-K dalton proteins did not also appear as IE proteins and were, therefore, presumed to be authentic E proteins. Seventeen presumptive L proteins were detected (Table 3). As noted by Lopetegui et al.[87], there are several potential difficulties in interpretation of the data: (1) the results depend upon the efficacy and specificity of the immune monkey serum. The lanes of immunoprecipitated mock-infected cell proteins show many bands; and (2) it is very difficult to coordinate VZV infection. Are the IE, E, and L 44-K dalton proteins, for example, the same protein or different proteins of indistinguishable mobilities? Nevertheless, Lopetegui et al.[87] have made a start at an experimentally difficult but important line of investigation.

The appearance of new enzymatic activities in VZV-infected cells led to the prediction of VZV-coded enzymatic activities. Several investigators reported the presence of a new thymidine kinase (TK) activity in VZV-infected cells.[88-91] Yamanishi et al.[92] reported the biochemical transformation of mouse LTK$^-$ cells to the TK$^+$ phenotype after co-cultivation with VZV-infected cells, and Lopetegui et al.[93] identified a 35-K dalton (denatured) protein specific to the VZV-biochemically transformed mouse LTK$^+$ cells. Purified, randomly sheared VZV DNA transfected into mouse LTK$^-$ cells produced the TK$^+$ phenotype.[94] Sawyer et al.[95] employed overlapping recombinant VZV DNAs in transfection experiments and localized the VZV TK transforming activity (and, hence, the VZV TK gene) to map position 0.50—0.52 (Figure 2). This assignment of the physical position of the VZV TK gene on VZV DNA is in agreement with the Davison[7] sequence data, which show a low level of homology to the HSV TK gene at the same VZV map position.

A novel high salt-resistant, phosphonoacetic acid (PAA)-sensitive DNA-dependent DNA polymerase activity has been found in VZV-infected cells.[96-98] These properties are entirely analogous to HSV-coded DNA polymerase, even to the selection of VZV PAA^R mutants.[99] A VZV DNA sequence with homology to the HSV-1 DNA polymerase has been identified,[7] and a nonstructural, antigenic protein of appropriate size has been immunoprecipitated from VZV-infected cells by Grose and Friedrichs.[100]

DNA-binding proteins are of interest because of their possible role in gene regulation. VZV-specific (or VZV-coded) DNA-binding proteins are difficult to distinguish from host-coded DNA-binding proteins for the technical reasons discussed at length previously. In an attempt to overcome some of the technical difficulties, Shiraki et al.[101] applied [^{35}S]methionine-labeled DNA-binding proteins from VZV-infected cells to an affinity column coupled to anti-VZV serum, and then analyzed the affinity-bound proteins by PAGE and fluorography. One prominent VZV-specific, DNA-binding protein of size 34 K daltons was observed.[101] Facing the same technical difficulties, Roberts et al.[102] used [^{35}S]methionine to radiolabel VZV-infected cells demonstrating 25 to 30% through 90 to 95% cytopathic effect (approximately 60 hr). Roberts et al.[102] detected seven prominent DNA-binding proteins. A 40-K dalton protein was closest in size to the DNA-binding protein previously identified by Shiraki et al.[101] Roberts et al.[102] tentatively identified a 155-K dalton VZV capsid protein, a 175-K dalton tegument protein, and a 125-K dalton major DNA-binding protein. The latter may correspond to the late, nuclear protein previously identified by a MAb.[85]

The proteins of the VZV virion have been enumerated by three groups of investigators.[78,103,104] The technical problems in distinguishing virion protein from cellular contaminants are formidable given the highly cell-associated replication of VZV in cell culture (see Chapter 1). Nevertheless, several generalizations emerge from the data. The VZV virion is composed of 33 proteins ranging in size from 16 to 244 K daltons,[103] 32 proteins ranging in size from 21 to 280 K daltons;[78] and/or 30 proteins ranging in size from 30 to 200 K daltons.[104] The most prominent VZV virion gps are gpI (VP7, VP8, VP14, VP19;[103] VP9, VP10, VP11, VP12, VP14, VP18, VP22,[78] 98K, 62 K,[104] and gpIII (VP6;[103]; VP8;[78] 118K).[104] The minor VZV virion gp is gpII (VP5, VP12,[103] VP16,[78] gp140[104]); gpI appeared as multiple bands; gpII appeared in gels in both its disulfide-linked and reduced form; and gpIII appeared as a single band. Given the technical difficulties and the differences in PAGE conditions, there is a remarkable concurrence on the proteins comprising the VZV virion (Table 4). The most prominent nonglycosylated protein is the major capsid protein: VP5, 145 K;[78] 155 K,[104] which is also a DNA-binding protein.[102] All three groups of investigators detected two very high molecular weight (greater than 170 K) proteins. Grose et al.[104] found that one of these proteins was more readily radiolabeled with [^{14}C]amino acids than with [^{35}S]methionine. All three groups of investigators found numerous proteins close in molecular weight from 70 K to 40 K.[101,103,104] The gel conditions were such that relatively low molecular weight proteins ($\leq$20 K), if present in VZV virions, would probably have escaped detection.

Table 3
KINETIC CLASSES OF VZV PROTEINS IN VZV-INFECTED CELLS

IE[a]	E[b]	L[c]
180		180
145		145
135	135	135
118	118	118
		108
		100
		84
		81
	68	68
		62
		48
44	44	44
42	42	42
		38
		37
	35	35
33		33

Note: The data are taken from Lopetegui et al.[87] The proteins are listed in order of decreasing apparent molecular weight × 10^{-3}.

[a] IE = immediate early proteins; VZV protein synthesis independent.
[b] E = early proteins; VZV DNA replication independent.
[c] L = late proteins; proteins radiolabeled late in the VZV infectious cycle.

Table 4
NONGLYCOSYLATED VZV VIRION PROTEINS AS DETERMINED BY THREE GROUPS OF INVESTIGATORS

Investigator		
Shemer[a]	Shiraki[b]	Grose[c]
244	280	200
222	220	174
180[d]	180	160
153	155	155[d]
77	145[d]	126
72.5	130	88
70.5	118	86
65	71	80
60	61	76
58	57	71
57	54	66
55	51	59
49.5	48	55
47	42	53
45.5[e]	38	52
43.5	35	51
42	33.5	50
38	32	49
36	30	48
34.5	28	45[e]
32	25	43
29.5	23.5	40
28	21.5	37
19.25		36
16.75		32
16		30

Note: The nonglycosylated virion proteins are listed in order of decreasing apparent molecular weight × 10^{-3}.

[a] Data from Shemer et al.[103]
[b] Data from Shiraki et al.[78]
[c] Data from Grose et al.[104]
[d] Apparent major virion capsid protein
[e] Comigrates with actin, a cellular protein contaminant.

Davison[7,16] and Davison and Scott[17] have identified 70 ORFs, or 67 potential genes (because of the IR_S/TR_S) from VZV DNA sequence analysis. There is an absence of knowledge concerning possible VZV RNA splicing and/or 3′ co-terminal nested sets of VZV mRNAs. In addition, Haarr et al.[105] have demonstrated that one HSV-1 mRNA is translated starting at three separate in phase AUG codons; three distinct, overlapping proteins were detected. Therefore, a minimum of 67 potential VZV-coded proteins can be predicted. Last, Gelb et al.[106] have presented preliminary evidence that VZV, like all other tested herpesviruses,[107] can transform normal cells in culture to oncogenic cells.

VII. MOLECULAR PATHOLOGY

The first in a series of related, fundamental questions to be asked concerning the VZV genome during long-term, inapparent infection in the human host is: in which cell type(s) is the VZV DNA? Experimental approaches have provided a first-step answer to this question.

Hyman et al.[108] employed *in situ* hybridization to look for VZV RNA in sections of human trigeminal ganglia. These experiments must be carefully controlled, because HSV also is found in human trigeminal ganglia,[109,110] and HSV DNA hybridizes with VZV DNA under relaxed hybridization conditions.[2] Figure 8 presents some of the control experiments. A radiolabeled VZV DNA probe detected VZV RNA in VZV-infected cells (observed as silver grains; Figure 8a), but not in uninfected cells (Figure 8b). A control HSV DNA probe was negative for VZV-infected cells (Figure 8d), while correctly positive for HSV-infected cells (Figure 8e). As a control for nonspecific binding, a λ DNA probe was negative for VZV-infected cells (Figure 8c) and HSV-infected cells (Figure 8f). This series of control experiments employed a 3-day exposure time. The signal was clearly present (Figure 8a), and the background was low (Figure 8b through d). In an attempt to increase the signal while not increasing the background, a 6-day exposure was employed. Figure 9 shows examples of the results of these control *in situ* hybridization experiments. The signal was definitely increased when a VZV DNA probe was hybridized to VZV-infected cells (Figure 9a). However, the background also increased, as seen when a λ DNA probe (Figure 9b) or a HSV DNA probe (Figure 9c) was hybridized to VZV-infected cells. Therefore, a 3-day exposure was chosen for the key experiment of VZV DNA probe hybridized to sections of human trigeminal ganglia. Of five individuals tested, neurons (and only neurons) were positive for VZV RNA. Examples of positive neurons are given in Figure 10. The percent of positive neurons ranged from 0 to 0.3%.[108] This result suggests strongly that the neuron is one site of VZV latency. However, an inherent problem with *in situ* hybridization is that the minimum detection limit is difficult, sometimes impossible, to determine. Therefore, other cell types might have appeared negative falsely. It might also be questioned whether the detected VZV RNA represented VZV transcription during true latency or, alternatively, during VZV reactivation caused by the patient's death. None of the patients had a history of recent herpes zoster, and the trigeminal ganglia were collected 7 to 11 hr after death. Because removal of the trigeminal ganglia before death is impossible, the question is experimentally moot, but the conclusion remains unchanged in either case: at least one cell type where VZV is latent is the neuron.

Gilden et al.[111] reported the apparent detection by blot hybridization of VZV DNA in preparations of total human ganglia DNA. One patient died while experiencing acute herpes zoster. The sacral ganglion DNA from that patient showed considerable VZV DNA, essentially a positive control. The trigeminal ganglia DNAs from 10 cadavers were examined for VZV DNA; 3 of the 10 were stated to be positive for VZV DNA.[111] However, there are serious problems in the interpretation of the data, especially as only one DNA band in each of two lanes (out of 18) representing one of the three positive human trigeminal ganglia DNAs was shown.[111]

1. Only two recombinant VZV DNAs, representing a small fraction of the VZV genome, were employed as probes.
2. For one of the two probes, the HSV DNA control lane was not clean, raising the possibility of cross-hybridization with HSV DNA.[2]
3. It was stated that the vector pBR322 DNA alone as probe did not hybridize to control DNAs, but no data were shown. It has been established that under certain

FIGURES 8a and 8b

FIGURE 8. *In situ* hybridization experiment controls. A VZV DNA probe (a mixture of recombinant VZV DNAs covering the VZV genome) was hybridized to (a) VZV-infected cells, the positive control; and (b) uninfected cells, the negative control. A λ DNA probe, as a control for nonspecific binding, was hybridized to (c) VZV-infected cells and (f) HSV-infected cells. A HSV DNA probe (a representative mixture of recombinant HSV DNAs) was hybridized to (e) HSV-infected cells, the positive control, and (d) VZV-infected cells, the control to demonstrate the absence of cross-hybridization under these conditions. Positive hybridization was visualized as silver grains (a and e). Exposure time was 3 days. (Panels a through d are reproduced from Hyman, R. W., Ecker, J. R., and Tenser, R. B., *Lancet*, 2, 814, 1983. With permission. Panels e and f are from Ecker and Hyman.[23])

FIGURES 8c and 8d

FIGURES 8e and 8f

FIGURE 9. Increasing the signal in *in situ* hybridization. A VZV DNA probe (a), a λ DNA probe (b), and an HSV DNA probe (c) were hybridized to VZV-infected cells. Exposure time was 6 days. (From Ecker and Hyman[23].)

FIGURE 9c.

hybridization conditions pBR322 DNA shows (presumably artifactual) hybridization to human DNA.[112,113]

4. Reconstruction experiments were not shown and, apparently, were not undertaken. Rather, a type of quantitation from densitometer traces of what may have been dot blots was stated but not shown.
5. The numbers derived from the quantitation yielded an average of 0.28 to 1 copies of VZV DNA per cell in the three positive human trigeminal ganglia. These numbers are dramatically higher than the results of careful quantitation of HSV DNA in the human trigeminal ganglia, as determined by Efstathiou et al.[114]: 0 to 0.1 HSV DNA copies per cell, with most positive samples yielding approximately 0.01 HSV DNA copies per cell.

For all of these reasons, the data of Gilden et al.[111] are difficult to assess.

Alternatively, Vlazny et al.[115] included the appropriate reconstruction and control experiments while using recombinant VZV DNAs as probes to attempt to detect VZV DNA as part of total human trigeminal ganglia DNA (Figure 11). The absence of hybridization to human embryo lung (HEL) cell DNA established a true negative background (Figure 11, lane 3). The absence of hybridization to HSV-infected cell DNA demonstrated no unwanted cross-hybridization under the stringent hybridization conditions employed (Figure 11, lane 11). The inclusion of sonicated, denatured pBR322

FIGURE 10. *In situ* hybridization to sections of human trigeminal ganglia. A VZV DNA probe was hybridized to flash-frozen sections of human trigeminal ganglia cut with a cryostat. Exposure time was 3 days. The photographs were taken of the silver grains in the emulsion layer above the cells, so that the cells are somewhat out of focus. (From Hyman, R. W., Ecker, J. R., and Tenser, R. B., *Lancet*, 2, 814, 1983. With permission.)

DNA as part of the carrier DNA removed the possibility of artifactual bands (Figure 11, lane 3). The reconstruction experiments determined a minimum detection limit of approximately 0.1 VZV DNA copies per haploid cell genome (Figure 11, lanes 1, 2), which is below the copy number determined by Gilden et al.[111] With these controls and

FIGURE 10 (continued)

under these hybridization conditions, all six total human trigeminal ganglia DNA samples were negative for VZV DNA (Figure 11, lanes 4 to 9). If there were an average of 0.01 VZV DNA copy per haploid cell genome, as HSV DNA in human trigeminal ganglia,[114] that would have been below the minimum detection limit. Therefore, one or more human trigeminal ganglia DNA may have appeared negative falsely. For purposes of argument, assuming there were an average of 0.1 VZV DNA copy per haploid cell genome (which should have been detected, but was not), what average copy num-

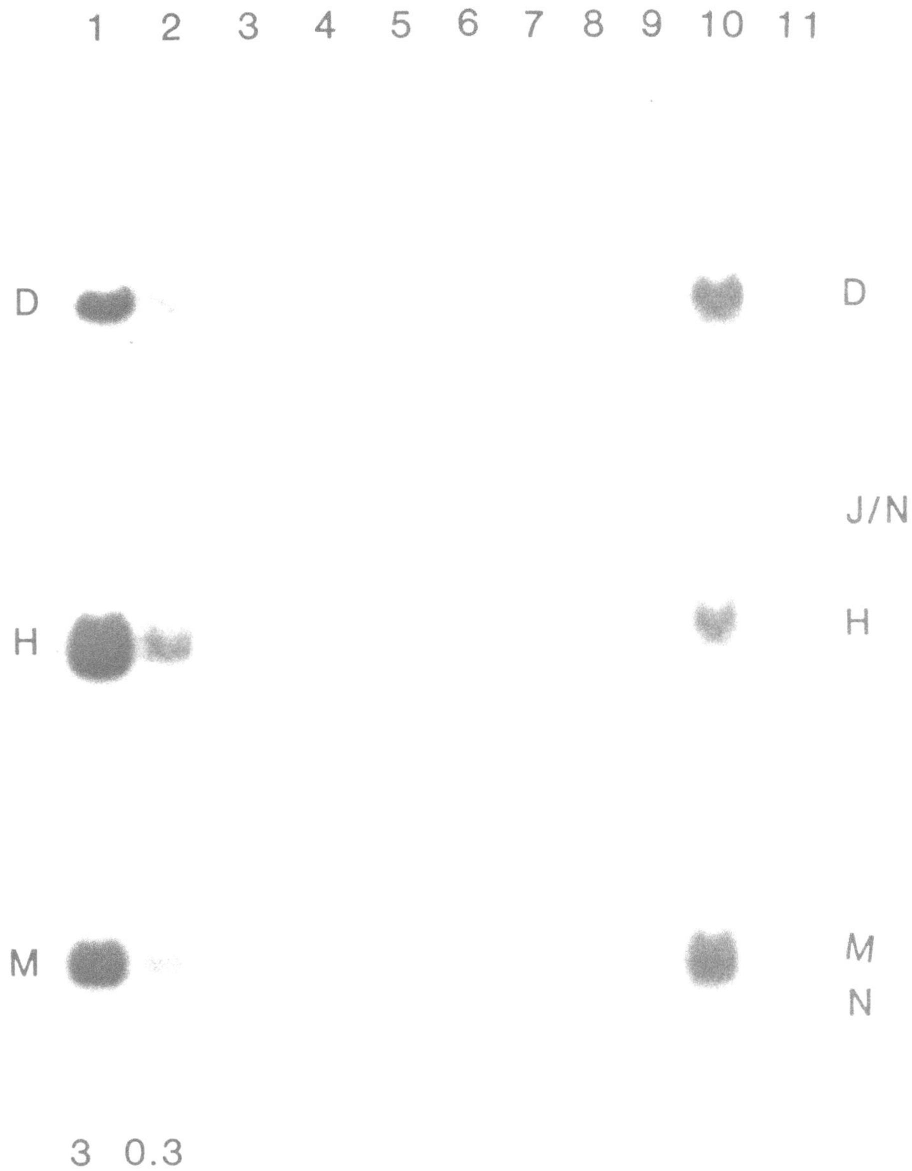

FIGURE 11. Blot hybridization of VZV DNA to human trigeminal ganglia DNA. Each lane contained 20 μg of DNA digested by *Hin*dIII. The DNA probe was a 1:2:2 mixture of pVHD (the VZV DNA *Hin*dIII-D fragment; Figure 2c), pVHH (the VZV DNA *Hin*dIII-H fragment, Figure 2c), and pVHM (the VZV DNA *Hin*dIII-M fragment; Figure 2c). Lanes 1 and 2 are reconstruction experiments mixing recombinant VZV DNAs with human embryo lung (HEL) cell DNA: pVHD at 3 (lane 1) and 0.3 (lane 2) copies per haploid cell genome (noted at the bottom of the lanes) and pVHH and pVHM at twice that concentration. Fragment designations are to the left. Lane 3 contains HEL DNA, the negative control. Lanes 4 to 9 contain DNA from each of six human trigeminal ganglia. Lane 10 contains total DNA from VZV-infected HEL cells. Hybridizing fragments are designated to the right. In VZV-infected cells, the circular and concatemeric VZV DNAs would yield the VZV DNA *Hin*dIII-J/N fusion fragment (Figure 2c), which would be present in much higher abundance than J_{IL}. The *Hin*dIII-M fragment is present at 2 *M* (Table 1) amounts relative to other fragments (Figure 2c). pVHH hybridizes to the VZV DNA *Hin*dIII-H, -N and -J/N fragments (Figure 2c). Lane 11 contains total DNA from HSV-infected HEL cells, the control for the absence of cross-hybridization. The prehybridization and hybridization mixture contained sonicated, denatured pBR322 DNA (2.25 μg/m*l*), as well as the standard ingredients.[113] The efficacy of the presence of unlabeled pBR322 DNA was demonstrated by the nonappearance of the vector band in lanes 1 and 2, as well as the absence of any artifactual bands in lane 3. (From Vlazny et al.[115])

ber would that be per neuron? There are 1.5×10^5 neurons in a human trigeminal ganglion out of (approximately) 2×10^6 cells.[116] Therefore, neurons are 7.5% of the cell population. If all the VZV DNA is confined to neurons during inapparent infection, then there is an average of (approximately) 3 VZV DNA copies per neuron. Of course, this is an average; the copy number in individual neurons could vary widely. These calculations are easily adjusted for any starting number for VZV DNA molecules in the human trigeminal ganglia.

VIII. PRESSING QUESTIONS

Despite the great leap forward over the last 10 years in our knowledge of VZV, VZV infection, and VZV DNA, there are immediate important questions still to be answered. What VZV gene products are expressed during viremia in the patient, a process assumed as a first approximation to be equivalent to VZV infection in cell culture? What VZV gene products are expressed for the virus to leave the blood stream (viremia) and enter the nervous system (neurotropism)? What VZV gene products, if any, are expressed during long-term inapparent infection in the human host? What is the physical form of the VZV DNA during inapparent infection: unit length linear, prototype arrangement, inverted L arrangement, circular, episomal, concatemeric, or integrated? What VZV genes must be triggered for the virus to leave the latent state and reemerge as herpes zoster? It is hoped that the techniques of contemporary molecular genetics, coupled with the incisive work of investigators, will provide the answers to these fundamental questions concerning the natural history of VZV infection.

IX. ACKNOWLEDGMENTS

I have been blessed over a 10-year period with the participation in my VZV research by intelligent and hard working technicians, graduate students, and postdoctoral fellows. I thank them. I am also grateful for the generosity of many colleagues who sent me virus isolates, monoclonal antibodies, recombinant DNAs, and other reagents. This work was supported by Grant CA 16498, National Cancer Institute, National Institutes of Health, U.S. Public Health Service.

REFERENCES

1. Schmidt, N., Further evidence for common antigens in herpes simplex and varicella-zoster viruses, *J. Med. Virol.*, 9, 27, 1982.
2. Davison, A. and Wilkie, N., Location and orientation of homologous sequences in the genomes of five herpesviruses, *J. Gen. Virol.*, 64, 1927, 1983.
3. Ludwig, H., Haines, H. G., Biswal, N., and Benyesh-Melnick, M., The characterization of varicella-zoster virus DNA, *J. Gen. Virol.*, 14, 111, 1972.
4. Iltis, J. P., Oakes, J. E., Hyman, R. W., and Rapp, F., Comparison of the DNAs of varicella-zoster viruses isolated from clinical cases of varicella and herpes zoster, *Virology*, 82, 345, 1977.
5. Richards, J. C., Hyman, R. W., and Rapp, F., Analysis of the DNAs from seven varicella-zoster virus isolates, *J. Virol.*, 32, 812, 1979.
6. Schildkraut, C. L., Marmur, J., and Doty, P., Determination of the base composition of deoxyribonucleic acid from its buoyant density in CsCl, *J. Mol. Biol.*, 4, 430, 1962.
7. Davison, A., Genetic blueprint for varicella-zoster virus, *Herpesvirus Workshop Abstr.*, 10, 3, 1985; Davison, A. and Scott, J., The complete DNA sequence of varicella-zoster virus, *J. Gen. Virol.*, 67, 1759, 1986.

8. Kieff, E., Bachenheimer, S. L., and Roizman, B., Size, composition and structure of the DNA of subtypes 1 and 2 herpes simplex virus, *J. Virol.*, 8, 125, 1971.
9. Dumas, A., Geelen, J., Maris, W., and van der Noordaa, J., Infectivity and molecular weight of varicella-zoster virus DNA, *J. Gen. Virol.*, 47, 233, 1980.
10. Ecker, J. R. and Hyman, R. W., Varicella-zoster virus DNA exists as two isomers, *Proc. Natl. Acad. Sci. U.S.A.*, 79, 156, 1982.
11. Ecker, J. R. and Hyman, R. W., Analysis of interruptions in the phosphodiester backbone of herpes simplex virus DNA, *Virology*, 110, 213, 1981.
12. Rapp, F., Iltis, J., Oakes, J., and Hyman, R. W., A novel approach to study the DNA of herpes zoster virus, *Intervirology*, 8, 272, 1977.
13. Dumas, A. M., Geelen, J. L. M., Weststrate, M., Wertheim, P., and van der Noordaa, J., *XbaI*, *PstI*, and *BglII* restriction enzyme maps of the two orientations of the varicella-zoster virus genome, *J. Virol.*, 39, 390, 1981.
14. Straus, S., Owens, J., Ruyechan, W., Takiff, H., Casey, T., Vande Woude, G., and Hay, J., Molecular cloning and physical mapping of varicella-zoster virus DNA, *Proc. Natl. Acad. Sci. U.S.A.*, 79, 993, 1982.
15. Roizman, B., The organization of the herpes simplex virus genomes, *Annu. Rev. Genet.*, 13, 25, 1979.
16. Davison, A., DNA sequence of the U_s component of the varicella-zoster virus genome, *EMBO J.*, 2, 2203, 1983.
17. Davison, A. and Scott, J., DNA sequence of the major inverted repeat in the varicella-zoster virus genome, *J. Gen. Virol.*, 66, 207, 1985.
18. Gilden, D., Shtram, Y., Friedman, A., Wellish, M., Devlin, M., Fraser, N., and Becker, Y., The internal organization of the varicella-zoster virus genome, *J. Gen. Virol.*, 60, 371, 1982.
19. Ecker, J. R., Molecular Cloning and Structural Determination of Varicella-Zoster Virus DNA, Ph.D. thesis, Pennsylvania State University, University Park, 1982.
20. Davison, A. and Scott, J., Molecular cloning of the varicella-zoster virus genome and derivation of six restriction endonuclease maps, *J. Gen. Virol.*, 64, 1811, 1983.
21. Ecker, J. R., Kudler, L., and Hyman, R. W., Variation in the structure of varicella-zoster virus DNA, *Intervirology*, 21, 25, 1984.
22. Vlazny, D. A. and Hyman, R. W., Errant processing and structural alterations of genomes present in a varicella-zoster virus vaccine, *J. Virol.*, 56, 92, 1985.
23. Ecker, J. R. and Hyman, R. W., unpublished data, 1982.
24. Davison, A., Structure of the genomic termini of varicella-zoster virus, *J. Gen. Virol.*, 65, 1969, 1984.
25. Kinchington, P., Reinhold, W., Casey, T., Straus, S., Hay, J., and Ruyechan, W., Inversion and circularization of the varicella-zoster virus genome, *J. Virol.*, 56, 194, 1985.
26. Mocarski, E., Post, L., and Roizman, B., Molecular engineering of the herpes simplex virus genome: insertion of a second L-S junction into the genome causes additional genome inversions, *Cell*, 22, 243, 1980.
27. Mocarski, E. and Roizman, B., Site-specific inversion sequence of the herpes simplex virus genome: domain and structural features, *Proc. Natl. Acad. Sci. U.S.A.*, 78, 7047, 1981.
28. Smiley, J. R., Fong, B. S., and Leung, W., Construction of a double-jointed herpes simplex viral DNA molecule: inverted repeats are required for segment inversion and direct repeats promote deletions, *Virology*, 113, 345, 1981.
29. Locker, H. and Frenkel, N., Structure and origin of defective genomes contained in serially passaged herpes simplex virus type 1 (Justin), *J. Virol.*, 29, 1065, 1979.
30. Chou, J. and Roizman, B., Isomerization of herpes simplex virus 1 genome: identification of the *cis*-acting and recombination sites within the domain of the *a* sequence, *Cell*, 41, 803, 1985.
31. Davison, A. and Wilkie, N., Nucleotide sequences of the joint between the L and S segments of herpes simplex virus types 1 and 2, *J. Gen. Virol.*, 55, 315, 1981.
32. Kilpatrick, B. and Huang, E.-S., Human cytomegalovirus genome: partial denaturation map and organization of genome sequences, *J. Virol.*, 24, 261, 1977.
33. Thomsen, D. and Stinski, M., Cloning of the human cytomegalovirus genome as endonuclease *XbaI* fragments, *Gene*, 16, 207, 1981.
34. Spaete, R. and Mocarski, E., The *a* sequence of the cytomegalovirus genome functions as a cleavage/packaging signal for herpes simplex virus defective genomes, *J. Virol.*, 54, 817, 1985.
35. Matsuo, T., Heller, M., Petti, L., O'Shiro, E., and Kieff, E., Epstein-Barr virus genome integrated into human lymphocyte DNA, *Science*, 226, 1322, 1984.
36. Albrecht, M., Darai, G., and Flugel, R., Analysis of the genomic termini of tupaia herpesvirus DNA by restriction mapping and nucleotide sequencing, *J. Virol.*, 56, 466, 1985.

37. Ben-Porat, T., Rixon, F., and Blankenship, M., Analysis of the structure of the genome of pseudorabies virus, *Virology*, 95, 285, 1979.
38. Lomniczi, B., Blankenship, M., and Ben-Porat, T., Deletions in the genomes of pseudorabies virus vaccine strains and existence of four isomers of the genomes, *J. Virol.*, 49, 970, 1984.
39. Straus, S., Aulakh, H., Ruyechan, W., Hay, J., Casey, T., Vande Woude, G., Owens, J., and Smith, H., Structure of varicella-zoster virus DNA, *J. Virol.*, 40, 516, 1981.
40. Jacob, R. and Roizman, B., Anatomy of herpes simplex virus DNA. VIII. Properties of the replicating DNA, *J. Virol.*, 23, 394, 1977.
41. Ben-Porat, T., Kaplan, A., Stehn, B., and Rubenstein, A., Concatemeric forms of intracellular herpesvirus DNA, *Virology*, 69, 547, 1976.
42. Hayward, G. S., Frenkel, N., and Roizman, B., Anatomy of herpes simplex virus DNA: strain differences and heterogeneity in the locations of restriction endonuclease cleavage sites, *Proc. Natl. Acad. Sci. U.S.A.*, 72, 1768, 1975.
43. Skare, J., Summers, W., and Summers, W., Structure and function of herpesvirus genomes. I. Comparison of five HSV-1 and two HSV-2 strains by cleavage of their DNA with *Eco*RI restriction endonuclease, *J. Virol.*, 15, 726, 1975.
44. Buchman, T., Roizman, B., Adams, G., and Stover, H., Restriction endonuclease fingerprinting of herpes simplex DNA: a novel epidemiologic tool applied to a nosocomial outbreak, *J. Infect. Dis.*, 138, 488, 1978.
45. Oakes, J. E., Iltis, J. P., Hyman, R. W., and Rapp, F., Analysis by restriction enzyme cleavage of human varicella-zoster virus DNAs, *Virology*, 82, 353, 1977.
46. Wilkie, N., Physical maps for herpes simplex virus type 1 DNA for restriction endonuclease *Hin*dIII, *Hpa*I, and *Xba*I, *J. Virol.*, 20, 222, 1976.
47. Skare, J., and Summers, W., Structure and function of herpesvirus genomes. II. *Eco*RI, *Xba*I and *Hin*dIII endonuclease cleavage sites on herpes simplex virus type 1 DNA, *Virology*, 76, 581, 1977.
48. Ecker, J. R., and Hyman, R. W., Varicella-zoster virus vaccine DNA differs from the parental virus DNA, *J. Virol.*, 40, 314, 1981.
49. Pichini, B., Ecker, J. R., Grose, C., and Hyman, R. W., DNA mapping of paired varicella-zoster virus isolates from patients with shingles, *Lancet*, 2, 1223, 1983.
50. Gershon, A., Steinberg, S., and Gelb, L., Clinical reinfection with varicella-zoster virus, *J. Infect. Dis.*, 149, 137, 1984.
51. Straus, S., Reinhold, W., Smith, H., Ruyechan, W., Henderson, D., Blaese, R., and Hay, J., Endonuclease analysis of viral DNA from varicella and subsequent zoster infections in the same patient, *N. Engl. J. Med.*, 311, 1362, 1984.
52. Williams, D., Gershon, A., Gelb, L., Spraker, M., Steinberg, S., and Ragab, A., Herpes zoster following varicella vaccine in a child with acute lymphocytic leukemia, *J. Pediatr.*, 106, 259, 1985.
53. Landry, M., Berkovits, N., Summers, W., Booss, J., Hsiung, G. D., and Summers, W., Herpes simplex encephalitis: analysis of a cluster of cases by restriction endonuclease mapping of virus isolates, *Neurology*, 33, 831, 1983.
54. Straus, S., Hay, J., Smith, H., and Owens, J., Genome differences among varicella-zoster virus isolates, *J. Gen. Virol.*, 64, 1031, 1983.
55. Wagner, M., and Summers, W., Structure of the joint region and the termini of the DNA of herpes simplex virus type 1, *J. Virol.*, 27, 374, 1978.
56. Post, L., Conley, A., Mocarski, E., and Roizman, B., Cloning of the reiterated and non-reiterated herpes simplex virus 1 sequence as *Bam*HI fragments, *Proc. Natl. Acad. Sci. U.S.A.*, 77, 4201, 1980.
57. Russell, R., Kudler, L., Miller, R., and Hyman, R. W., Stability of the cloned "joint region" of herpes simplex virus type 1 DNA, *Intervirology*, 18, 98, 1982.
58. Kinchington, P., Casey, T., Remenick, J., Ostrove, J., Ruyechan, W., Straus, S., and Hay, J., A 42 bp repeat sequence of variable copy number in DNAs of different strains, *Herpesvirus Workshop Abstr.*, 10, 25, 1985; Kinchington, P., Remenick, J., Ostrove, J., Straus, S., Ruyechan, W., and Hay, J., Putative glycoprotein gene of varicella-zoster virus with variable copy number of a 42-base pair repeat sequence has homology to herpes simplex virus glycoprotein C., *J. Virol.*, 59, 660, 1986.
59. Rixon, F. and McGeoch, D., Detailed analysis of the mRNAs mapping in the short unique region of herpes simplex virus type 1, *Nucleic Acids Res.*, 13, 953, 1985.
60. Ostrove, J., Reinhold, W., Fan, C.-M., Zorn, S., Hay, J., and Straus, S., Transcription mapping of the varicella-zoster virus genome, *J. Virol.*, 56, 600, 1985.
61. Anderson, K., Costa, R., Holland, L., and Wagner, E., Characterization of HSV-1 RNA present in the absence of *de novo* protein synthesis, *J. Virol.*, 34, 9, 1980.
62. Wagner, E. K., HSV transcription patterns, *Adv. Viral Oncol.*, 3, 239, 1983.
63. Maguire, H. and Hyman, R. W., unpublished observations, 1985.
64. Wolff, M., The proteins of varicella-zoster virus, *Med. Microbiol. Immunol.*, 166, 21, 1978.

65. Asano, Y. and Takahashi, M., Studies on the polypeptides of varicella-zoster (V-Z) virus. I. Detection of varicella-zoster virus polypeptides in infected cells, *Biken J.*, 22, 81, 1979.
66. Grose, C., The synthesis of glycoproteins in human melanoma cells infected with varicella-zoster virus, *Virology*, 101, 1, 1980.
67. Grose, C., Edwards, D., Friedrichs, W., Weigle, K., and McGuire, W., Monoclonal antibodies against three major glycoproteins of varicella-zoster virus, *Infect. Immun.*, 40, 381, 1983.
68. Grose, C., Edwards, D., Weigle, K., Friedrichs, W., and McGuire, W., Varicella-zoster virus-specific gp140: a highly immunogenic and disulfide-linked structural glycoprotein, *Virology*, 132, 138, 1984.
69. Forghani, B., Schmidt, N., Myoraku, C., and Gallo, D., Serological reactivity of some monoclonal antibodies to varicella-zoster virus, *Arch. Virol.*, 73, 311, 1982.
70. Forghani, B., Dupuis, K., and Schmidt, N., Varicella-zoster viral glycoproteins analyzed with monoclonal antibodies, *J. Virol.*, 52, 55, 1984.
71. Okuno, T., Yamanishi, K., Shiraki, K., and Takahashi, M., Synthesis and processing of glycoproteins of varicella-zoster virus (VZV) as studied with monoclonal antibodies to VZV antigens, *Virology*, 129, 357, 1983.
72. Keller, P., Neff, B., and Ellis, R., Three major glycoprotein genes of varicella-zoster virus whose products have neutralizing epitopes, *J. Virol.*, 52, 293, 1984.
73. Edson, C., Hosler, B., Poodry, C., Schooley, R., Waters, D., and Thorley-Lawson, D., Varicella-zoster virus envelope glycoproteins: biochemical characterization and identification in clinical material, *Virology*, 145, 62, 1985.
74. Davison, A., Edson, C., Ellis, R., Forghani, B., Gilden, D., Grose, C., Keller, P., Vafai, A., Wroblewska, Z., and Yamanishi, K., A new common nomenclature for the glycoprotein genes of varicella-zoster virus and their glycosylated products, *J. Virol.*, 57, 1195, 1986.
75. Namazue, J., Campo-Vera, H., Kitamura, K., Okuno, T., and Yamanishi, K., Processing of virus-specific glycoproteins of varicella-zoster virus, *Virology*, 143, 252, 1985.
76. Ellis, R., Keller, P., Lowe, R., and Zivin, R., Use of a bacterial expression vector to map the varicella-zoster virus major glycoprotein gene, gC, *J. Virol.*, 53, 81, 1985.
77. Keller, P., Davison, A., Lowe, R., Bennett, C., and Ellis, R., Identification and structure of the gene encoding gpII, a major glycoprotein of varicella-zoster virus, *Virology*, 152, 181, 1986.
78. Shiraki, K., Okuno, T., Yamanishi, K., and Takahashi, M., Polypeptides of varicella-zoster virus (VZV) and immunological relationship of VZV and herpes simplex virus (HSV), *J. Gen. Virol.*, 61, 255, 1982.
79. Edson, C., Hosler, B., Respess, R., Waters, D., and Thorley-Lawson, D., Cross-reactivity between herpes simplex virus glycoprotein B and a 63,000-dalton varicella-zoster virus envelope glycoprotein, *J. Virol.*, 56, 333, 1985.
80. Sarmiento, M., Haffey, M., and Spear, P., Membrane proteins specified by herpes simplex virus. III. Role of glycoprotein VP7 (B_2) in virion infectivity, *J. Virol.*, 29, 1149, 1979.
81. Little, S., Jofre, J., Courtney, R., and Schaffer, P., A virion associated glycoprotein essential for infectivity of herpes simplex virus type 1, *Virology*, 115, 149, 1981.
82. Weigle, K. and Grose, C., Molecular dissection of the humoral immune response to individual varicella-zoster viral proteins during chickenpox, quiescence, reinfection, and reactivation, *J. Infect. Dis.*, 149, 741, 1984.
83. Friedrichs, W. and Grose, C., Glycoprotein gp118 of varicella-zoster virus: purification by serial affinity chromatography, *J. Virol.*, 49, 992, 1984.
84. Davison, A., Waters, D., and Edson, C., Identification of the products of a varicella-zoster virus glycoprotein gene, *J. Gen. Virol.*, 66, 2237, 1985.
85. Vafai, A., Wroblewska, Z., Wellish, M., Green, M., and Gilden, D., Analysis of three late varicella-zoster virus proteins, a 125,000-molecular-weight protein and gp1 and gp3, *J. Virol.*, 52, 953, 1984.
86. Wroblewska, Z., Gilden, D., Green, M., Devlin, M., and Vafai, A., Affinity-purified varicella-zoster virus glycoprotein gp1/gp3 stimulates the production of neutralizing antibody, *J. Gen. Virol.*, 66, 1795, 1985.
87. Lopetegui, P., Campo-Vera, H., and Yamanishi, K., Varicella-zoster virus (VZV)-specific polypeptides detected in cells treated with metabolic inhibitors, *Microbiol. Immunol.*, 29, 569, 1985.
88. Dobersen, M., Jerkofsky, M., and Greer, S., Enzymatic basis for the selective inhibition of varicella-zoster virus by 5-halogenated analogues of deoxycytidine, *J. Virol.*, 20, 478, 1976.
89. Ogino, T., Otsuka, T., and Takahashi, M., Induction of deoxypyrimidine kinase activity in human embryonic lung cells infected with varicella-zoster virus, *J. Virol.*, 21, 1232, 1977.
90. Ogino, T., Lopetegui, P., and Yamanishi, K., Deoxythymidine kinases in varicella-zoster virus infected and biochemically transformed cells, *Biken J.*, 25, 149, 1982.
91. Hackstadt, T. and Mallavia, L., Deoxypyrimidine nucleoside metabolism in varicella-zoster-virus-infected cells, *J. Virol.*, 25, 510, 1978.

92. Yamanishi, K., Matsunaga, Y., Ogino, T., and Lopetegui, P., Biochemical transformation of mouse cells by varicella-zoster virus, *J. Gen. Virol.*, 56, 421, 1981.
93. Lopetegui, P., Matsunaga, Y., Okuno, T., Ogino, T., and Yamanishi, K., Expression of varicella-zoster virus-related antigens in biochemically transformed cells, *J. Gen. Virol.*, 64, 1181, 1983.
94. Eizuru, Y. and Hyman, R. W., unpublished data, 1982.
95. Sawyer, M., Ostrove, J., Felser, J., and Straus, S., Mapping of the varicella-zoster virus deoxypyrimidine kinase gene and preliminary identification of its transcript, *Virology*, 149, 1, 1986.
96. May, D., Miller, R., and Rapp, F., The effect of phosphonoacetic acid on the in vitro replication of varicella-zoster virus, *Intervirology*, 8, 83, 1977.
97. Miller, R. and Rapp, F., Varicella-zoster virus-induced DNA polymerase, *J. Gen. Virol.*, 36, 515, 1977.
98. Mar, E.-C., Huang, Y.-S., and Huang, E.-S., Purification and characterization of varicella-zoster virus-induced DNA polymerase, *J. Virol.*, 26, 249, 1978.
99. Shiraki, K., Ogino, T., Yamanishi, K., and Takahashi, M., Isolation of drug resistant mutants of varicella-zoster virus: cross resistance of acyclovir resistant mutants with phosphonoacetic acid and bromodeoxyuridine, *Biken J.*, 26, 17, 1983.
100. Grose, C. and Friedrichs, W., Immunoprecipitable polypeptides specified by varicella-zoster virus, *Virology*, 118, 86, 1982.
101. Shiraki, K., Yamamoto, T., Yamanishi, K., and Takahashi, M., DNA binding proteins induced by varicella-zoster virus in human cells, *Biken J.*, 25, 185, 1982.
102. Roberts, C., Weir, A., Hay, J., Straus, S., and Ruyechan, W., DNA-binding proteins present in varicella-zoster virus-infected cells, *J. Virol.*, 55, 45, 1985.
103. Shemer, Y., Leventon-Kriss, S., and Sarov, I., Isolation and polypeptide characterization of varicella-zoster virus, *Virology*, 106, 133, 1980.
104. Grose, C., Friedrichs, W., and Smith, G. C., Purification and molecular anatomy of the varicella-zoster virion, *Biken J.*, 26, 1, 1983.
105. Haarr, L., Marsden, H., Preston, C., Smiley, J., Summers, W. C., and Summers, W. P., Utilization of internal AUG codons for initiation of protein synthesis directed by mRNAs from normal and mutant genes encoding herpes simplex virus-specified thymidine kinase, *J. Virol.*, 56, 512, 1985.
106. Gelb, L., Huang, J., and Wellinghoff, W., Varicella-zoster virus transformation of hamster embryo cells, *J. Gen. Virol.*, 51, 171, 1980.
107. Rapp, F., *Oncogenic Herpesviruses*, Vol. 1 and 2, CRC Press, Boca Raton, Fla., 1980.
108. Hyman, R. W., Ecker, J. R., and Tenser, R. B., Varicella-zoster virus RNA in human trigeminal ganglia, *Lancet*, 2, 814, 1983.
109. Bastian, F., Rabson, A., Lee, C., and Tralka, T., Herpesvirus hominis: isolation from human trigeminal ganglion, *Science*, 178, 306, 1972.
110. Baringer, J. R. and Swoveland, P., Recovery of herpes simplex virus from human trigeminal ganglions, *N. Engl. J. Med.*, 288, 648, 1973.
111. Gilden, D., Vafai, A., Shtram, Y., Becker, Y., Devlin, M., and Wellish, M., Varicella-zoster virus DNA in human sensory ganglia, *Nature (London)*, 306, 478, 1983.
112. Cameron, I., Park, M., Dutia, B., Orr, A., and Macnab, J., Herpes simplex virus sequences involved in the initiation of oncogenic morphological transformation of rat cells are not required for maintenance of the transformed state, *J. Gen. Virol.*, 66, 517, 1985.
113. Jones, T. R., Parks, C. L., Spector, D. J., and Hyman, R. W., Hybridization of herpes simplex virus DNA and human ribosomal DNA and RNA, *Virology*, 144, 384, 1985; Jones, T. R. and Hyman, R. W., Sequences in the proximal IRL of herpes simplex virus DNA hybridize to human DNA, *Virus Research*, 4, 369, 1986.
114. Efstathiou, S., Minson, A. C., Field, H., Anderson, J., and Wildy, P., Detection of herpes simplex virus specific DNA sequences in latently infected mice and in humans, *J. Virol.*, 57, 446—455, 1986.
115. Vlazny, D., Tenser, R. B., and Hyman, R. W., unpublished data, 1985.
116. Klein, R. J., Initiation and maintenance of latent herpes simplex virus infections: the paradox of perpetual immobility and continuous movement, *Rev. Infect. Dis.*, 7, 21, 1985.

Chapter 4

THERAPY OF VARICELLA-ZOSTER VIRUS INFECTIONS*

Richard J. Whitley

TABLE OF CONTENTS

* Studies performed by the author were initiated under and supported by a contract (NO1-A1-62554) with the Development and Applications Branch of the National Institute of Allergy and Infectious Diseases, and by grants from the National Cancer Insitutue (CA-13148), the General Clinical Research Centers Programs (RR-032) of the National Institutes of Health, and the State of Alabama.

I. INTRODUCTION

The past decade has witnessed significant advances in the therapy of human viral infections, particularly those caused by herpes simplex and varicella-zoster viruses. These advances were once only a basis for discussion among clinical and biomedical investigators; however, they have now been applied by the practicing physician to the management of viral diseases, resulting in improved patient care. For the most part, these therapeutic improvements were first achieved in treatment of adult viral infections. More recently, therapy has been proven efficacious for herpesvirus and certain respiratory tract viral infections in children. Advances in treatment were adopted, at least at the outset, for severe and life-threatening diseases, but, more recently, less severe mucocutaneous herpes simplex virus infections have been found amenable to therapy.

The development of specific and selective inhibitors of viral replication is especially important in the field of treatment of herpesvirus infections. As described in prior chapters, herpesviruses have enzymes which are unique for their replication and, as a consequence, are targets for antiviral drugs. The ability of drugs to distinguish enzymes unique for viral replication from those enzymes necessary for cellular replication has led to the development of compounds with low toxicity and broader clinical applicability. Controlled clinical studies, utilizing appropriate biostatistical methodology, have defined viral infections for which treatment is of not just laboratory but also of clinical value. During the conduct of these controlled studies, the natural history, pathogenesis, need for prompt diagnosis, and expedient therapeutic intervention have been reiterated. Much of what has been learned from studies of varicella-zoster virus (VZV) infections has provided a basis for more promising interventive strategies which will need to be tested over the next decade.

The treatment of VZV infections in both the immunocompromised and normal host has been attempted for approximately 20 years. Because of the nature of the vesicular rash of both primary and recurrent infection, diagnosis can be made promptly, shortly after the appearance of skin lesions and, therefore, therapy can be instituted in a timely manner. As has been well described in prior chapters, VZV is responsible for two clinically distinct diseases. Primary infection results in chickenpox, the consequence of exposure of a susceptible or seronegative individual to VZV. It is ubiquitous, extremely contagious, and, for the most part, a benign illness of childhood characterized by a generalized exanthematous vesicular rash. Infection occurs most frequently in an epidemic fashion in the late winter and early spring. With recurrence, infection results in the localized clinical phenomenon of herpes zoster or shingles.

Either primary or recurrent infection in the immunocompromised host is more severe than in the normal host. Infection in these populations is characterized by a period of prolonged viral replication, vesicle formation, delayed healing, and a high frequency of complications. It is for these reasons that VZV infections in such high-risk patient populations have been targets for antiviral therapy.

The epidemiology, pathogenesis, and natural history of VZV infections have been reviewed in prior chapters. The treatment of these infections, particularly as it relates to the administration of immune globulin products for prevention or the administration of either interferon or antivirals for treatment, will be the subject of this chapter. While it is not the intent of this section to review the clinical epidemiology and natural history of VZV infections, salient issues, as they relate to intervention, will be briefly reviewed in order to put prevention and therapy in the proper perspective. Prevention by utilization of vaccines is the subject of a separate chapter.

II. CHICKENPOX

A. Clinical Findings of Relevance to Therapy

The onset of chickenpox begins with a rash, low-grade fever, and malaise. In the immunocompetent child, fever ranges from 100 to 103°F and lasts only 3 to 5 days.[1-3] Other constitutional symptoms include pruritis, anorexia, and listlessness. The lesions are highly suggestive of the diagnosis. Beginning as small erythematous macules on the face and trunk, lesions spread rapidly centripetally to involve other areas of the body and become vesicular, pustular, or scabbed. While they remain small in the immunocompetent child (ranging from 5 to 15 mm in diameter), lesions are both more numerous and larger, even bullous, in the immunocompromised child. The erythematous base of lesions in the normal host is often replaced by one which is hemorrhagic in the immunocompromised child. The immunocompetent child develops successive crops of lesions over a period of 2 to 4 days. In contrast, the period of lesion formation in the immunocompromised child is prolonged, likely dependent upon the extent of immune impairment. The median duration of lesion formation in the immunocompromised child is approximately 10 days, with fever persisting a median of 11 days and time to total scabbing nearly 12 days. Most immunocompromised patients are not completely scabbed until 2 weeks into the course of their illness. Such healing represents a two- to threefold increase over that in the normal host.

While chickenpox in the normal child is usually a benign illness, this infection in the adult or immunocompromised individual is of increased severity, as evidenced by enhanced morbidity and mortality. Several studies have summarized the severity of chickenpox in the immunocompromised child.[2,4,5] For children receiving cancer chemotherapy, nearly one third have progressive involvement of multiple organs, including lungs, liver, and the central nervous system. The majority of children with visceral disease develop pneumonitis within the first week after the onset of infection. The overall mortality of chickenpox in the immunocompromised host has been reported to be 8 to 10%, but in the absence of therapy it is likely higher. It is probably that visceral organ involvement in patients suffering such complications is the consequence of viremia. In a series of antiviral studies performed in immunocompromised children, an even higher percentage of patients on contemporary anticancer chemotherapy regimens developed life-threatening disease with a higher mortality in those patients who did not receive antiviral therapy.[4-8]

The most serious complication of chickenpox is pneumonitis which is characterized by progressive tachypnea and cough. The chest roentgenogram demonstrates a diffuse nodular infiltrate. In the immunocompromised child, respiratory distress increases rapidly as a consequence of hemorrhagic necrosis of the lungs and frequently necessitates ventilatory support.

When the central nervous system is involved, an event even more common than varicella pneumonitis, the clinical outcome is less severe. It is estimated that between 0.1 and 0.2% of individuals with chickenpox have encephalitic findings, characterized by altered levels of consciousness with progressive headaches, vomiting, and seizures. Cerebellar ataxia appears between 7 and 14 days after the onset of the rash, and infrequently occurs during the preeruptive phase of the disease. The most prominent clinical finding is ataxia with meningeal irritation. The cerebrospinal fluid contains lymphocytes and elevated levels of protein.

The most common infectious complication of varicella is secondary bacterial infection of the skin, which is usually caused by *Streptococcus pyogenes* or *Staphylococcus aureus*. These infections may result from excoriation of skin lesions, the consequence of pruritis. A Gram stain of the skin lesion should help clarify the etiology of unusually erythematous and pustulated lesions.

B. Clinical Management of Chickenpox

1. Prevention

As with other infectious diseases, prophylaxis is the more acceptable form of management. Such an approach can best be accomplished by immunization, or alternatively by the administration of immune serum products at a time when they can be most beneficial. For the normal host, prophylaxis and treatment are of no importance since the disease is usually benign. However, as noted above, the immunocompromised child is at risk for developing progressive varicella. A series of studies have been performed to determine the effectiveness of the administration of prophylactic immune serum globulin for disease prevention. Ross suggested the clinical benefit of the administration of varying dosages of serum gamma globulin administered to 242 children within 72 hr of exposure to chickenpox.[9] These results were compared with data derived from 209 children who served as untreated but also unmatched controls. Higher doses of immune serum globulin were associated with significantly decreased total vesicular counts. Children who formed fewer than 20 lesions accounted for approximately 40% of the immune serum globulin recipients as compared with only 1% of those individuals in the control group.

This observation provided a basis for more detailed attempts at altering the natural history of chickenpox with either specific zoster immune globulin (ZIG), derived from several patients with herpes zoster, or varicella-zoster immune globulin (VZIG), particularly in the immunocompromised child. An alternative approach is the administration of zoster immune plasma (ZIP). In a preliminary study, the administration of ZIG was clearly beneficial.[10] More detailed studies have since demonstrated safety and tolerance following administration of ZIG, ZIP, and VZIG to high-risk household contacts of children with chickenpox. High-risk children have been defined as patients immunocompromised by underlying disease, such as acute lymphoblastic leukemia, or requiring cytotoxic and cytolytic medication for control of their disease. Both ameliorative and protective effects have been demonstrated.[11,12] The administration of these products does not unequivocally prevent disease but simply modifies its course.[13]

Table 1 summarizes the indications for administration of ZIG or VZIG. Varicella-zoster immune globulin should be given to an immunodeficient child who is less than 15 years of age, who has a negative or unknown history of chickenpox, who has not been vaccinated against VZV, and who has had contact in a household or with a playmate for more than 1 hr indoors or in a shared hospital room. Furthermore, it should be administered within 72 hr of exposure to be most effective. It can be given as late as 96 hours. An additional indication for administration of VZIG in the prevention of chickenpox is to the newborn whose mother had an onset of chickenpox less than 5 days before delivery or 48 hr post partum. The use of VZIG for susceptible individuals over 15 years of age must be evaluated on an individual basis. There is no evidence that VZIG is useful for adults with chickenpox, including pregnant women.

2. Treatment

Medical management of chickenpox in the normal host is directed toward prevention of avoidable complications. Secondary bacterial infection of the skin can be avoided by meticulous care, particularly the close cropping of fingernails. Good hygiene, including daily bathing and astringent soaks, is important for avoiding this complication. Pruritis can be decreased with topical dressings or the administration of such appropriate antipruritics as Benadryl® or Atarax®.

Infrequently, visceral complications appear in the normal child or adult. Patients with varicella pneumonitis require excellent supportive nursing care. This should include the removal of bronchial secretions, supplementary oxygen, and ventilatory sup-

Table 1
INDICATIONS FOR ADMINISTRATION OF ZOSTER IMMUNE GLOBULIN

I. Exposure:
  A. Both exposure to person with chickenpox or zoster as:
    1. Continuous household contact
    2. Playmate of >1 hr duration indoors
    3. Hospital contact (same room or prolonged face-to-face)
    4. Newborn exposure whereby mother had onset of chickenpox 5 days before delivery or 48 hr post partum
  B. And time elapsed ≤96 hr (preferably sooner)
II. Candidates should:
  A. Have significant exposure (see I above)
  B. Be susceptible to VZV infection
  C. Be <15 years of age (older immunocompromised patients require individual decision)
  D. Have one of the following conditions:
    1. Leukemia or lymphoma
    2. Congenital or acquired immune deficiency
    3. Immunosuppressive treatment
    4. Newborn defined above (see IA4)

port, as required. At this time no other form of interventive antiviral therapy has been proven efficacious for the management of varicella or complications of varicella in the normal host.

Chickenpox in the immunocompromised host is a disease which often requires treatment. Several antiviral therapeutic approaches have been considered. Of the reported studies, three warrant consideration. A successful trial performed by the National Institute of Allergy and Infectious Diseases (NIAID), Collaborative Antiviral Study Group utilized vidarabine for the management of chickenpox in the immunocompromised child.[6]

Vidarabine (9-beta-D-arabinofuranosyladenine) is a nucleoside analog which was developed in the early 1960s, first as an anticancer agent, but subsequently was evaluated for the treatment of herpesvirus infections. While its mechanism of action is not precisely known, it is phosphorylated to its triphosphate derivative which acts as a competitive inhibitor of VZV DNA polymerase.[14] The drug acts at the level of viral polymerase, among other sites. Importantly, the triphosphate derivative is also an inhibitor of cellular DNA polymerase. As a consequence, its therapeutic index or ratio of efficacy to toxicity in vitro is lower than that of other antiviral compounds. When vidarabine is administered to humans, it is rapidly deaminated to arabinosyl hypoxanthine, a compound which has antiviral activity but significantly less than the parent compound. This drug is currently the only licensed compound for treatment of VZV infection in the immunocompromised host at this time, although others will likely soon be approved.

Immunocompromised patients suffering from chickenpox were treated at a dosage of 10 mg/kg per day with vidarabine. They ceased forming new lesions and defervesced more rapidly than their placebo counterparts as displayed in Figure 1.[6] Most important, the incidence of life-threatening complications which developed after the onset of treatment was significantly lower in drug recipients than in placebo counterparts, 53% vs. 5%, $p < 0.01$. Thus, when therapy was instituted within 72 hr of the onset of chickenpox in these high-risk patients, it beneficially influenced the outcome of disease.

A similar study was performed to evaluate acyclovir. Acyclovir, 9-[(2-hydroxy-

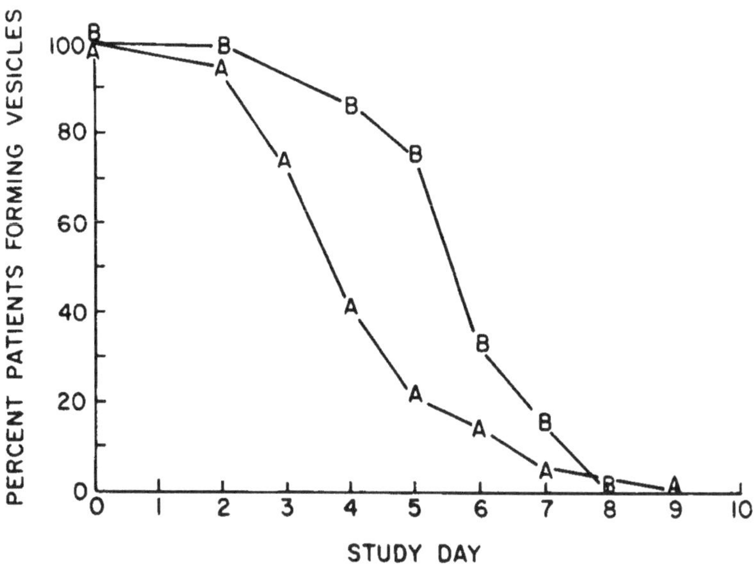

FIGURE 1. Percentage of patients forming new vesicles on each day of study. The response of vidarabine (A, N = 19) compared to that for placebo (B, N = 15) recipients for new vesicle formation: $p = 0.015$. (From Whitley, R. J., Hilty, M., Haynes, R., Bryson, Y., Connor, J. D., Soong, S. J., Alford, C. A., Jr., and the NIAID Collaborative Antiviral Study Group, *J. Pediatr.*, 101, 125, 1982. With permission.)

ethoxy)methyl]quanine, was developed and synthesized by the Burroughs Wellcome Company. It has been shown to be specifically activated by viral thymidine kinase induced by herpes simplex and VZV.[15,16] As a consequence of selective phosphorylation of acyclovir by viral enzymes, the drug is considered a specific inhibitor of viral replication.

Following conversion of acyclovir to its monophosphate derivative, the drug is further phosphorylated to a triphosphate derivative by both cellular and viral enzymes. As a triphosphate derivative, it is a competitive inhibitor and chain terminator of viral DNA polymerase.[16,17] Because of the nature of the selectivity of action, the compound has a higher therapeutic index than other antiviral compounds which have been clinically evaluated, including vidarabine. Drug concentrations required for 50% inhibition of VZV replication are higher by over a factor of 10 than that necessary to achieve a similar endpoint with herpes simplex virus. Acyclovir levels of 5 to 8 μM are necessary to inhibit VZV replication in tissue culture as compared to 0.1 to 0.4 μM for herpes simplex virus replication.[18,19] This difference in the sensitivity of VZV isolates as compared to those of herpes simplex is important when considering the likelihood that a therapeutic agent will achieve appropriate plasma levels required for successful therapy.

A therapeutic trial of acyclovir for the management of chickenpox in the immunocompromised host was conducted in 20 patients.[7] Prober and colleagues found that therapy had no demonstrable effect on the events of cutaneous healing or fever; however, treatment significantly decreased the development of pneumonitis in treated patients from 45 to 0%. Both acyclovir and vidarabine were without significant laboratory or clinical toxicity in the patient populations evaluated. Thus, the therapeutic index for the utilization of these compounds in the management of chickenpox in high-risk populations appears adequate. At this time, a comparative trial of vidarabine and acyclovir has not been conducted.

Another therapeutic modality used for the treatment of chickenpox is the administration of large doses of interferon early in the disease course.[8] In a controlled study, Arvin and colleagues found that 30 million units of interferon administered intramuscularly for 5 days decreased the frequency of visceral complications and prevented progressive disease in children with varicella. Interferon for this study was produced by Dr. Kari Cantel from the Finnish Blood Bank. It was derived from leucocytes harvested and challenged for interferon production. To date, no published studies have evaluated recombinant DNA interferons alone or in combination with antiviral agents for the management of chickenpox in high-risk patients.

Interferons are a series of naturally occurring low-molecular weight proteins that, among other functions, act as a host defense to viral infection. High-titered interferon has been found in vesicle fluid from patients with varicella at the termination of illness, suggesting that interferon may play a role in the control of this disease.[20] Because of the difficulty of producing interferon from leukocytes, the availability of interferon for further clinical trials and, ultimately, licensure, will be dependent upon recombinant DNA technology. Even so, if efficacy can be demonstrated in utilizing recombinant DNA interferon for the management of chickenpox in the high-risk patients, therapeutic benefit will have to be weighed against the high frequency of adverse effects encountered following interferon administration

3. *New Drugs*

New compounds have been developed and evaluated for the management of VZV infections in the normal and immunocompromised host. Since none of these compounds have been evaluated for the treatment of chickenpox (at least at the present time), a discussion of these agents will be deferred to the consideration of the treatment of herpes zoster.

C. Conclusion

While chickenpox is benign in the normal host, the propensity for severe illness in adults and immunocompromised patients is striking. Current therapeutic research efforts have focused on both the prevention of VZV infections through the development of a live, attenuated vaccine and improved therapeutic modalities. The existing antiviral drugs, vidarabine and acyclovir, demonstrate that hospital management with intravenous administration of medication can significantly alter outcome. Effective orally administered medications would be particularly desirable for even the high risk patients. Research efforts both in academia and industry will continue to focus on the development of such therapeutic approaches.

III. HERPES ZOSTER

Herpes zoster, as has been previously identified, is a sporadic disease which is the consequence of reactivation of virus latent in the dorsal root ganglia. While this disease occurs at all ages, it is particularly common among the elderly, with an incidence which ranges between 5 to 10 cases per 1000 persons in the 6th through 8th decades of life.[21] Approximately 2% of patients who suffer from herpes zoster and who do not receive immunosuppressive therapy will develop a second episode of infection. This figure is at least fivefold higher in immunocompromised individuals. Since the characteristics of this infection, namely, the events of reactivation, the clinical course, and the complications, have been previously identified, only those salient complications in need of therapy are worthy of reiteration.

The most debilitating complication of herpes zoster both in the normal and the immunocompromised host is acute neuritis and postherpetic neuralgia. While postherpetic neuralgia is uncommon in younger patients, at least 50% of afflicted individuals over 60 years of age will suffer from persistent pain. The mean duration of pain is variable but generally last 6 months in the majority of patients. Approximately 5 to 10% of the patients will return to the physician with recurrent complaints, particularly the persistence of pain, which continues to persist 6 months after the onset of infection. This complication has been subject to a variety of studies.

Involvement of the central nervous system in the presence of cutaneous herpes zoster is probably more common than has been recognized. Invariably, patients who undergo cerebrospinal fluid examination for other reasons during episodes of shingles are found to have evidence of pleocytosis, but without proteinosis. These patients are frequently without signs of meningeal irritation and infrequently complain of headaches. Symptomatic meningoencephalitis is common in many patients. Onset begins with headache, fever, photophobia, and meningeal irritation. A rare manifestation of herpes zoster infection of the central nervous system is granulomatis arteritis. This complication can be diagnosed by arteriography and is associated with contralateral hemiparesis.

Involvement of the ophthalmic branch of the fifth cranial nerve occurs in approximately 7% of all cases of herpes zoster and can be identified as zoster ophthalmicus. Inflammation can lead to cicatricial lid retraction, ptosis, conjunctivitis, scleritis, uveitis, chorioretinitis, and even blindness.

As with chickenpox in the immunocompromised host, herpes zoster poses unique therapeutic challenges in the presence of an underlying malignancy. Disease is exaggerated with prolonged periods of vesicle formation and delayed pustulation. The median time to total cessation of new vesicle formation, pustulation, and scabbing has been defined as 8, 9, and 18 days, respectively, in immunocompromised patients who do not receive therapy. Because of the prolonged duration of cutaneous disease in these patients, which can be at least a week greater than that in the normal host, a therapeutic interval exists for intervention with appropriate antiviral agents.

Those patients at greatest risk for developing herpes zoster suffer from Hodgkin's disease or non-Hodgkin's lymphoma and other lymphoproliferative malignancies. This group of patients is at great risk for cutaneous dissemination and visceral involvement. Logistic regression analyses have demonstrated that patients with Hodgkin's disease, non-Hodgkin's lymphomia, leukemia, and oat cell carcinoma of the lung had a 40% risk of cutaneous dissemination following the development of localized herpes zoster. Furthermore, there was a direct correlation between cutaneous dissemination and the appearance of visceral complication. No other factor, including preceding and concurrent corticosteroid and cytotoxic therapy, age, sex, or duration of disease, predicted visceral complications other than the underlying malignancy.[22] Thus, these findings indicate a subset of patients for whom therapy must be developed to prevent progressive and life-threatening infection. Mortality from herpes zoster in the immunocompromised host has been found to vary from 5 to 8%.[1]

A. Clinical Management

1. Supportive Care

At the present time there is no specific therapy for herpes zoster in the normal host. The optimal approach is supportive management with the appropriate administration of analgesics for control of pain. Cleansing of the lesions with Domebro® soaks, which are applied to the involved skin, is soothing and will help prevent secondary bacterial infection. With severe neuritis, it has been suggested that amitriptyline hydrochloride and fluphenazine hydrochloride alone or in combination are useful for the amelioration of acute neuritis and may decrease the frequency of postherpetic neuralgia. In addition, the administration of steroids has been suggested to be useful in small studies.[23] Utilization of orally administered steroids, however, may delay healing in the primary dermatome and has been reported anecdotally to be associated with an increased frequency of cutaneous dissemination. Large-scale studies are lacking to confirm or refute the utility or adverse clinical effects of steroid administration to these patients.

2. Prevention

At the present time there are no preventive approaches to herpes zoster in individuals at risk for the development of this infection. The utilization of ZIG or ZIP in high-risk patient populations has not been of demonstrated utility.

3. Antiviral Therapy

Therapy has been proven effective for the management of herpes zoster only in the immunocompromised host. Several studies have reported the utilization of acyclovir both in an intravenous and oral formulation for the management of herpes zoster in normal host; however, the reported studies cannot distinguish between the antiviral effect of acyclovir and clinical improvement.[24-29] It should be recognized that intravenous treatment of localized herpes zoster in the normal host is associated with acceleration of clearance of virus from lesions and a trend toward accelerated lesion healing. There is no evidence, however, that treatment decreases either acute neuritis or postherpetic neuralgia, the real determinants of therapeutic benefit.

Because herpes zoster in the normal host is not a severe and life-threatening disease, intravenous therapy for these patients appears less than optimal, especially with the expense of the medication, its delivery, and hospitalization. As a consequence, oral formulations of acyclovir have been evaluated in the normal host. Oral medication has been administered at 200 and 400 mg 5 times a day for periods of 7 to 14 days. These therapeutic trials have demonstrated trends of effectiveness for accelerating the events of healing, but not of such significance as to indicate treatment with acyclovir of localized herpes zoster in the normal host.[25-29] In all likelihood, the inability to demonstrate a more marked therapeutic effect on the healing of herpes zoster is the consequence of the nature of the disease. For instance, with the infection being of short duration, as evidenced by a relatively brief period of vesicle formation and virus positivity, the likelihood that an antiviral could be effective for recurrent infection is less than with primary disease.

Infection in the immunocompromised host, however, introduces alternative variables. Because of the prolonged period of vesicle formation, as noted above, a therapeutic window exists for the introduction of appropriate antiviral therapy. Specific therapeutic measures have been studied for nearly two decades in the treatment of herpes zoster in the immunocompromised host. Studies utilizing cytosine arabinoside, while initially encouraging, later proved that the therapeutic index was unsatisfactory.[30-37] This compound is no longer being evaluated as an antiviral for any viral infection.

Two compounds have been demonstrated to have unequivocal activity for treatment of herpes zoster in the immunocompromised host. Both drugs have been evaluated in large, multicenter, double-blind, placebo-controlled studies. In two studies, vidarabine has been shown to accelerate the events of cutaneous healing, namely, the cessation of new vesicle formation, time to total pustulation, and time to total scabbing, as shown in Figure 2.[22,38] Therapy which was initiated within 72 hr of the onset of lesion formation prevented progression to cutaneous dissemination. Furthermore, there was a gradual regression of disease within the dermatome. Most important, treatment decreased

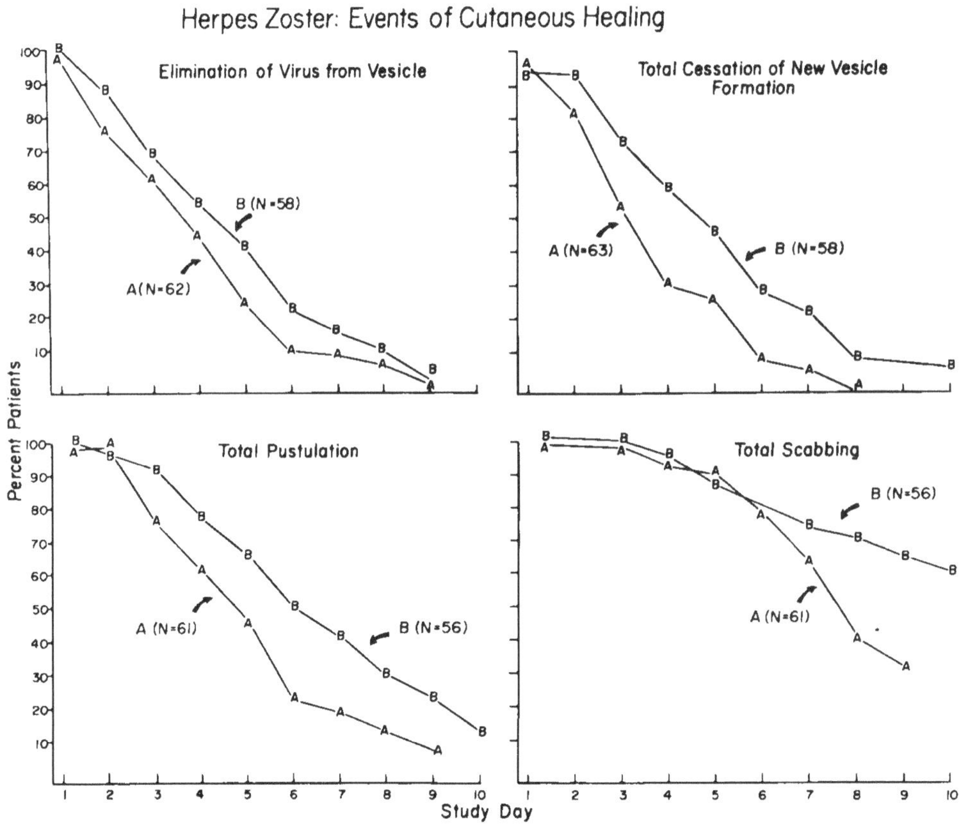

FIGURE 2. Resolution of cutaneous disease in relation to therapy. A represents vidarabine recipients, and B represents placebo recipients. For elimination of virus from vesicles, $p = 0.104$ by generalized Wilcoxon test and $p = 0.162$ by log-rank test; for cessation of lesion formation, $p = 0.002$ by Wilcoxon test and $p = 0.011$ by log-rank test; for time to total pustulation, $p = 0.003$ by Wilcoxon test and $p = 0.0068$ by log-rank test; for time to total scabbing, $p = 0.059$ by Wilcoxon test and $p = 0.0068$ by log-rank test. (From Whitley, R. J., Soong, S. J., Dolin, R., Bretts, R., Linneman, C., Jr., Alford, C. A., Jr., and the NIAID Collaborative Antiviral Study Group, reprinted by permission of the *N. Engl. J. Med.*, 307, 971, 1981.)

the frequency of visceral complications from 19 to 5%. Thus, vidarabine therapy administered within 72 hr of disease onset at a dosage of 10 mg/kg/day in standard intravenous solutions (concentration of 0.5 mg/mℓ) over 12 hr for 5 days was beneficial.

The frequency of adverse effects following vidarabine administration was 18%, overall, in all studies. The most frequent complications included nausea, vomiting, and diarrhea, which appeared in approximately 15% of the study population. Jitters, hallucinations, and myoclonus have been reported as a consequence of vidarabine administration in small numbers of patients receiving this compound; however, it has not been possible to demonstrate these complications in the controlled studies evaluating this medication. Furthermore, bone marrow suppression can be a consequence of administration of drug but has not been routinely encountered, even in the immunocompromised host.

Data are currently available from a large controlled study of acyclovir, administered at dosages of 5 mg/kg every 8 hr for a period of 7 days to immunocompromised patients.[39] While this study failed to demonstrate an effect of treatment on the events of cutaneous healing, it did reveal that therapy decreased progressive disease both within the dermatome and cutaneous dissemination. Of equal importance, there was a lower incidence of visceral complications in treated as compared to placebo counterparts. In the performance of the acyclovir studies, adverse effects were infrequently encountered.

The problem with the treatment of herpes zoster, utilizing either vidarabine or acyclovir in current formulations, is the necessity for in-hospital treatment. Vidarabine is a relatively insoluble drug that requires continuous administration over 12 hr. Thus, therapy can be associated with a risk of nosocomial infection because of the need for hospitalization. Since many immunocompromised patients heal spontaneously, it would be desirable to identify methods by which only patients at risk for cutaneous dissemination and life-threatening complications could be identified and their therapy provided in-hopsital. This latter approach remains to be proven efficacious in high-risk patients.

Only recently have data become available comparing vidarabine with acyclovir for treatment of localized zoster in the immunocompromised host.[40] This study was performed in 22 patients who had undergone bone marrow transplantation. From the comparative evaluations, cutaneous dissemination did not occur in the acyclovir recipients, while it developed in 50% of those who received vidarabine ($p = 0.016$). Furthermore, acyclovir treatment significantly decreased the median period of positive viral cultures and new lesion formation, and the time until the first decrease in pain, pustulation, and crusting of lesions. These data would indicate, then, that acyclovir is superior to vidarabine for the treatment of localized herpes zoster in this patient population. The intravenous formulation of acyclovir will probably replace vidarabine as the modality of choice for treatment of herpes zoster in these patients.

An alternative therapeutic modality is the administration of interferon, a strategy which has been successfully evaluated in a dose escalating study of 90 patients suffering from herpes zoster and having a underlying malignancy.[41] Merigan and colleagues reported that administration of leukocyte interferon at a dosage of 1 to 2 million units/kg intramuscularly resulted in decreased progressive vesicle formation in the primary dermatome and a lower frequency of cutaneous dissemination. The interferon-treated patients did not show a significant decrease in total healing time, but they did have significant amelioration of postherpetic neuraglia. Toxicity was encountered in this trial as manifest by fever, malaise, myalgia, and depressed leukocyte counts.

The usefulness of leukocyte interferon for treatment of herpes zoster infections in the immunocompromised host was recently reevaluated, utilizing recombinant DNA-produced interferon. With the availability of larger quantities of interferon, this study reported that interferon A decreased the frequency of cutaneous dissemination and progressive disease, findings similar to that reported by Merigan with Cantel leukocyte interferon.[42] Nevertheless, interferon administered in such a fashion failed to alter the events of cutaneous healing or pain associated with acute neuritis and postherpetic neuralgia. In addition, toxicity similar to that encountered in earlier studies was reported in a high frequency of patients. Thus, the acceptability of interferon as a single treatment of herpes zoster in the immunocompromised host is unlikely. It is conceivable that this compound could be administered with other drugs, particularly for localized herpes zoster in the immunocompromised host, on an ambulatory care basis in hope of the acceleration of events of healing. Side effects in the later study were reported in as many as 40 to 45% of patients entered into that trial. These side effects were mainly "flu-like" symptoms, including fever to 102 to 104°F.

At the present time, no data are available to indicate whether therapy is effective for disseminated or visceral complications associated with VZV infections. Clearly, the goal of any form of therapeutic should be the introduction of treatment before such complications occur. Thus, the availability of oral formulations for treatment would avoid the necessity for in-hospital management of a patient with these infections.

Zoster ophthalmicus, both in the normal and immunocompromised host, requires immediate medical attention. The identification of a patient with ophthalmic branch involvement should lead to prompt referral to an ophthalmologist. Therapy consists of administration of analgesics for severe pain and the use of atropine. The role of parenteral administration of antivirals for the management of zoster ophthalmicus remains unknown at this time, although both acyclovir and vidarabine have been used.

B. Future Therapeutic Approaches to the Management of Herpes Zoster

As with any field of evolving antimicrobial therapy, new drugs are constantly being synthesized in hopes of improving the therapeutic outcome. Varicella zoster virus infections are no exception. Several compounds have attracted particular attention as they relate to the management of VZV infections. These include bromovinyl deoxyuridine (BVDU), dihydroxypropoxymethylguanine (DHPG), 6-deoxyacyclovir, and a series of fluorinated arabinosyl nucleoside derivatives, particularly 2′-fluoro-2′-deoxy-5-methylarabinosyl uracil, FMAU. Another compound considered for treatment of herpes zoster is phosphonoformate. While each of these compounds offers unique potential as a therapeutic, all suffer from significant drawbacks.

1. BVDH

Bromovinyl deoxyuridine is particularly effective in vitro against VZV. In cell culture, BVDU inhibits VZV at concentrations less than 0.01 μg/m*l*, being the most active of currently available antiviral compounds against this virus.[19] As with the other currently available antivirals, BVDU is phosphorylated to a 5′-monophosphate derivative and, subsequently, a 5′-diphosphate derivative. Further phosphorylation involves a cellular kinase, leading to BVDU triphosphate, which can act either as an inhibitor or substrate for viral DNA polymerase.[43] Both in animals and humans, BVDU is rapidly cleaved to bromovinyl uracil in the plasma which is cleared more slowly by the kidney. This compound has been used in treatment of VZV infections. Oral therapy of ophthalmic and localized zoster as well as chickenpox in high-risk patient populations has been reported.[44,45] The studies which have been performed to date have not been placebo controlled and, as a consequence, must be interpreted with caution. Nevertheless, it has been reported that oral therapy of children with VZV infection as well as adults with localized or disseminated zoster led to a rapid cessation of lesion formation and the failure to develop significant complications. Therapeutic benefit was also reported for the management of zoster ophthalmicus. The investigators did not report toxic side effects attributable to BVDU administration. As indicated, controlled studies will be required to validate the utility of BVDU for human treatment. Unfortunately, the identification of toxicity in animal models performed by an American pharmaceutical firm have negated the further development of this compound in the U.S. Until issues of potential toxicity are resolved, it is unlikely that BVDU will be studied extensively in the U.S.

2. DHPG and 6-Deoxyacyclovir

Two compounds similar to acyclovir are attracting considerable attention for treatment of severe and life-threatening herpesvirus infections. These are DHPG and the 6-deoxyacyclovir derivative, BW-A515U. The former, DHPG, has significant activity against all herpesviruses, including VZV and cytomegalovirus. However, the toxicity profile of the compound indicates potential for bone marrow suppression in humans and animals as well as gonadal toxicity in animal models. It is likely that this drug will only be studied for severe and life-threatening herpesvirus infections, particularly cytomegalovirus infections in immunocompromised patients, or where other forms of antiviral therapy remain to be developed.

3. BW-A515U

A more exciting compound, particularly from the vantage point of utilization for management of VZV infections, is BW-A515U. This compound is a prodrug of acyclovir and results in high plasma levels following oral administration. It is cleaved by xanthine oxidase leading to plasma levels of acyclovir in humans which are approximately fourfold in excess of in vitro inhibitory levels (plasma steady state levels of 25.8 ± 6.0 μM). These levels can be achieved at a dosage of 250 mg of BW-A515U administered every 6 hr for 5 days. In the conduct of phase one pharmacology studies, no adverse clinical effects of significance were noted.

4. Other

A series of compounds developed by Dr. Jack Fox at Memorial Sloan Kettering have been evaluated in a limited number of patients suffering from herpesvirus infections. One compound, FMAU, was compared with vidarabine for treatment of localized zoster in immunocompromised patients. This study suggested superiority of FMAU for resolution of events of cutaneous healing when compared to vidarabine; however, the data remained to be published in a peer-reviewed journal. The availability of this compound for treatment of patients with herpes zoster remains to be established. Of further note, the phase one pharmacologic studies performed with this compound indicated toxicity at dosages of 16 and 32 mg/m^2.[46]

IV. CONCLUSION

It is apparent that significant advances have been achieved in the treatment of VZV infections in the immunocompromised host over the past decade. Significant advances in the management of localized zoster in the normal host remain to be achieved. With the availability of a variety of medications for management of infection, treatment, at least at this time, rests on administration of drug in a hospital environment; therefore, the risks of nosocomial infection as well as the cost of treatment remain higher than are desirable.

Future research and development efforts will likely lead to the development of compounds which can be administered on an ambulatory care basis. Such compounds with likelihood of success in the clinical arena include BW515U, a compound which, when cleaved, has a similar mechanism of action to acyclovir. Identification of patients at high-risk for progressive disease for in-hospital therapy will also be of importance in future therapeutic studies.

REFERENCES

1. Weller, T. H., Varicella and herpes zoster: changing concepts of the natural history, control, and importance of a not-so-benign virus, *N. Engl. J. Med.*, 309, 1362, 1983.
2. Whitley, R. J., Varicella-zoster virus infections, in *Antiviral Agents and Viral Diseases of Man*, 2nd ed., Galasso, G. J., Ed., Raven Press, New York, 1984.
3. Whitley, R. J., Varicella-zoster virus infections, in *Harrison's Principles of Internal Medicine*, McGraw-Hill, New York, in press.

4. Whitley, R. J., Chickenpox in the immunocompromised host, in *Advances in Antiviral Therapy of Herpesvirus Infections in Immunocompromised Hosts,* Park Row Publishers, New York, in press.
5. Feldman, S., Huges, W. T., and Kim, H. Y., Herpes zoster in children with cancer, *Am. J. Dis. Child.,* 126, 178, 1973.
6. Whitley, R. J., Hilty, M., Haynes, R., Bryson, Y., Connor, J. D., Soong, S. J., Alford, C. A., Jr., and NIAID Collaborative Antiviral Study Group; Vidarabine therapy of varicella in immunosuppressed patients, *J. Pediatr.,* 101, 125, 1982.
7. Prober, C. G., Kirk, L. E., and Keeney, R. E., Acyclovir therapy of chickenpox in immunosuppressed patients, *J. Pediatr.,* 101, 622, 1982.
8. Arvin, A. M., Kushner, J. H., Feldman, S., Buchner, R. L., Hammond, D., and Merigan, T. C., Human leukocyte interferon for treatment of varicella in children with cancer, *N. Engl. J. Med.,* 306, 761, 1982.
9. Ross, A. H., Modification of chickenpox in family contacts by administration of gamma globulin, *N. Engl. J. Med.,* 267, 369, 1962.
10. Brunell, P. A., Ross, A., Miller, L., and Kuo, B., Prevention of varicella by zoster immune globulin, *N. Engl. J. Med.,* 280, 1191, 1969.
11. Brunell, P. A., Gershon, A. A., Hughes, W. T., Riley, H. D., and Smith, J., Prevention of varicella in high-risk children: a collaborative study, *Pediatrics,* 50, 718, 1972.
12. Brunell, P. A. and Gershon, A. A., Passive immunization against varicella-zoster infections and other modes of therapy, *J. Infect. Dis.,* 127, 415, 1973.
13. Gershon, A. A., Steinberg, S., and Brunell, P. A., Zoster immune globulin. A further assessment, *N. Engl. J. Med.,* 290, 243, 1974.
14. Cohen, S. S., Introduction to the Biochemistry of D-arabinosyl nucleosides, in *Progress in Nucleic Acid Research and Molecular Biology,* Vol. 5, Davidson, J. N. and Cohn, W. E., Eds., Academic Press, New York, 1966, 1.
15. Schaeffer, H. J., Beauchamp, L., deMiranda, P., Elion, G. B., Baver, D. J., and Collins, P., 9(2-hydroxyethoxymethyl) guanine activity against viruses in the herpes group, *Nature (London),* 272, 583, 1978.
16. Elion, G. B., Furman, P. A., Fyfe, J. E., deMiranda, P., Beauchamp, L., and Schaeffer, H. J., Selectivity of action of an anti-herpetic agent, 9(2-hydroxyethoxymethyl) guanine, *Proc. Natl. Acad. Sci. U.S.A.,* 74, 5716, 1977.
17. Miller, W. H. and Miller, R. L., Phosphorylation of acyclovir (acycloguanosine) monophosphate by GMP kinase, *J. Biol. Chem.,* 255, 7204, 1980.
18. Biron, K. K. and Elion, G. B., In vitro susceptibility of varicella-zoster virus to acyclovir, *Antimicrob. Agents Chemother.,* 18, 443, 1980.
19. DeClercq, E., Descamps, J., Verhelst, G., et al., Comparative efficacy of antiherpes drugs against different strains of herpes simplex virus, *J. Infect. Dis.,* 141, 563, 1980.
20. Stevens, D. A. and Merigan, T. C., Interferon, antibody, and other host factors, in herpes zoster, *J. Clin. Invest.,* 51, 1170, 1972.
21. Hope-Simpson, R. E., The nature of herpes zoster: a long-term study and a new hypothesis, *Proc. R. Soc. Med.,* 58, 9, 1965.
22. Whitley, R. J., Soong, S. J., Dolin, R., Betts, R., Linnemann, C., Jr., Alford, C. A., Jr., and the NIAID Collaborative Antiviral Study Group, Early vidarabine therapy to control the complications of herpes zoster in immunosuppressed patients, *N. Engl. J. Med.,* 307, 971, 1982.
23. Eaglstein, W. H., Hatz, R., and Brown, J. A., The effects of early corticosteroid therapy on the skin eruption and pain of herpes zoster, *JAMA,* 211, 1681, 1970.
24. Bean, B., Braun, C., and Balfour, H. H., Jr., Acyclovir therapy for acute herpes zoster, *Lancet,* 2, 118, 1982.
25. McKendrick, M. W., Care, C., Burke, C., Hickmott, E., and McKendrick, G. D. W., Oral acyclovir in herpes zoster, *J. Antimicrob. Chemother.,* 14, 661, 1984.
26. Peterslund, N. A., Esmann, V., Ipsen, J., Christensen, K. D., and Petersen, C. M., Oral and intravenous acyclovir are equally effective in herpes zoster, *J. Antimicrob. Chemother.,* 14, 185, 1984.
27. McKendrick, M. W., Oral acyclovir in herpes zoster, *Scand. J. Infect. Dis.,* 47, 76, 1985.
28. Peterslund, N. A., The treatment of herpes zoster infections, *Scand. J. Infect. Dis.,* 47, 80, 1985.
29. McKendrick, M. W., McGill, J. J., Bell, A. M., Hickmott, E., and Burke, C., Oral acyclovir for herpes zoster, *Lancet,* 2, 925, 1984.
30. Hall, T. C., Wilfert, C., Jaffe, N., et al., Treatment of varicella-zoster with cytosine arabinoside, *Trans. Assoc. Am. Physicians,* 82, 201, 1969.
31. McKelvey, E. M. and Kwaan, H. C., Cytosine arabinoside therapy for disseminated herpes zoster in a patient with IgG pyroglobulinemia, *Blood,* 34, 706, 1969.
32. Prager, D., Bruder, M., and Sawitsky, A., Disseminated varicella in a patient with acute myelogenous leukemia: treatment with cytosine arabinoside, *J. Pediatr.,* 78, 321, 1971.

33. Chow, A. W., Forester, J., and Hryniuk, W., Cytosine arabinoside therapy for herpesvirus infections, *Antimicrob. Agents Chemother.*, 10, 214, 1970.
34. Stevens, D. A., Jordan, G. W., Waddell, T. F., et al., Adverse effect of cytosine arabinoside on disseminated zoster in a controlled trial, *N. Engl. J. Med.*, 289, 873, 1973.
35. Schimpff, S. C., Fortner, C. L., Greene, W. H., et al., Cytosine arabinoside for localized herpes zoster in patients with cancer failure in a controlled trial, *J. Infect. Dis.*, 130, 673, 1974.
36. Davis, C. M., VanDersari, J. V., and Coltman, C. A., Jr., Failure of cytarabine in varicella-zoster infections, *JAMA*, 224, 122, 1973.
37. Betts, R. F., Zaky, D. A., Douglas, R. G., Jr., et al., Ineffectivenesss of subcutaneous cytosine arabinoside in localized herpes zoster, *Ann. Intern. Med.*, 82, 778, 1975.
38. Whitley, J., Chien, L. T., Dolin, R., Galasso, G. J., Alford, C. A., Jr., and the Collaborative Antiviral Study Group, Adenine arabinoside therapy of herpes zoster in the immunosuppressed, *N. Engl. J. Med.*, 294, 1193, 1976.
39. Balfour, H. H., Laskin, O. L., Meyers, J. D., Zaia, J. A., Kirk, L. E., Bean, B., Ambinder, F., Wade, J. C., Appeli, D., and Segreti, A. C., Acyclovir halts progression of herpes zoster in immunocompromised patients, *N. Engl. J. Med.*, 308(24), 1448, 1983.
40. Shepp, D., Dandliker, P., and Myers, J., Treatment of varicella-zoster infection in severely immunocompromised patients: a randomized comparison of acyclovir and vidarabine, *N. Engl. J. Med.*, 314(4), 208, 1986.
41. Merigan, T. C., Rand, K. H., and Pollard, R. B., Human leukocyte interferon for the treatment of herpes zoster in patients with cancer, *N. Engl. J. Med.*, 298, 981, 1978.
42. Winston, D. J., Eron, L., Ho, M., Pazin, G., Kessler, H., Sartiano, G., Gallagher, J., Ho, E. G., Champlin, R. E., Bernhardt, L., Bigley, J., Nadler, P., and Hoffmann-LaRoche Herpes Zoster Study Group, Treatment of herpes zoster in immunocompromised cancer patients with recombinant leukocyte A interferon (rINF-A0), in 25th Interscience Conf. on Antimicrob. Agents and Chemother., Minneapolis, September 29 to October 2, 1985.
43. DeClercq, E., Pyrimidine nucleoside analogues as antiviral agents, in *Targets for the Design of Antiviral Agents*, DeClercq, E. and Walker, R. T., Eds., NATO ASI Series, Vol. 73, Geneva, 1984, 203.
44. Maudgal, P. C., DeClercq, E., and Missottin, L., Treatment of ocular herpes by new agents, in *Herpesviruses and Virus Chemotherapy: Pharmacological and Clinical Approaches*, Kono, R. and Nakajima, A., Eds., International Congress Series, Vol. 667, Excerpta Medica, Amsterdam, 1985, 95.
45. DeClercq, E., Benoit, Y., Laureys, G., and Delbeka, M. J., Clinical potentials of bromovinyldeoxyuridine (BVDU): in particular for oral treatment of varicella-zoster virus infections in children with cancer, in *Herpesviruseses and Virus Chemotherapy: Pharmacological and Clinical Approaches*, Kono, R. and Nakajima, A., Eds., International Congress Series, Vol. 667, Excerpta Medica, Amsterdam, 1985, 49.
46. Fanucchi, M. P., Leyland-Jones, B., Young, C. W., Burchenal, J. H., Watanabe, K. A., and Fox, J. J., Phase 1 trial of 1-(2′-deoxy-2′-fluoro-1-β-D-arabinofuranosyl)-5 methyluracil (FMAU), *Cancer Treat. Rep.*, Vol. 69(1), 1985.

Chapter 5

A VACCINE TO PREVENT CHICKENPOX

Michiaki Takahashi

TABLE OF CONTENTS

I. INTRODUCTION

Varicella has been considered a usually mild disease in normal children. However, it is recognized that this disease can be severe or even fatal in immunocompromised patients and such risk is increased by the use of steroid for treatment of various diseases.

Until recently, there has been little effort to develop an attenuated live varicella vaccine. Technical difficulties with varicella-zoster virus (VZV), such as the poor virus yield due to its cell-associated nature and the labile infectivity of the virus under usual conditions, have hampered the study of this virus.

Previously, Greenthal used vesicle fluid from chickenpox and zoster patients on 36 persons, of whom 19 had a local reaction.[1] These vaccinated persons did not develop chickenpox after exposure to the diseases, whereas some of the unvaccinated persons did. However, it is difficult to assess Greenthal's results because serological tests were not then available.

Recently we have developed a live varicella vaccine (Oka strain).[2-7] This vaccine has been extensively tested in Japan, and subsequently tested in the U.S. and Europe. Its safety and effectiveness have been recognized for immunocompromised children as well as normal children.

II. DEVELOPMENT OF A LIVE VARICELLA VACCINE

Vesicular fluid was taken from vesicles of a 3-year-old boy (named Kenji Oka, Osaka) who had typical chickenpox but was otherwise healthy, and varicella-zoster virus (VZV, Oka strain) was isolated in a primary human embryo lung cell culture. Attempts to obtain an attenuated strain were made by serially cultivating the VZV Oka strain in human embryonic lung (HEL) cells at 34°C and then passaging it in guinea

Table 1
TEMPERATURE SENSITIVITY TEST OF THE VACCINE AND WILD-TYPE STRAINS

Strain	Passage no. in HEL cell	Infectivity PFU/0.2 ml 39°C	37°C	34°C	Ratio 39°C:34°C
Oka vaccine		4.0×10^1	1.5×10^3	1.5×10^3	1:38
Oka parental	7	3.0×10^2	8.5×10^2	9.5×10^2	1:3.2
Izawa	7	7.0×10^1	4.0×10^2	4.0×10^2	1:5.7
Inoue	5	3.0×10^2	1.0×10^3	1.0×10^3	1:3.3
Kawaguchi	12	2.0×10^2	6.5×10^2	7.0×10^2	1:3.5

From Hayakawa Y., Torigoe, S., Shiraki, K., Yaminishi, K., and Takahashi, M., *J. Infect. Dis.*, 149, 956, 1984. Copyright by The University of Chicago Press, 1984. With permission.

pig embryonic fibroblasts (GPEF). As the latter cells were found to be the only non-primate cells susceptible to VZV, they were thought suitable for obtaining a host-dependent VZV variant. The virus was then cloned and passaged in human diploid cells (WI-38 and MRC-5) several times to prepare an experimental vaccine. For vaccine preparation, infected tissue cultures were washed well with phosphate-buffered saline (PBS) and infected cells were harvested with EDTA treatment. Then the cell suspension in the vaccine medium was sonicated to obtain cell-free virus; a dose of 500 to 1500 plaque-forming units (PFU) was usually used. Safety testing of the vaccine included lack of pathogenicity by inoculation (including intracerebral) of small animals and monkeys. The absence of C-type particles and of latent viruses was also confirmed morphologically and biochemically.

III. BIOLOGICAL AND BIOPHYSICAL CHARACTER OF THE OKA VACCINE STRAIN

The vaccine virus has been analyzed in order to find biological or biophysical markers that may be used to distinguish it from wild viruses.

A. Temperature Sensitivities of the Vaccine and Wild-Type Strains

The vaccine strain was found to be slightly temperature sensitive at 39°C, unlike the wild-type strains.[8] The foci of the vaccine strains were also smaller than those of wild-type strains at high temperature but were similar in size to the latter at lower temperatures (Table 1).

B. Difference in Infectivities on GPEF and HuEF of the Vaccine and Wild-type Strains

The infectivities of the vaccine strain and wild-type strains were assayed by plaque titration on GPEF and human embryo fibroblasts (HuEF). The vaccine strain showed a higher ratio of infectivity in GPEF to that in HuEF than any of the wild-type strains[8] (Table 2).

The immunogenicity of the vaccine virus is also far better than other wild viruses in guinea pigs.[9] This is probably related to the difference in the capacity of the viruses to replicate in cultured guinea pig cells. In fact, the growth of the vaccine virus is better than the original Oka strain and other strains in guinea pig embryo cultures as determined by infectious center assays.[9] In addition, adsorption rate of the vaccine virus to cultured guinea pig cells assessed by infectious center assay was higher than other strains.[9]

Table 2
INFECTIVITY OF THE VACCINE AND WILD-TYPE STRAINS OF VZV IN GPEF AND HuEF CELLS

Strain	Source	Passage no. in HEL cells	Titer on GPEF cells(a) (PFU/0.2 m*l*)	Titer on HuEF cells(b) (PFU/0.2 m*l*)	Ratio a/b
Oka (vaccine)			7.0×10^3	8.5×10^3	0.82
Oka (parental)	Varicella	10	2.8×10^2	6.8×10^3	0.041
Tsuchiyama	Varicella	5	5.2×10^1	1.5×10^3	0.035
Inoue	Varicella	7	6.6×10^2	1.3×10^4	0.051
Watanabe	Varicella	6	5.8×10^1	1.6×10^3	0.036
Wada	Varicella	8	3.8×10^2	6.2×10^3	0.061
Terada	Varicella	9	2.2×10^2	1.7×10^4	0.013
Morita	Zoster	2	1.2×10^2	6.5×10^3	0.018
Kato	Zoster	4	3.8×10^1	1.2×10^3	0.032
Takenaka	Zoster	4	1.5×10^2	7.0×10^3	0.021
Yamashita	Zoster	5	5.8×10^1	1.3×10^3	0.045
Yamaguchi	Zoster	7	1.2×10^2	1.9×10^3	0.063
Ellen	Varicella		8.0×10^1	2.1×10^3	0.038
Mean					0.038 ± 0.015

From Hayakawa, Y., Torigoe, S., Shiraki, K., Yamanishi, K., and Takahashi, M., *J. Infect. Dis.*, 149, 956, 1984.

These results suggest that the vaccine virus is a variant of VZV, at least with respect to thermosensitivity and host range.

C. DNA Cleavage Profile of Vaccine Virus

Differences in the migration patterns of DNA of vaccine virus and other wild strains have been found after cleavage with restriction endonucleases. In a comparison of DNA from the vaccine virus and its parental strain by endonuclease enzyme cleavage, a significant difference was found in the cleavage patterns of vaccine strain (Oka) and other wild strains by *Hpa*I, *Bam*HI, and *Bgl*I enzymes.[10] The cleavage profiles of the vaccine strain, its possibly derivative strain, and wild-type strains were examined using restriction endonucleases, and one distinctive fragment (k) produced by *Hpa*I enzyme was detected which is unique to Oka strains (Figure 1).[8,11] When the clinical isolates from the vaccine recipients were later examined by this method, the results were in accordance with those from the test of infectivity ratio (GPEF/HuEF) and consistent with the clinical judgement which could be given to several clinical isolates (Table 3).

IV. DEVELOPMENT OF VARICELLA SKIN TEST TO MEASURE CELL-MEDIATED IMMUNITY (CMI) AND ITS CLINICAL APPLICATION

A varicella skin test has been developed in the study of cell-mediated immunity to VZV.[12,13] A VZ skin antigen was prepared by harvesting the VZ-infected human diploid cells, washing the cells with PBS, and sonicating the cells, followed by centrifugation at 3000 r/min for 20 min. The supernatant was used as skin antigen after heating at 56°C for 30 min, which destroyed infectivity but not antigenicity. The CF antigen titer of this material was in a 1:64 ratio. A control antigen was prepared similarly. A volume of 0.1 m*l* of viral or control antigen was injected intracutaneously into the forearm. The reaction was read usually after 48 hr by measuring the diameters of erythematous change. Using the criterion of 5 mm of erythema as a positive result, 50 of 53 normal children with a history of varicella and VZ-neutralizing antibody had positive skin reactions (Figure 2). Induration was observed in approximately one half of the reactions positive for erythema. In 22 children without a history of varicella and no neutralizing antibody to VZV, skin tests were negative. (See Figure 2.)

The skin test was applied prospectively to identify the susceptibles exposed to a varicella patient in an institution for mentally retarded children.[12] The skin test was immediately performed on 65 patients; 24 children with negative skin test results were vaccinated; those identified as skin test positive were observed. As a result, no spread of varicella was observed except for three vaccinated cases that developed varicella 2 weeks after the onset of the index case. These children could have been in the incubation period when the vaccine was given.

The VZ skin test was again utilized to identify varicella susceptibles rapidly when a case of varicella occurred in an institution housing 49 children.[13] The skin test was immediately administered. Among 20 children with negative results, 17 were vaccinated; the index case and two children with fever were not vaccinated. Only 1 of the 17 vaccinated children developed mild varicella within 2 weeks after vaccination; typical varicella developed in 2 unvaccinated children within 4 weeks after appearance of the index case. None of the 29 exposed children with positive skin tests developed varicella.

It was also found that when the VZ skin test was performed on vaccinees at sequential 5- to 7-day intervals, positive reactions were observed 5 days after vaccination, which was 7 to 9 days prior to detection of the neutralizing antibody.[13]

In natural varicella infection, skin test reactions in all children before the appearance of a rash were negative, while positive results were consistently obtained after the appearance of a rash. After vaccination, the skin reaction was positive as early as 4 days after vaccination in about half the children (Figure 3). It was also reported that positive lymphoproliferative (LPF) activity was observed a week after vaccination, which preceded appearance of neutralizing antibody by 1 to 3 weeks.[14] This early appearance of cellular immunity after vaccination seems to be closely related to the protective effect of prompt vaccination of susceptible children, even shortly after their contact with varicella patients.

An improved VZ skin test antigen, free from viral particles and cell debris, mainly composed of viral glycoproteins, was later prepared by using culture fluid of infected cells as starting material.[15,16] It is more purified than the original crude VZ skin antigen, still retaining antigenicity comparable to the crude antigen. The improved skin antigen was found to be composed of two glycoproteins by SDS-gel electrophoresis.[17]

Recently it was reported by LaRussa et al.[18] that the skin test was evaluated in 16 adults known to be immune or susceptible to varicella and in 109 adults with no history of varicella, and that the skin test was well tolerated and compared favorably with established methods of determining immunity to varicella, such as FAMA, ELISA, and LPF assay. It accurately predicted which subjects would develop clinical varicella after dose exposure. That is, 23 subjects with no history of varicella had a close household exposure to the disease, and none of 15 adults with a positive skin test at the time of exposure developed clinical varicella, whereas five of seven adults with a negative skin test developed clinical disease. One subject, whose skin test was uninterpretable, did not develop clinical disease after exposure.

It was further reported that among normal individuals, the skin test when compared with the ELISA had a sensitivity of 85%, a specificity of 100%, and positive predictive value of 100%.[19] Thus the skin test, which represents delayed-type hypersensitive CMI, would be conveniently used to identify individuals susceptible to clinical varicella. Recently, the role of delayed-type hypersensitive T-cell in protection against HSV disease was documented.[20] Similar immune mechanisms may be working in protection against VZV infection.

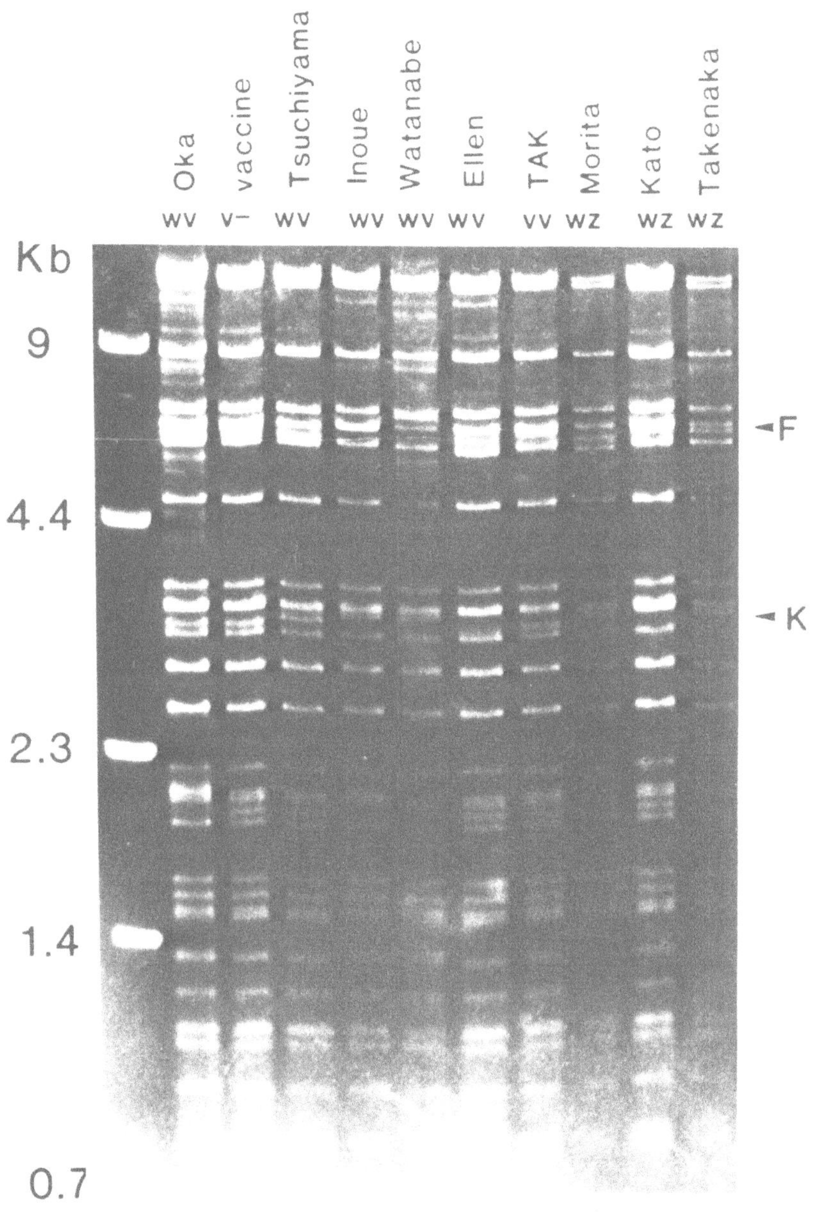

A

FIGURE 1. (A) Profiles of DNA cleavage by restriction endonuclease HpaI. DNAs from ten strains of VZV are shown: seven wild-type strains, the Oka parental strain, the Oka vaccine strain, and one strain (TAK) recovered from a vaccine recipient. (B) HpaI cleavage profiles of DNAs from nine strains of VZV recovered from vaccine recipients (KIT, MEV, MEZ, TAK, ABE, OSA, FMO, KAK, and OYA), the Oka parental strain, the Oka vaccine strain, and a wild-type strain (Kawaguchi). WV, a wild-type VZV isolated from a varicella patient; WZ, a wild-type VZV isolated from a zoster patient; V−, the vaccine strain, VV, vaccine-derived strain isolated from the vaccine recipient who developed varicella symptoms; VZ, vaccine-derived strain isolated from the vaccine recipient who developed zoster symptoms. (From Hayakawa, Y., Torigoe, S., Shiraki, K., Yamanishi, K., and Takahashi, M., *J. Infect. Dis.*, 149, 956, 1984. Copyright by The University of Chicago Press, 1984. With permission.)

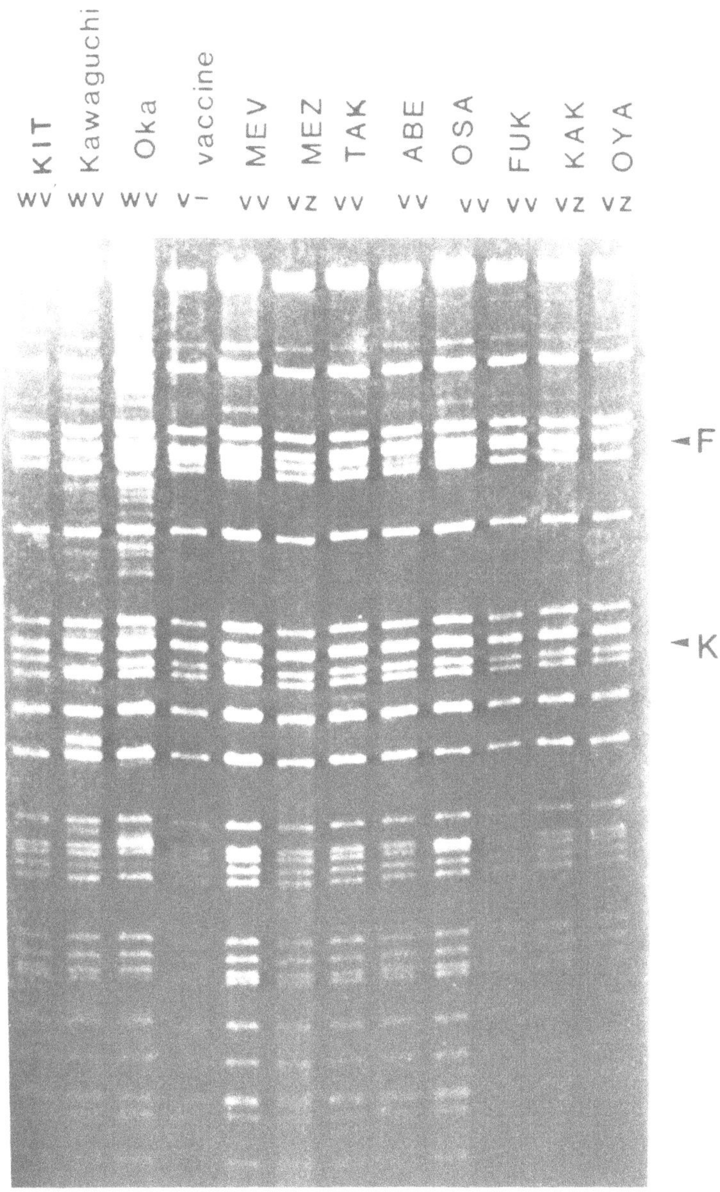

FIGURE 1B

V. CLINICAL STUDIES

A. Initial Clinical Trial: Vaccination of Normal Children and Protective Effect of Immediate Vaccination after Household Exposure

As there is no suitable small animal for experimental infection of VZV, varicella vaccine was tested carefully stepwise from normal children to immunocompromised children.

The vaccine was first given to 70 normal children, with the virus dose ranging from 100 to 2000 PFU. Clinical reactions were noticed in none of the vaccines, whereas antibody responses were detected in all the vaccines who received a dose of 200 PFU or more.[2]

Table 3
RESULTS OF BIOLOGICAL AND BIOPHYSICAL MARKER TEST ON THE CLINICAL VZV ISOLATES FROM THE VACCINE RECIPIENTS

Case no.	Sex	Age at time of vaccination	Underlying diseases	Symptom	Period after vaccination	Passage no. of VZV in HEL cells	Contact with varicella patients	Clinical judgement	Infectivity ratio in GPEF/HuEF	Judgement	DNA pattern
1 (KIT)	M	4	ALL[a]	Varicella	8 mo	3	+	Wild	0.097	Wild ($p < 0.05$)	Wild
2 (MEV)[b]	M	4	ALL	Varicella	27 days	6	–	Vaccine	0.65	Vaccine	Vaccine
3 (MEV)[b]	M	4	ALL	Zoster[a]	12 mo	5	–	Vaccine	0.76	Vaccine	Vaccine
4 (TAK)	M	9	ML[d]	Varicella	27 days	3	–	Vaccine	0.67	Vaccine	Vaccine
5 (ABE)	M	2	ALL	Varicella	18 days	2	+	?	0.32	Vaccine	Vaccine
6 (OSA)	M	6	ML	Varicella	15 days	7	+	?	0.81	Vaccine	Vaccine
7 (FMO)	F	7	ML	Varicella	20 days	5	+	?	0.45	Vaccine	Vaccine
8 (KAK)	F	5	ML	Varicella	11 days	3	+	?	0.36	Vaccine	Vaccine
9 (OYA)[c]	M	3	ALL	Zoster	4 mo	2	–	Vaccine	0.21	Vaccine ($p < 0.01$)	Vaccine
Oka (vaccine)									0.59		Vaccine
Kawaguchi						12			0.014		Wild
Ellen									0.039		Wild

Note: Sources of all atrains were vesicles.

[a] Acute lymphocytic leukemia (ALL).
[b] MEV and MEZ strains were recovered from the same patient.
[c] Case 9 (OYA) showed a symptom of mild varicells 20 days after vaccination.
[d] Malignant Lymphoma (ML).

From Hayakawa, Y., Torigoe, S., Shiraki, K., Yamauishi, K., and Takahashi, M., *J. Infect. Dis.*, 149, 956, 1984. Copyright by The University of Chicago Press, 1984. With permission.

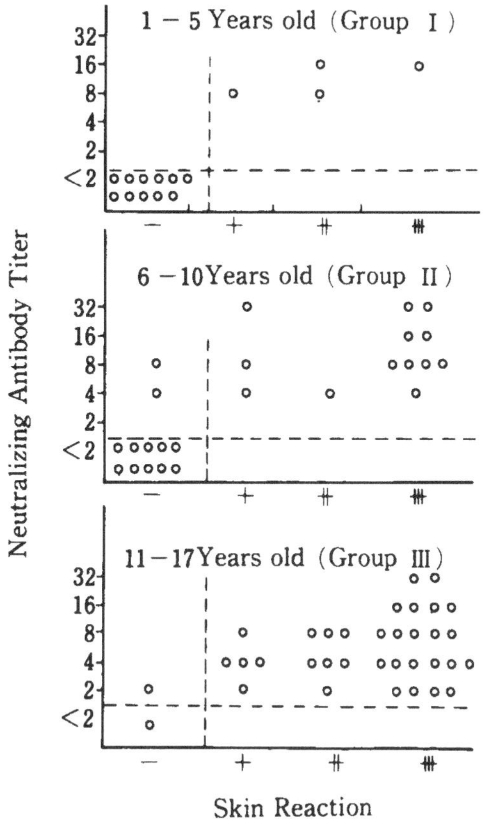

FIGURE 2. Correlation between varicella skin test reaction in children and reciprocal titer of neutralizing antibody. The skin test reaction was graded according to the diameter of the erythematous change: –, <5 mm, +, ≥ 5 mm but <10 mm, ++, ≥ 10 mm but <15 mm, +++, ≥ 15 mm. (From Kamiya, H., Ihara, T., Hattori, A., Iwasa, T., Sakurai, M., Izawa, T., Yamada, A., and Takahashi, M., *J. Infect. Dis.*, 136, 784, 1977. Copyright by The University of Chicago Press, 1977. With permission.)

To examine the protective efficacy of the vaccination, vaccine was given to the susceptible household contacts immediately after exposure.[21] Of 21 families, 26 contacts were vaccinated, mostly within 3 days after exposure to the index cases. None of the vaccinated children developed clinical symptoms of varicella. On the other hand, all of the 19 unvaccinated contacts in 15 families, who served as controls, exhibited typical varicella symptoms 10 to 20 days after the onset of the index cases. In three families, in which only one sibling contact received vaccine and the other was an unvaccinated control, none of the vaccinated children showed any clinical symptoms, whereas unvaccinated controls exhibited typical varicella symptoms (Table 4). In general, the antibody titers after clinical varicella were higher than those following immunization. This study clearly demonstrated that vaccination soon after exposure could protect against clinical varicella.

In a subsequent study, the relationship of viral dose, interval between exposure and vaccination, and protective effect were examined in household contacts.[22] As shown in Figure 4, when approximately 1000 PFU was administered within 3 days after expo-

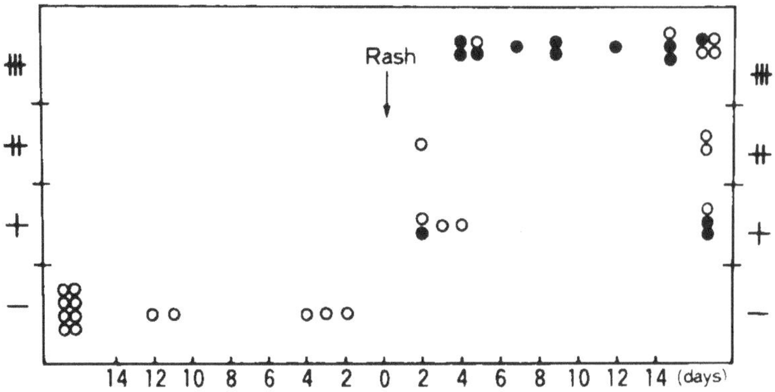

A

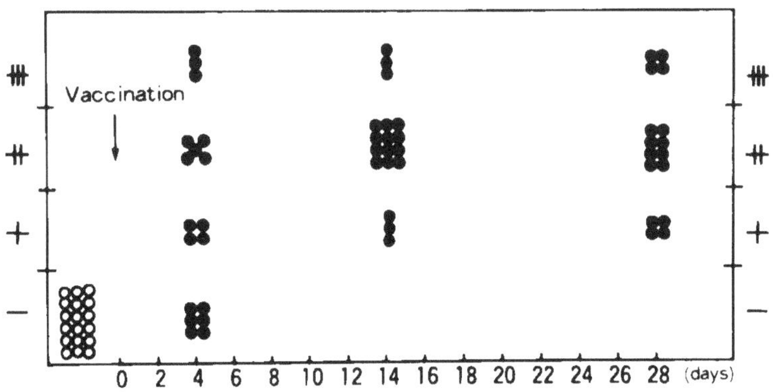

B

FIGURE 3. Comparison of the time conversion of varicella skin reaction in children with natural varicella and with vaccination; indicates the first skin test; • indicates the 2nd, 3rd, and 4th skin tests. (A) Skin test on 22 children with natural varicella; (B) test done 4 times on each of 18 vaccinated children. (From Takahashi, M. and Baba, K., *Medical Virology*, vol. 3, de la Maza, L. and Peterson, E. M., Eds., Elsevier, New York, 255, 1984. With permission.)

sure, protection against varicella was observed in all the vaccinees. Even in the failure cases that received a smaller dose of vaccine, clinical symptoms were mild, suggesting that natural varicella was modified by vaccination.

B. Use of VZ Vaccine to Terminate the Spread of Varicella in Hospitals and Institutions

The first clinical trial of this vaccine for hospitalized children was done to terminate the spread of varicella in a children's ward.

In the children's ward of a hospital, 23 children with no history of varicella and no detectable antibody were vaccinated immediately after the occurrence of a case of varicella.[1,22] These children suffered from such diseases as the nephrotic syndrome, nephritis, purulent meningitis, and hepatitis. Twelve of them had been receiving corticosteroid therapy. An antibody response was demonstrated in all of the vaccinated children; mild fever occurred in six and a mild rash in two of the six patients within 10 to 14 days after vaccination. It was uncertain whether these reactions were due to

Table 4
PREVENTION OF SPREAD OF VARICELLA IN FAMILIES BY IMMEDIATE VACCINATION

No. of family	No of index case	No. of vaccinated contacts	Clinical varicella (in contact)[a]
21	21	26/26	0/26
15	15	0/19	19/19
3	3	3/6	3/6

Note: Vaccination was given to the contacts within 3 days after onset of the index cases in families.

[a] Unvaccinated only.

From Yazaki, T., Ito, S., Isuzuki, K., Kato, R., Hirose, S., and Takahashi, M., *Pediatrics,* 59, 3, 1977. Reproduced by permission.)

vaccination or to naturally acquired infection modified by vaccination. No other clinical reactions or abnormalities of the blood or urine were detected. Thus, spread of varicella infection was prevented, with the exception of one severe case who was unvaccinated because his mother mistakenly believed that he had a history of varicella. This study offered the first proof that the vaccine was well tolerated in patients receiving immunosuppressive therapy.

A follow-up was made of 13 of the original vaccinees and 5 newly vaccinated children.[24] Many of these children were receiving steroid therapy for renal disease. These vaccines were reexposed to varicella or zoster several times in a hospital setting for up to 9 months after vaccination. None developed clinical varicella, although exposure boosted antibody responses in several exposed vaccinees.

In the following studies, it was also proved that prompt immunization of susceptible children exposed to varicella in hospitals or institutions could prevent epidemics.[25-27]

In an institution for children less than 2 years old, a similar protective effect of prompt vaccination was observed.[13] Varicella developed in an 11 month-old infant on a ward for 86 children. A total of 33 children over 11 months of age were vaccinated and 43 children less than 10 months of age were unvaccinated, partly because they were expected to still possess maternal antibody. A small viral dose of 80 PFU per person was purposely used to find the minimum effective dose for immunization.

In the vaccinated group of 33 children, mild rash appeared in 8, and a temperature less than 38°C was detected in one of the 8 children 2 to 4 weeks after vaccination. In contrast, typical varicella developed in all of the 43 unvaccinated group during the 10 weeks following onset of the index case. Symptoms were severe in 16, with confluent vesicles and high fever; skin scars were left in 13 of the 16 children after recovery. The relationship between the severity of clinical symptoms and the patient's age by month is shown in Table 5. All of the four 1-month-old children had symptoms in a mild form, but clinical symptoms were generally more severe in the 2- to 11-month-old infants than in those aged 12 to 14 months old. These results suggest that vaccination with as small a dose as 80 PFU was still effective to stop, although not completely, the spread of varicella in children in close contact, that young infants tend to develop severe varicella, as compared with older ones.

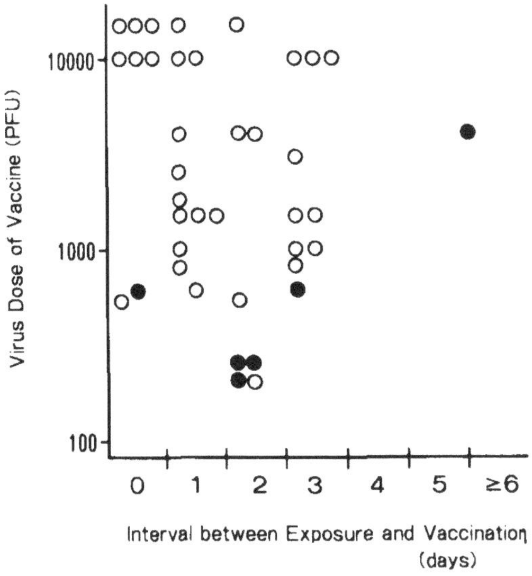

FIGURE 4. Protective effect of immediate varicella vaccination against household contacts in relation to viral dose, interval between exposure and vaccination; O, no varicella symtoms; ● with varicella symptoms. (From Asano, Y., Hirose, S., Iwasa, A., Miyata, T., Yazaki, T., and Takahashi, M., *Biken, J.*, 25. 43. 1982. With permission.)

C. Vaccination of Children with Malignant Diseases

The first vaccination trials on children with malignant diseases were done with suspension of chemotherapy from 1 week before to 1 week after vaccination.[28,29] Of 12 children with acute leukemia, 10 who had been in remission for 6 months or less were immunized. The remaining two patients had been off therapy for 9 and 48 months, respectively. Of these children, four were leukemic with less than 3000 WBC/mm, but most had positive skin test reactions with dinitrochlorobenzene (DNCB), purified protein derivative (PPD), or phytohemagglutinin (PHA). Chemotherapy, usually 6-mercaptopurine (6-MP) (2 to 3 mg/kg/day orally) and methotrexate (30 mg/m^2 once a week, intravenously) as maintenance therapy, and intermittent intrathecal methotrexate, vincristine, and steroids as reinforcement therapy, was suspended 1 week prior to and 1 week after vaccination. The patients were vaccinated with virus doses of 200, 500, or 1500 PFU. Three of the 12 developed a mild rash with 13, 30, and 25 papular or incomplete vesicles; a fever of 39°C for 1 day was noted in one of the children about 3 weeks after vaccination.

Six children with solid tumors were also vaccinated, with suspension of chemotherapy for 2 weeks. They had no symptoms after vaccination and all the vaccinees developed low but detectable neutralizing antibodies.[29]

Subsequently, the same research group immunized leukemic children with continuation of 6-MP daily treatments.[4] Of the 22 children immunized, only three developed mild fever and/or rash; a serological response was observed in all. This method of immunization, therefore, appears suitable for routine vaccination of children with malignancies. Another research group immunized 15 children with leukemia and 12 with solid tumors, without suspension of anticancer therapy.[30] Low-grade fever and/or papules (from 3 to 30) appeared in 12 of the 15 leukemic children and in 1 of the 12 children with solid tumors.

Table 5
RELATIONSHIP BETWEEN SEVERITY OF SYMPTOMS AND AGE OF CHILDREN AT OUTBREAK OF VARICELLA

Age (mo.)	No. of cases	Clinical symptoms: Mild	Moderate	Severe
		Unvaccinated		
1	4	4	0	0
2	4	1	0	2
4	1	1	1	0
5	3	0	0	2
6	4	1	1	2
7	6	0	1	2
8	5	1	4	1
9	1	1	3	0
10	2	0	0	2
11	3	0	0	3
12—24	10	3	5	2
Total	43	12	15	16
		Vaccinated		
11	2	0	0	0
12—27	31	8	0	0
Total	33	8	0	0

From Baba, K., Yabuuchi, H, Okuni, H., and Takahashi, M., *Pediatrics*, 61, 550, 1978. Reproduced by permission.

D. Summarized Results of Clinical Trials in Japan

Results of clinical trials in Japan for 1974 to 1983 are summarized as follows:[31,32]

1. Vaccination of Acute Leukemic Children

By the end of 1983, a total of 330 children with acute leukemia had been immunized with the varicella vaccine. Of these patients, 251 were vaccinated while in remission. They showed a normal CMI, as assessed with phytohemagglutinin (PHA) or other reagents, and their anticancer therapy, except for 6-MP, was suspended from 1 week before to 1 week after vaccination. Of the 251 patients who received vaccination under the above conditions, 46 (18.3%) showed mostly mild, but sometimes moderate, clinical reactions. By contrast, out of the 79 patients who received either emergency vaccination or vaccination without suspension of chemotherapy, 37 (46.8%) showed clinical reactions. One patient with myelocytic leukemia from the latter group developed uncountable vesicles with high fever (40°C). Immune responses, assessed by the Immune Adherence Hemagglutination (IAHA) or the fluorescent antibody to membrane antigen (FAMA) test, were detected in 91 and 96% of them, respectively (Table 6).

The viral dose and serological and clinical reaction in acute leukemic children are shown in Table 7. There is no significant difference in serological and clinical reaction with varying dose of vaccine (300 to 2500 PFU/dose) administered. Similar results were observed in children with nonmalignant underlying diseases and normal children, though serological responses (average antibody titer after vaccination) are a little higher in normal children than in other groups of children.

Table 6
IMMUNIZATION OF ACUTE LEUKEMIC CHILDREN WITH LIVE VARICELLA VACCINE (Oka STRAIN)

	No. of vaccinated children	Clinical reaction	Serological reaction	Clinical varicella infection	Zoster
With suspension of chemotherapy (2 weeks)	251	46 (18;3%)	229/251 (91.2%)	43 (17.1%)[a]	12
Without suspension of chemotherapy	79	37 (46.8%)[b]	76/79 (96.2%)	9 (11.3%)	6

Note: As reported by 44 hospitals and the Collaborative Study Group of Varicella Vaccine, Ministry of Health and Welfare, Japan, 1974 through 1984.

[a] Two severe cases.
[b] One severe case.

The relationship between the number of lymphocytes and clinical reaction due to vaccination in acute leukemic children is presented in Table 8. In patients with lymphocytes below 1000, the frequency of clinical reaction is higher (30 to 50%), as compared to those with lymphocytes above 1000 (approximately 20%).

Incidence of zoster in acute leukemic children with and without clinical symptoms after vaccination is compared. As shown in Table 9, zoster occurred far more frequently in cases that exhibited rashes than those without rash after vaccination. With regard to pathogenesis of zoster, it has generally been believed that VZV in skin vesicles travel up the sensory nerve to the posterior ganglion, where it persists. When immunity to VZV declines, virus is reactivated and travels down the sensory nerve to the skin, where its replication occurs, leading to formation of vesicles. Thus, it was reasonable that incidence of zoster was far higher in acute leukemic children who developed rashes after vaccination than in those who did not. As no rash usually appears in normal children after vaccination as described below, it is expected that incidence of zoster would be far less in the vaccinated children in their later lives than in those who had acquired natural varicella.

Exogenous varicella infection was observed in 52 out of 330 (15.8%) vaccinees in follow-up of them. Symptoms were usually mild, but two cases showed severe symptoms, with one vaccinee contracting varicella during meningeal leukemia, and the other while receiving intensive chemotherapy.

In one of the major study groups where 52 children with acute leukemia were vaccinated,[33] mild varicella developed in 3 cases after a total of 48 occasions of contact with varicella patients, two of whom were among 11 household contracts (protective effect, 81.8%). The incidences of relapse in vaccinated and nonvaccinated leukemic children who had been receiving the same kind of chemotherapy were compared. There was no significant difference between the rates of relapse in the vaccinated group (12/35 = 34.3%) and the nonvaccinated group (13/39 = 33.3%), values being comparable to those observed in other hospitals. Thus the results gave no indication that vaccination had an adverse effect on the prognosis of the underlying acute leukemia.

2. Vaccination of Children with Malignant Lymphoma and Solid Tumors

Twenty children with malignant lymphoma were vaccinated. Clinical symptoms developed in eight of them, with four showing severe symptoms. These children had received vaccine during continuation of chemotherapy. Immune response was detected

Table 7
VIRAL DOSE, SEROLOGICAL AND CLINICAL REACTIONS IN VACCINATED ACUTE LEUKEMIC CHILDREN[a]

Viral dose (PFU/dose)	No. of vaccinees	Serological reaction (IAHA)	Mean antibody titer (IAHA)	Clinical reaction			
				Mild[b]	Moderate[c]	Severe[d]	Rate %
300	4	4 (100%)	$2^{2.8}$	0	0	0	0
500	31	29 (93.5%)	$2^{3.9}$	8	1	0	29.0
600	41	39 (95.1%)	$2^{3.6}$	7	1	0	19.5
750	46	41 (89.1%)	$2^{2.8}$	5	2	0	15.2
1500	31	28 (90.3%)	$2^{3.2}$	1	2	0	9.7
2500	3	3 (100%)	$2^{3.0}$	0	0	0	0
Total	156	144 (92.3%)		21	6	0	7.3

Note: As reported by 41 hospitals in Japan and the Collaborative Study Group of Varicella Vaccine, Ministry of Health and Welfare, Japan, 1980 through 1982. Vaccine from single batch 7900 was used.

[a] Vaccinated with suspension of chemotherapy.
[b] Mild:severe rash, fever <38°C.
[c] Moderate:countable rash, fever <40°C.
[d] Severe:uncountable rash, fever ≥40°C.

in 18 (90%) of them. Of the 54 vaccinated children with malignant solid tumors (neuroblastoma, hepatoblastoma, rhabdomyosarcoma, Wilms' tumor), 6 (11.1%) developed mild clinical reactions with 49/54 (90.7%) showing seroconversion.

3. Vaccination of Children with Nonmalignant Underlying Diseases

Approximately 1500 children with nonmalignant underlying diseases including neurological disorders, heart diseases, allergic dispositions, congenital diseases, nephrotic diseases, and endocrine and metabolic diseases were vaccinated. Mild clinical reactions were observed in only about 3% and an antibody response in approximately 96%. Of these vaccinated children, 171 had the nephrotic syndrome and 106 were on steroid therapy. Eleven patients (6.4%) with the nephrotic syndrome developed mild clinical symptoms, and an immune response was detected in almost all of them. Vaccination was usually carried out under the following conditions: (1) the patients were in remission; (2) the vaccine was not given when the patients were in the early stage of the disease with evident clinical symptoms or during steroid therapy (>1 mg/kg/day), but was usually administered 1 or 2 weeks after suspension of steroid therapy; and (3) a skin test or in vitro test with PHA or other reagents showed normal CMI. However, emergency vaccination was performed without full consideration of these criteria; nevertheless, no serious adverse reaction was observed.

In another major study group,[34] 70 to 80 children with underlying diseases were vaccinated every year from 1977, and a total of 607 vaccinated seronegative children were followed until 1984. Seroconversion, skin test with varicella antigen, and clinical follow-up results are presented in Tables 10 and 11. A seroconversion rate of 94.1% and a conversion rate of 84.5% in the skin tests were detected, with few clinical reactions (3/541 = 0.6%) except for children with malignancies. A clinical follow-up for 1 to 7 years showed that only 23 out of 607 vaccinated children developed mild symptoms of natural varicella. In 60 known household exposures to varicella patients, 5 children acquired mild clinical varicella (a protection rate of 91.7%). Mild zoster occurred in 4, all of whom had malignancies.

Table 8
LYMPHOCYTE NUMBER AND CLINICAL REACTION IN ACUTE LEUKEMIC CHILDREN AFTER VACCINATION

Lymphocyte no.	No. of vaccinees	Clinical reaction	
		Fever	Rash
<500	4	2 (50%)	2 (50%)
500—1000	26	8 (30.7%)	6 (23.0%)
1000—2000	77	12 (15.6%)	13 (16.9%)
2000—3000	44	10 (22.7%)	8 (18.1%)
3000—4000	99	1 (11.1%)	2 (22.2%)
4000—5000	1	0 (0%)	0 (0%)
>5000	4	0 (0%)	0 (0%)
Total	165	33 (20.0%)	31 (18.8%)

Note: As reported by the Collaborative Study Group of Varicella Vaccine, Ministry of Health and Welfare, Japan, 1980 through 1983.

4. Vaccination of Normal Children

In approximately 4000 normal vaccinated children, hardly any clinical reaction was observed and a high rate (> 95%) of antibody response was recorded.

In a study group,[35] 1531 healthy children were vaccinated between 1980 and 1985. Following vaccination, fever and a slight rash were observed in 10, and a serological response was noted in 99.4% of them. Of the vaccinees, only 15 developed clinical varicella with extremely mild or mild symptoms, while at least 600 vaccinated children came in contact with varicella patients during the 5-year observation period (Table 12). Mild zoster developed in one child who was admitted to a hospital with mycoplasma pneumonia. A similar result was obtained in another study group,[36] indicating that vaccination conferred solid immunity in normal children.

E. Long-Term (7 to 10 years) Follow-Up Studies of the Vaccinees

A long-term (7 to 10 years) follow-up was carried out for 106 vaccinated children, of whom 40 had underlying diseases and 14 had been receiving steroid therapy at the time of vaccination.[37] Despite a total of 147 occasions of contact with varicella patients (26 household, 66 school, and 55 neighborhood contacts), only 5 children (1 household, 4 school contacts) developed varicella. Four children contracted the disease within a year after vaccination and their symptoms were very mild, while one exhibited typical varicella symptoms 3 years after vaccination. No case of zoster was reported. For examination of long-term immunity to VZV, blood was drawn from 38 of the 106 vaccinees. As controls, 29 children having had natural varicella within the previous 10 years were included. FAMA antibody (> 1:4) was detected in 37/38 (97.4%) of the vaccinees with a geometric mean titer (GMT) of $2^{3.1}$, while 29/29 (100%) of the natural varicella group were FAMA positive with a GMT of $2^{3.5}$ (Table 13).

Using the VZV skin test, 37/38 (97.4%) of the vaccinees were positive, while 28/29 (96.6%) of the natural varicella group were positive. These results indicate that immunization with live varicella vaccine conferred long-lasting (7 to 10 years) immunity against VZV.

F. Conclusions of Clinical Trials of Live Varicella Vaccine (Oka Strain) in Japan

The results of extensive clinical use of the Oka strain live varicella vaccine for 10

Table 9
COMPARISON OF INCIDENCE OF ZOSTER IN ACUTE LEUKEMIC CHILDREN WITH OR WITHOUT CLINICAL REACTION FOLLOWING VACCINATION

	No. of cases that developed zoster		
Years after vaccination	Rash (+) after vaccination	Rash (−) after vaccination	Total
—1	7	3	10
1—2	5	2	7
2—3	0	0	0
3—4	0	0	0
4—5	1	0	1
Total:	13/83 (15.7%)	5/247 (2.0%)	18/330 (5.5%)

Note: As reported by the Collaborative Study Group of Varicella Vaccine, Ministry of Healthy and Welfare, Japan, 1974 through 1983.

years in Japan indicate that children with malignancies can be immunized safely and effectively, provided that these patients are in remission, their CMI is normal (using the PHA test or other methods), and chemotherapy other than 6-MP is suspended from 1 week before to 1 week after vaccination. An immune response was observed in a high percentage of the vaccinees. Some vaccinees (52/330, 15.8%) acquired exogenous varicella infection, but their symptoms were mostly mild. As patients with T-cell leukemia or malignant lymphoma tend to develop more severe clinical rections, a careful immunological examination should be carried out for such patients prior to vaccination. Booster revaccination is recommended in children with malignancies, because immunity against VZV declines in those receiving continuous chemotherapy.

Since it is important that vaccination should not aggravate the underlying diseases, it was noteworthy that no significant different in relapse rate was found in vaccinated and unvaccinated leukemic children, indicating that vaccination did not have an adverse effect on the prognosis of acute leukemia. In high-risk children, other than those with malignancies such as children with nephrosis, autoimmune diseases, and bronchial asthma receiving immunosuppressive drugs, vaccination resulted in few clinical reactions but a good immune response, when carried out with the same or fewer precautions than those outlined for leukemic patients. Of more than 200 vaccinated high-risk patients of this type, only one patient with infantile spasms, receiving a large amount of ACTH exhibited an atypical clinical reaction with numerous vesicular lesions and had a high fever, but this patient recovered completely. In this group, the decrease in immunity was less and the protective effect was greater than the group with malignancies. Consequently, no revaccination appears necessary to maintain VZV immunity.

In non-high-risk diseased children, the precautions necessary for vaccination of high-risk children were not needed, although children with high fever or an extremely poor state of nutrition were excluded for vaccination. After vaccination, clinical reactions were rare (<1%) and the immune response was good. The rate of exogenous varicella infection was low (<10%) with mild clinical symptoms.

On vaccination of normal children, few clinical reactions, if any, were observed. In accordance with a high immune response (>95%), the protective effect of the vaccine was also high (>95%).

Table 10
CLINICAL AND IMMUNOLOGICAL RESPONSES OF CHILDREN WITH VARIOUS UNDERLYING DISEASES AFTER RECEIVING A LIVE VARICELLA VACCINE

Disease	No. of vaccinees	Clinical reaction	Seroconversion (IAHA or FAMA)	Skin reaction (Varicella antigen)
Neurological	194	0	181 (93.2%)	107/125 (85.6%)
Cardiovascular	115	0	109 (94.8%)	96/110 (87.2%)
Neoplasms	66	26 (34.4%)	63 (95.7%)	49/59 (83.1%)
Blood, immunity, or allergy	132	2 (1.5%)	122 (93.4%)	102/125 (81.6%)
Malformation, digestive, or prenatal	58	0	55 (94.0%)	45/51 (88.2%)
Kidneys, endocrine, or metabolism	42	1 (2.3%)	41 (97.6%)	37/41 (90.2%)
Total	607	29 (4.8%)	571 (94.1%)	496/587 (84.5%)

Note: Department of Pediatrics, Osaka University, 1975 to 1984. Abstracted from Reference 34.

G. Clinical Trials of the Varicella Vaccine in the U.S. and Europe

In the U.S., 99 healthy children were first immunized with the Oka strain of varicella vaccine, and seroconversion rates of over 94% were induced with minimal clinical reactions.[38] A doubleblind, placebo-controlled efficacy trial of the live attenuated Oka/Merck varicella vaccine among 956 children was then conducted.[39] Of 914 children, 468 received vaccine and 446 received placebo. The vaccine produced few clinical reactions and no evidences of viral spread from vaccinated children to sibling controls. Approximately 8 weeks after vaccination, 94% of initially seronegative children who received vaccine had detectable antibody to varicella. During the 9-month surveillance period, 39 clinically diagnosed cases of varicella (38 of which were confirmed by laboratory tests) occurred among study participants. All 39 cases occurred in placebo recipients; no child who received vaccine contracted varicella. Thus the vaccine was reported as 100% effective in preventing varicella in this population of healthy children.

Lately, a clinical trial among 137 healthy children, ages 1 to 12 years was conducted to titrate the dose of live Oka/Merck varicella vaccine and determine the minimum effective dose, induction time of measurable antibody, and clinical reactions.[40] Doses of approximately 43, 435, 870, and 4350 PFU were administered. The vaccine was well tolerated with no significant difference in the rate of clinical symptoms by dose. The frequency of varicella-like rash was 3% (4/137); all rashes were mild. Among the 99 initially seronegative children who received doses of 435 PFU or greater, 94% assayed at 2 weeks and 100% assayed at 4 or 6 weeks seroconverted. Children receiving the lowest dose (43 PFU) seroconverted more slowly than those who received a higher dose. The geometric mean titers were similar for all four doses at 6 weeks. All 29 children tested for IgG responded; 5 of 29 (17%) demonstrated an IgA response with no relation to the vaccine dose. SGOT and SGPT values of 26 children tested were essentially unchanged after vaccination. Minor variations found in platelet counts after vaccination were not associated with any bleeding, bruising, or clotting.

As for vaccination of acute leukemic children, Brunell et al. first reported that live varicella vaccine was given to 23 children with lymphoreticular malignancies: 12 whose chemotherapy was complete, and 11 whose therapy was continuing.[41] Seroconversion was observed in all 23 children, only one of whom lost his vaccine-induced antibodies. Out of 13 children, 8 experienced exposure to varicella, including 4 continuing exposures in their household. Varicella, manifested by the appearance of seven vesicular

Table 11
PROTECTIVE EFFICACY AGAINST VARICELLA INFECTION: INCIDENCE OF VARICELLA AND ZOSTER IN CHILDREN AFTER VZV VACCINATION

Diseases	No. of vaccinees	Occasion of contact: Family	Neighbors	Classmates	Unknown	Developed varicella	Developed zoster
Neuromuscular	194	15	44 (3)	34 (3)	—	6	0
Cardiovascular	115	10 (1)	36	16 (1)	—	2	0
Neoplasms	66	10 (1)	10 (2)	11 (1)	— (4)	8	4
Blood, immunity, or allergy	132	12	17	16 (1)	— (1)	2	0
Malformation, digestive, or prenatal disturbances	58	6 (2)	12	11 (1)	—	3	0
Glomerulus, endocrine, or metabolism	42	7 (1)	12	6 (1)	—	2	0
Total	607	60 (5)	131 (5)	94 (8)	— (5)	23 (3.8%)	4 (0.7%)

Note: In parentheses, no. of cases that developed varicella. Department of Pediatrics, Osaka University, 1975 to 1984. Abstracted from Reference 34.

lesions, developed only in one. In all but two of the children, an in vitro blastogenic response to VZV antigen developed; both children had a biphasic rash after immunization. A sibling of one of these children seroconverted without clinical evidence of varicella, presumably because of infection with vaccine virus. None of the other household contacts had significant rises in VZV antibody.

The National Institute of Allergy and Infectious Diseases (NIAID) Varicella Vaccine Collaborative Study Group conducted studies on the efficacy of the vaccine in children with leukemia in remission: 191 susceptible children with leukemia in remission were immunized with varicella vaccine.[42] There was serological evidence of an immune response in approximately 80% after one dose and in more than 90% after several doses. The major side effect was a mild to moderate rash, seen especially in children with maintenance chemotherapy suspended for 1 week before and 1 week after vaccination. Mild to moderate clinical reactions were observed in 2 out of 53 (4%) vaccinees who had completed chemotherapy and 49 out of 138 (36%) vaccinees who had been receiving maintenance chemotherapy. Children with a rash had higher antibody titers than those without a rash, but those with a rash were also at risk (10%) to transmit vaccine virus to others. Subsequently, 22 vaccinees had household exposure to varicella or zoster. The attack rate of clinical varicella in these vaccinees was 18% (4/22), significantly lower than the attack rate of approximately 90% in varicella-susceptible persons with household exposures. All cases of clinical illness were extremely mild, with an average of about 50 vesicles. The mild character of the illness was clearly different from varicella in unimmunized children receiving chemotherapy for leukemia. It was thus concluded that varicella vaccine was approximately 80% effective in preventing clinical varicella in children with leukemia and completely effective in preventing severe varicella in this high-risk group.

In Europe, clinical trials with the varicella vaccine (Oka strain) have been conducted in high-risk as well as normal children, and similar results to those in Japan and the U.S. have been reported.[43] It was recently reported that 34 nurses who had no previous history and were seronegative to VZV were immunized with Oka-RIT vaccine and were followed for up to 36 months.[44] No major vaccine reactions were observed. At 5 months and at 1 year, 94% of the nurses had seroconverted, but at 3 years only 66% had detectable antibody. Lymphocyte transformation to VZV antigen tested in seven nurses in whom antibody was no longer detectable gave positive stimulation indices.

Table 12
VACCINATION OF HEALTHY CHILDREN AND RESULT OF CLINICAL FOLLOW-UP (1980—1985)

Year of vaccination	No. of vaccinated children	Serological reaction	Clinical reinfection	Type of contact
1980	160	160	0	
1981	213	209	3	2 Nursery 1 Playmate
1982	380	380	2	2 Nursery
1983	227	222	4	2 Nursery 2 Playmate
1984	200	200	4	2 Nursery 1 Playmate 1 Unknown
1985	351	351	2	2 Nursery
Total	1531	1522 (99.4%)	15 (1.0%)	

Note: Ten children showed extremely mild reactions; five children showed mild reactions (with a few rashes and fever <38°C). Number having contact with varicella patients was at least 600; zoster patient: one (mycoplasma pneumonia mild).

From Horiuchi, K., *Biken, J.*, 27, 37, 1984, and personal communication. With permission.

Of 13 vaccinated nurses who looked after children with chickenpox, 2 became infected but 1 of them had not seroconverted after vaccination. Chickenpox developed in six out of seven unvaccinated seronegative nurses who looked after children with chickenpox. This study shows that the varicella vaccine is safe and immunogenic in adults and that both cell-mediated and antibody tests are needed for long-term assessment of immunity of VZV. The varicella vaccine of Oka strain was licensed in several countries of Europe in 1984 and in Japan in 1986.

H. Possible Use of the Vaccine to Prevent Zoster in the Aged Group

It was observed that approximately 20% of 55- to 65- year-old healthy people with confirmed history of chickenpox and humoral antibodies to VZV show a negative LPF immune response to VZV antigen.[45,46] With a view to the possible prevention of herpes zoster, 33 subjects having a reduced LPF immune response to VZV were given varicella vaccine.[47] In 28 (85%) of them, the vaccination induced a change from a negative to positive cell-mediated immune response to VZV antigen in vitro. In all subjects, humoral antibodies to VZV were present prior to administration of the booster vaccination. An increase in antibody titer was observed in >75% of the vaccinees, suggesting at least some VZV replication, despite preexisting humoral antibodies. Retesting of a few of the vaccinees 1 year after vaccination showed that the boosting effect remained. It is uncertain whether such an improvement in the LPF immune response to VZV does in fact give protection against reactivation of VZV, leading to herpes zoster, but the result has given the hope that zoster in the older person will be prevented by varicella vaccine. Stimulation of CMI to VZV by vaccination in adults was also reported by Baba et al.[48] Thus in the future the vaccination may be applicable to an older person with decreased CMI response to VZV, after the clinical evaluation of a large number of vaccinees over a period of several years. In addition, as discussed in section V.D.1.,

Table 13
A LONG-TERM (7 TO 10 YEARS) CLINICAL AND IMMUNOLOGICAL FOLLOW-UP OF THE VACCINATED CHILDREN

No. of vaccinee	Occasions of contact with varicella patients		Cases of clinical varicella
106*	Family	24	4 mild
	School	64	1 moderate
	Playmate	58	
	Total	146	5

Subjects	Age at vaccination or natural varicella (years)	Age at immunological test (years)	Humoral immunity: FAMA test (+)	Humoral immunity: Mean antibody	Cell-mediated immunity: Varicella skin test	Cell-mediated immunity: Mean diameter of erythema (mm)
Vaccinated group	4.2 ± 2.1	12.1 ± 3.0	37/38 (97.4%)	$2^{3.1}$	37/38 (97.4%)	13.4 ± 4.9
Natural varicella group	4.3 ± 2.1	13.3 ± 3.3	29/29 (100%)	$2^{3.5}$	28/29 (96.6%)	12.9 ± 5.7

* No cases of zoster were reported in any of the follow-up studies.

Data from Reference 37.

it is expected that vaccination of susceptible children would greatly decrease the incidence of zoster in their later lives.

VI. ISOLATION OF VZV FROM THE BLOOD OF NATURALLY INFECTED AND VACCINATED CHILDREN

VZV could be recovered from blood mononuclear cells of nonimmunocompromised patients for several days before and after onset of the disease (Table 14).[49,50,51] In contrast, no VZV could be recovered from a total of 27 children, 4 to 14 days after vaccination at a dose of 5000 PFU (Table 15).[51] It is generally believed that at the time of primary VZV infection, the virus multiplies on the respiratory mucosa and in the regional lymph nodes, and this multiplication leads to a primary viremia, during which the virus is delivered to the viscera, where further multiplication ensues. A secondary viremia, greater in magnitude than the first, then occurs and delivers virus to the skin, leading to the appearance of a rash. The above results suggest that the magnitude of replication of the vaccine virus in the susceptible viscera is far less than that of wild-type VZV, but sufficient to induce an immune response. Thus it seems that viremia may be a marker of the virulence of VZV for the host, and the vaccine virus may be attenuated to the degree that it lacks the capacity to cause a viremia, except possibly in rare instances.

VII. TRANSMISSIBILITY OF VIRUS FROM VACCINE RECIPIENTS

No contact infection was observed clinically and serologically from the vaccine recipients to the susceptibles in an institution for children.[52] It was recently reported that

Table 14
VIRAL ISOLATION FROM MONONUCLEAR CELLS AND ANTIBODY RESPONSES AFTER CLOSE CONTACT WITH PATIENTS WITH VARICELLA

Day of testing after onset of varicella	Viral isolation from mononuclear cells	Detectable antibodies*
−11	0/3 (0)	ND
−7	0/4 (0)	ND
−6	0/1 (0)	0/1 (0)
−5	1/2 (50)	ND
−4	1/3 (33)	0/2 (0)
−3	ND	0/1 (0)
−2	4/4 (100)	0/4 (0)
−1	4/5 (80)	0/5 (0)
0	4/17 (24)	0/13 (0)
1	7/32 (22)	0/28 (0)
2	0/14 (0)	0/12 (0)
3	0/3 (0)	4/12 (33)
4	0/1 (0)	9/18 (50)
≥5	0/3 (0)	14/14 (100)

Note: Data are no. of positive subjects/no. tested (%). ND = not done.

* Measured by the assay for fluorescent antibody to membrane antigen.

From Asano, Y., Itakura, N., Hiroishi, Y., Hirose, S., Ozaki, T., Kuno, K., Nagai, T., Yazaki, T., Yamanishi, K., and Takahashi, M., *J. Infect. Dis.*, 152, 863, 1985. Copyright by The University of Chicago Press, 1986. With permission.

there was evidence of spread of VZV from the vaccinated children with acute leukemia to a susceptible sibling in four families, two subclinically and two clinically.[42] Each case occurred only when the vaccinee had experienced rash, and the incidence of spread to siblings from vaccinees with rash was four (10%) out of 40. However, it has been observed that few clinical reactions occurred in children successfully vaccinated by the inhalation method.[53] Considering the low virulence of vaccine virus, spread of vaccine virus may be limited, if it occurs at all, and it is unlikely that the vaccine virus spreads in succession to a wide range of susceptibles.

VIII. DURATION OF IMMUNITY INDUCED BY LIVE VARICELLA VACCINE: REINFECTION WITH VZV IN RELATION TO PERSISTENCE OF IMMUNITY AFTER VACCINATION

It has been reported that humoral antibody and protective immunity lasted for 2 to 5 years after vaccination.[54-57]

A recent follow-up study of 26 immunized healthy children revealed that 100% had VZV-neutralizing antibody, and 96% had VZV FAMA antibody detectable 5 years after immunization.[55] None of these vaccines developed varicella during that interval

Table 15
ISOLATION OF VZV FROM CHILDREN INOCULATED WITH LIVE VIRUS VACCINE (STRAIN Oka)

	Viral isolation source			
Day of testing after vaccination	Mononuclear cells	Throat	Detection antibodies[a]	Positive skin reaction
0	ND	0/3 (0)	0/28 (0)	0/22 (0)
3	0/11 (0)	0/8 (0)	0/8 (0)	0/10 (0)
4—5	0/14 (0)	0/13 (0)	0/11 (0)	1/11 (9)
6—7	0/17 (0)	0/16 (0)	2/8 (25)	8/11 (73)
8—9	0/6 (0)	0/6 (0)	2/5 (40)	4/5 (80)
10—14	0/11 (0)	ND	8/8 (100)	6/7 (86)
30—60	ND	ND	28/28 (100)	17/20 (85)

Note: Data are no. of positive vaccinees/no. tested (%). ND = not done.

[a] Measured by the assay for fluorescent antibody to membrane antigen (FAMA).

From Asano. Y., Itakura, N., Hiroishi, Y., Hirose, S., Ozaki, T., Kuno, K., Nagai, T., Yazaki, T., Yamanishi, K., and Takahashi, M., *J. Infect. Dis.*, 152, 863, 1985. Copyright by The University of Chicago Press, 1985. With permission.

although many had been exposed to the virus. Because no decrease in antibody was observed during this time and because of the many exposures, it can be postulated that subclinical reinfections may have occurred, causing boosts in VZV antibody titer. In addition, a long-term follow-up of vaccinees indicates ≥95% protection for children immunized 7 to 10 years previously and that average antibody titer (FAMA) of the immunized group was comparable to the matched control group who experienced natural varicella.[37] As average antibody titer following immunization is usually one fourth to one eighth that of those who acquired natural varicella,[21] the above results suggest that boost in antibody titer had been caused by subclinical reinfection.

Other investigators have also observed the phenomenon of subclinical reinfection in recipients of VZV vaccine.[41,42] Subclinical reinfection with VZV has also been observed after natural varicella.[58-59] In addition, occasional mild clinical reinfections with VZV have been observed in normal persons after natural varicella.[58] Whether these secondary booster responses are important in maintaining long-term immunity to VZV is an extremely important subject for further research and may have significant implications for the future use of varicella vaccine.

Varicella is a more severe illness in adults than it is in children; therefore, if the vaccine does not confer long-term protection, its routine use in children could be dangerous if it serves only to postpone varicella until adult life. For these problems, however, it might be very difficult to see the immune status of the vaccinees for a long period in the closed condition where no exogenous varicella infection could occur. In a semiclosed institution, the incidence of varicella outbreaks was not reduced despite an average rate of immune individuals in the population of more than 70%.

In contrast, active immunization of children more than 1 year of age with measles vaccine resulted in a significant reduction of measles epidemics in the same institution (Figure 5). Even after introduction of varicella vaccine, the occurrence of varicella outbreaks among subjects less than 6 months of age was not uncommon.[60] Thus, if it is not possible to actively immunize the 6-months-or-younger age group, there will always be a pool of susceptibles to VZV infection. It has also been observed that in-

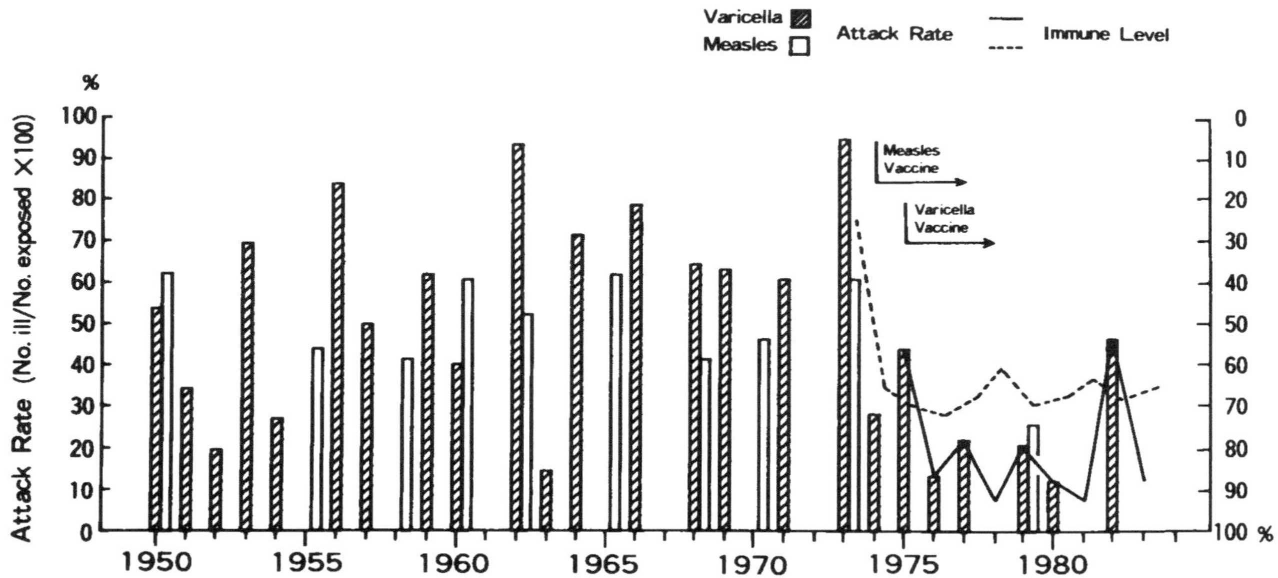

FIGURE 5. Clinical attack rate and immune rate against varicella and measles in an institution for children from 1950 to 1983. (Immune rate: number of prior infected + prior immunized/number of exposed). (From Baba, K., Yabuuchi, H., Takahashi, M., and Ogra, P. L., Immunologic and academic aspects of varicella acquired during infancy and early childhood, *J. Pediatr.*, 100, 881, 1982. With permission.)

Table 16
THE EFFECT OF PREEXISTING TRANSPLACENTALLY ACQUIRED MATERNAL ANTIBODY ON THE OUTCOME AND SEVERITY OF CLINICAL VARICELLA

Preexisting antibody titer (FAMA)	No. Subjects and severity of disease			
	Total	Mild	Moderate	Severe
<4	7	0	3	4
4	5	3	1	1
8	4	2	1	1
16	2	1	1	0

From Baba, K., Yabuuchi, H., Takahashi, M., and Ogra, P. L., Immunologic and academic aspects of varicella acquired during infancy and early childhood, *J. Pediatr.*, 100, 881, 1982. With permission.

fants who received vaccination at the age of less than 6 months are prone to lose immunity earlier, becoming susceptible to VZV infection. In addition, some of those who have experienced natural varicella may develop zoster in their later lives. This means that though zoster is less infectious than varicella, there will always be a considerable number of zoster patients who could be a source of varicella epidemic, at least for several decades in our communities. In fact, it was observed that outbreak of varicella was initiated from the zoster patients in 4 of 13 outbreaks in a semiclosed institution for children.[61] From these results, it may be postulated that epidemiologic modification of VZV infection in a community might not occur for a long time after introduction of varicella vaccine to normal children.

IX. ROLE OF CELL-MEDIATED IMMUNITY IN PREVENTION OF CLINICAL VARICELLA INFECTION

We have followed-up VZV infection in a population of infants under 1 year of age during three outbreaks of varicella in a semiclosed domiciliary institution for infants.[61] Over a period of 4 years, 252 residents ranging in age from 27 days to 32 months were tested for skin reaction with VZV antigen and VZV antibody before, during, and after each outbreak of varicella. Of these, 85 subjects developed clinical varicella, with an overall attack rate of 100% for all individuals with a negative skin test with VZV antigen before outbreak of varicella, including infants under 2 months of age. The clinical disease appeared to be milder in infants less than 2 months old. Severe disease occurred more frequently in subjects 2 to 11 months of age than in those over 12 months. Preexisting antibody did not prevent development of illness, though the severity of disease is somewhat milder in those with higher antibody titer (Table 16). Varicella vaccine was given to 18 susceptible children, including those below 6 months of age. All the vaccinated children examined showed conversion in skin test with VZV antigen, with little enhancement of preexisting humoral antibody. All of them were resistant to clinical varicella epidemic which occurred subsequently in that institution.[6]

These results suggest that cell-mediated immunity to VZV induced by vaccination is playing a major role in the prevention of clinical varicella and that the presence of humoral antibody itself may not be enough to prevent clinical varicella unless antibody titer is extremely high. It has been well known from in vitro experiments that VZV spreads from cell to cell, not via liberated virus from infected cells. Therefore, it is reasonable that humoral antibody is not effective to block the spread of the virus once infection is established, but cell-mediated immunity is effective in prevention of the disease by destroying the infected cells.

The importance of cell-mediated immunity to clinical varicella infection was also proved by the observation on the vaccinated Bruton-type agammaglobulinemia patients. Four Bruton-type agammaglobulinemia patients were vaccinated with no clinical symptoms. They exhibited a positive skin test reaction with varicella antigen, whereas no humoral antibody reactions (IAHA, FAMA) were detected after vaccination. In the follow-up for 5 to 9 years, three of them had evident contacts with varicella patients, and all of them have been free from any clinical symptom, with consistent positive skin test and negative humoral antibody.[71] These observations indicate that cell-mediated immunity is of vital importance in protection against clinical varicella.

X. LONG-TERM SAFETY OF THE VACCINE

Concern for long-term safety of the vaccine involves the incidence of zoster and malignancy.

A. Incidence of Zoster after Vaccination

A major question on live varicella vaccine has been whether the vaccine virus becomes latent, resulting in later development of zoster. A long-term follow-up of vaccinated normal children will be required to answer this question definitely. Children with acute leukemia tend to develop zoster early after natural infection. Careful comparative studies on groups of vaccinated and naturally infected children with acute leukemia are as follows.[30,31] In one study group,[33] the incidence of zoster in vaccinated and naturally infected children is 8/52 (15.4%) and 11/63 (17.5%), respectively, and in another study group[34] 4/44 (9.1%) and 8/37 (21.6%), respectively. The clinical symptoms of zoster in the vaccinated children were usually mild and clinically not troublesome, whereas two of the naturally infected children in the latter study group exhibited moderate or severe symtpoms. Some VZV isolated from zoster that developed in the vaccinated ALL patients were identified as vaccine-virus derived.[8,62]

Recently, Brunell et al. reported that 19 of 76 children with acute leukemia who had varicella developed zoster, while none of the 48 vaccinees did.[63] Adjusting for time of observation, removing the vaccinees who failed to have a sustained antibody response or developed chickenpox, the risk of zoster was still less in vaccinees ($p = 0.017$). It was thus concluded that there is no reason to suspect that recipients of varicella vaccine are more likely to develop zoster than children who get varicella.

B. Malignancy

It is difficult to rule out completely VZV as a cause of malignancy, but VZV has never been linked epidemiologically to any form of cancer. It has been found that VZV-infected human cells transformed mouse cells in vitro biochemically but not morphologically, although cell-free material (e.g., vaccine) did not.[64] This finding does not indicate that VZV has oncogenic potential but indicates that genes for viral thymidine kinase can be transferred under certain conditions to other cells deficient in that enzyme.

XI. STUDIES ON ANOTHER ATTENUATED VZV STRAIN (KMcC)

Results of clinical trials with another live varicella vaccine (KMcC strain) in healthy children were also reported (43 children were immunized: 26 with the 40th-passage vaccine and 17 with the 50th passage).[65,66] There was 100% seroconversion to the KMcC passage in 40th and 50th vaccines. Papular skin lesions occurred in 31% of the 40th-passage vaccinees but only in 6% of the 50th-passage vaccinees. However, somewhat low immunogenicity of the 50th passage of KMcC was also noted, and no following trials with KMcC strain in high-risk children have been reported.

It was further reported that Oka and KMcC strains of VZV vaccine were administered to 369 healthy seronegative children in a sequence of 10 comparative clinical trials.[67] Postimmunization clinical reactivity was minimal with the Oka vaccine but was unacceptably high (32%) with the KMcC passage-40 vaccine. Immunogenicity of 93 to 100% was noted by FAMA assay, and in vitro LPF responses. Follow-up studies demonstrated persistence of antibody and in vitro LPF responses and protection and modification of infection 9 to 48 months after immunization. Only five episodes of mild varicella occurred in children in whom seroconversion had occurred. These episodes were noted after at least 281 known varicella exposures. The KMcC passage-50 vaccine was concluded not acceptable because all the episodes of varicella occurred in children who had previously shown seroconversion, and the lesion count was higher in KMcC breakthrough cases. Vaccine virus reactivation as zoster had not occurred in any child. Better immunogenicity of attenuated Oka strain than KMcC strain was also observed in rhesus monkeys.[68]

XII. PROTECTIVE EFFECT OF VACCINATION

Despite the observation that varicella may occasionally occur after an exposure in vaccinees with specific antibody and cellular immunity (breakthrough cases), this has not been the usual outcome of exposed vaccinees, however, in any study so far. The vast majority of vacinees with close exposures to VZV have been protected from clinical chickenpox. The low clinical attack rate after vaccination is quite different than the attack rate of 80 to 100% seen in children with household exposures who have not been vaccinated.

XIII. FUTURE USE OF VARICELLA VACCINE

It has been generally accepted that the varicella vaccine is useful for immunocompromised children, i.e., children with acute leukemia and solid tumors, and children receiving steroids for nephrosis, asthma, and collagen and vascular diseases. This vaccine is also recognized as useful for termination of nosocomial varicella outbreaks. Varicella-susceptible adults, particularly medical staff and nurses,[44] will also benefit from varicella vaccine. Whether the vaccine ought to be administered to normal children on a routine basis is the pressing and controversial question. However, based on the discussion in Section VIII, there would be no definite reason for objection to using the vaccine on normal children, considering that there are some severe varicella cases even in normal children (Table 17). Recently, a measles-rubella-mumps vaccine combined with varicella vaccine was tested in clinical trials, and promising results have been obtained.[69,70] It is expected that in the near future children will be free by vaccination from the main exanthematous diseases.

Table 17
POSSIBLE SUBJECTS FOR VACCINATION

Immunosuppressed children
Acute leukemia
Solid tumor
Nephrosis, asthma, etc.
Siblings of high-risk children
Termination of nosocomial varicella outbreak
Seronegative adults
Medical staffs (doctors and nurses)
Prepregnant aged women
Normal children

REFERENCES

1. Greenthal, R. M., The prophylaxis of varicella with vesicle fluid, *Am. J. Dis. Child.*, 32, 851, 1926.
2. Takahashi, M., Otsuka, T., Okuno, Y., Asano, Y., Yazaki, T., and Isomura, S., Live vaccine used to prevent the spread of varicella in children in hospital, *Lancet*, 2, 1288, 1974.
3. Takahashi, M., Okuno, Y., Otsuka, T., Osame, J., Takamizawa, A., Sasada, T., and Kubo, T., Development of a live attenuated varicella vaccine, *Biken, J.*, 18, 25, 1975.
4. Takahashi, M., Asano, Y., Kamiya, H., Baba, K., and Yamanishi, K., Active immunization for varicella-zoster virus, in *Human Herpes Viruses*, Nahmias, A. J., Dowdle, W. R., and Schinazi, R. E., Eds., Elsevier, New York, 1981, 414.
5. Takahashi, M., Chickenpox virus, in *Advances in Virus Research*, Academic Press, Vol. 29, Lauffer, M. A. and Maramorosch, K., Eds., New York, 1983, 285.
6. Takahashi, M. and Baba, K., A live varicella vaccine; its protective effects and immunological aspects of varicella-zoster virus infection, in *Medical Virology*, Vol. 3, de la Maza, L. and Peterson, E. M., Eds., Elsevier, New York, 1984, 255.
7. Takahashi, M., Hayakawa, Y., Shiraki, K., Yamanishi, K., Asano, Y., and Ozaki, T., Attenuation and laboratory markers of the Oka strain varicella-zoster virus, *Postgrad. Med. J.*, London, 61 (Suppl. 4), 37, 1985.
8. Hayakawa, Y., Torigoe, S., Shiraki, K., Yamanishi, K., and Takahashi, M., Biological and biophysical markers of a varicella vaccine strain (Oka strain): identification of clinical isolates from vaccine recipients, *J. Infect. Dis.*, 149, 956, 1984.
9. Matsunaga, Y., Yamanishi, K., and Takahashi, M., Experimental infection and immune response of guinea pig with varicella-zoster virus, *Infect. Immun.*, 37, 407, 1982.
10. Martin, J. H., Dohner, D. E., Wellinghoff, W. J., and Gelb, L. D., Restriction endonuclease analysis of varicella zoster vaccine virus and wild-type DNAs, *J. Med. Virol.*, 9, 69, 1982.
11. Hayakawa, Y., Yamamoto, T., Yamanishi, K., and Takahashi, M., Analysis of varicella-zoster virus (VZV) DNAs of clinical isolates by endonuclease HpaI, *J. Gen. Virol.*, 67, 1817, 1986.
12. Kamiya, H., Ihara, T., Hattori, A., Iwasa, T., Sakurai, M., Izawa, T., Yamada, A., and Takahashi, M., Diagnostic skin test reactions with varicella virus antigen and clinical application of the test, *J. Infect. Dis.*, 136, 784, 1977.
13. Baba, K., Yabuchi, H., Okuni, H., and Takahashi, M., Studies with live varicella vaccine and inactivated skin test antigen: protective effect of the vaccine and clinical application of the skin test, *Pediatrics*, 61, 550, 1978.
14. Kumagai, T., Chiba, Y., Fujiwara, M., Hanazono, H., Nakamura, S., Chiba, S., Nakao, T., and Takahashi, M., Humoral and cellular immune response to varicella-zoster virus in children inoculated with live attenuated varicella vaccine, *Biken, J.*, 23, 135, 1980.
15. Asano, Y., Shiraki, K., Takahashi, M., Nagai, T., Ozaki, T., and Yazaki, T., Soluble skin test antigen of varicella-zoster virus prepared from the fluid of infected cultures, *J. Infect. Dis.*, 143, 684, 1981.
16. Shiraki, K., Yamanishi, K., and Takahashi, M., Biologic and immunologic characterization of the soluble skin-test antigen of varicella-zoster virus, *J. Infect. Dis.*, 149, 501, 1984.

17. Shiraki, K., and Takahashi, M., Virus particles and glycoproteins excreted from cultured cells infected with varicella-zoster virus, *J. Gen. Virol.*, 61, 271, 1982.
18. LaRussa, P., Steinberg, S. P., Seeman, M. D., and Gershon, A. A., Determination of immunity of varicella-zoster virus by means of an intradermal skin test, *J. Infect. Dis.*, 152, 869, 1985.
19. Florman, A. L., Umland, E. T., Ballow, D., Cushing, A. H., McLaren, L. C., Gribble, T. J., and Duncan, M. H., Evaluation of a skin test for chickenpox, *Infect. Control.*, 6, 314, 1985.
20. Schrier, R. D., Pizer, I. I., and Moorhead, J. W., Type-specific delayed hypersensitivity and protective immunity induced by isolated herpes simplex virus glycoprotein, *J. Immunol.*, 130, 1413, 1983.
21. Asano, Y., Nakayama, H., Tazaki, T., Kato, R., Hirose, S., Tsuzuki, K., Ito, S., and Takahashi, M., Protection against varicella in family contacts by immediate inoculation with live varicella vaccine, *Pediatrics*, 59, 3, 1977.
22. Asano, Y., Hirose, S., Iwasa, A., Miyata, T., Yazaki, T., and Takahashi, M., Protective effect of immediate inoculation of a live varicella vaccine in household contacts in relation to the viral dose and interval between exposure and vaccination, *Biken J.*, 25, 43, 1982.
23. Asano, Y., Yazaki, T., Miyata, T., Nakayama, H., Hirose, S., Ito, S., Tanaka, E., Isomura, S., Suzuki, H., and Takahashi, M., Application of a live varicella vaccine to hospitalized children and its protective effect on spread of varicella infection, *Biken J.*, 18, 35, 1975.
24. Asano, Y., Nakayama, H., Yazaki, T., Ito, S., Isomura, S., and Takahashi, M., Protective efficacy of vaccination in children in four episodes of natural varicella and zoster in the ward, *Pediatrics*, 59, 8, 1977.
25. Ueda, K., Yamada, I., Goto, M., Nanri, T., Fukuda, H., Katsuta, M., Otsuka, T., and Takahashi, M., Use of a live varicella vaccine to prevent the spread of varicella in handicapped or immunosuppressed children including MCLS (Muco-Cutaneous Lymphonode Syndrome) patients in hospital, *Biken, J.*, 20, 117, 1977.
26. Asano, Y., Iwayama, S., Miyata, T., Yazaki, T., Ozaki, T., Tsuzuki, K., Ito, S., and Takahashi, M., Spread of varicella in hospitalized children having no direct contact with an indicator zoster case and its prevention by a live vaccine, *Biken, J.*, 23, 157, 1980.
27. Katsushima, N., Yazaki, T., Sakamoto, M., Fujiyama, J., Nakagawa, M., Okuyama, Y., and Takahashi, M., Application of a live varicella vaccine to hopsitalized children and its follow-up study, *Biken, J.*, 25, 29, 1982.
28. Hattori, A., Ihara, T., Iwasa, T., Kamiya, H., Sakurai, M., Izawa, T., and Takahashi, M., Use of live varicella vaccine in children with acute leukemia or other malignancies, *Lancet*, 2, 210, 1976.
29. Izawa, T., Ihara, T., Hattori, A., Iwasa, T, Kamiya, H., Sakurai, M., and Takahashi, M., Application of a live varicella vaccine in children with acute leukemia or other malignant diseases, *Pediatrics*, 60, 805, 1977.
30. Ha, K., Baba, K., Ikeda, T., Nishida, M., Yabuchi, H., and Takahashi, M., Application of live varicella vaccine to children with acute leukemia or other malignancies without suspension of anticancer therapy, *Pediatrics*, 65, 346, 1980.
31. Takahashi, M., Asano, Y., Kamiya, H., and Baba, K., Varicella vaccine: case studies, *Microbiol. Sci.*, 2, 249, 1985.
32. Takahashi, M., Kamiya, H., Baba, K., Asano, Y., Ozaki, T., and Horiuchi, K., Clinical experience with Oka live varicella vaccine in Japan, *Postgrad. Med. J.*, London, 61 (Suppl. 2), 61, 1985.
33. Kamiya, H., Kato, R., Isaji, M., Oitani, K., Ito, M., Ihara, T., Sakurai, M., and Takahashi, M., Immunization of acute leukemic children with a live varicella vaccine (Oka strain), *Biken J.*, 27, 99, 1984.
34. Yabuuchi, H., Baba, K., Tsuda, N., Okada, S., Nose, O., Seino, Y., Timota, K., Ha, K., Mimaki, T., Ogawa, M., Kanesaki, T., Yoshida, A., and Takahashi, M., A live varicella vaccine in a pediatrics community, *Biken, J.*, 27, 43, 1984.
35. Horiuchi, K., Chickenpox vaccination of healthy children; immunological and clinical responses and protective effect in 1978—1982, *Biken, J.*, 27, 37, 1984.
36. Ozaki, T., Matsui, T., Ichikawa, T., Asano, Y., Yamanishi, K., and Takahashi, M., Clinical trial of the Oka strain of live attenuated varicella vaccine on healthy children, *Biken J.*, 27, 39, 1984.
37. Asano, Y., Nagai, T., Miyata, T., Tazaki, T., Ito, S., Yamanishi, K., and Takahashi, M., Long-term protective immunity of recipients of the OKA strain of varicella vaccine, *Pediatrics*, 75, 667, 1985.
38. Arbeter, A. M., Starr, S. E., Seibel, R. E., and Plotkin, S. A., Live attenuated varicella vaccine: immunization studies with the Oka strain in healthy children, *J. Pediatr.*, 100, 886, 1982.
39. Weibel, R. E., Neff, B. J., Kuter, B. J., Buess, H. A., Rothenberger, C. A., Fitzgerald, A. J., Connor, K. A., McLean, A. A., Hilleman, M. R., Buynak, E. B., Scolnick, E. M., Live attenuated varicella virus vaccine: efficacy trial in healthy children, *N. Engl. J. Med.*, 310, 1409, 1984.
40. Weibel, R. E., Kutter, B. J., Neff, B. J., Rothenberger, C. A., Fitzgerald, A. J., Connor, K. A., Morton, D., McLean, A. A., and Scolnick, E. M., Live Oka/Merck varicella vaccine in healthy children: further clinical and laboratory assessment, *JAMA*, 254, 2435, 1985.

41. Brunell, P. A., Shehab, Z., Geiser, C., Waugh, J. E., Administration of live varicella vaccine to children with leukemia, *Lancet*, 2, 1069, 1982.
42. Gershon, A. A., Steinberg, S. P., Gelb, L., Galasso, G., Borkowsky, W., LaRussa, P., Ferrara, A., and the National Institute of Allergy and Infectious Diseases Varicella Vaccine Collaborative Study Group, Live varicella vaccine: efficacy for children with leukemia in remission, *JAMA*, 252, 355, 1984.
43. Andre, F. E., Worldwide experience with Oka-strain live varicella vaccine, *Postgrad. Med. J.*, London, 61 (Suppl. 4), 113, 1985.
44. Ndumbe, P. M., Cradock-Watson, J. E., MacQueen, S., Dunn, H., Halzel, H., Andre, F. E., Davies, E. G., Dudgeon, J. A., Immunization of nurses with a live varicella vaccine, *Lancet*, 1, 1144, 1985.
45. Berger, R., Florent, G., Just, M., Decrease of the lymphoproliferative response to varicella-zoster virus antigen in the agent, *Infect. Immun.*, 32, 24, 1981.
46. Burke, B. L., Steele, R. W., Beard, O. W., Wood, J. S., Cain, T. D., and Marmer, D. L., Immune responses to varicella-zoster in the aged, *Arch. Inter. Med.*, 142, 291, 1982.
47. Berger, R., Lauscher, D., Just, M., Enhancement of varicella-zoster-specific immune responses in the elderly by boosting with varicella vaccine, *J. Infect. Dis.*, 149, 647, 1984.
48. Baba, K., Tsuda, N., Yabuuchi, H., Konishi, S., Fujinami, A., Tsuzuki, G., Takahashi, M., Prevention of varicella in urgent cases by passive transfer of vaccine-induced immunity, *Biken, J.*, 23, 89, 1980.
49. Ozaki, T., Nakayama, H., Yazaki, T., Kato, R., Hirose, S., Tsuzuki, K., Ito, S., and Takahashi, M., Viremic phase in nonimmunocompromised children with varicella, *J. Pediat.*, 104, 85, 1984.
50. Asano, Y., Itakura, N., Hiroishi, Y., Hirose, S., Nagai, T., Ozaki, T., Yazaki, T., Yamanishi, K., and Takahashi, M., Viremia is present in incubation period in nonimmunocompromised children with varicella, *J. Pediat.*, 106, 69, 1985.
51. Asano, Y., Itakura, N., Hiroishi, Y., Hirose, S., Ozaki, T., Kuno, K., Nagai, T., Yazaki, T., Yamanishi, K., and Takahashi, M., Viral replication and immunologic responses in children naturally infected with varicella-zoster virus and in varicella vaccine recipients, *J. Infect. Dis.*, 152, 863, 1985.
52. Asano, Y., Yazaki, T., Ito, S., Isomura, S., and Takahashi, M., Contact infection from live varicella vaccine recipients, *Lancet*, 1, 965, 1976.
53. Bogger-Goren, S., Baba, K., Hurley, P., Yabuuchi, H., Takahashi, M., and Ogra, P. L., Antibody response to varicella-zoster virus after natural or vaccine-induced infection, *J. Infect. Dis.*, 146, 260, 1982.
54. Asano, Y., and Takahashi, M., Clinical and serological testing of a live varicella vaccine and two year follow-up for immunity of the vaccinated children, *Pediatrics*, 60. 810, 1977.
55. Asano, Y., Albrecht, P., Vujcic, L. K., Quinnan, G. V., Kawakami, K., and Takahashi, M., Five-year follow-up study of recipients of live varicella vaccine using enhanced neutralization and fluorescent antibody membrane antigen assays, *Pediatrics*, 72, 291, 1983.
56. Katsushima, N., Yazaki, T., Sakamoto, M., Effect and follow-up study on varicella vaccine, *Biken J.*, 27, 51, 1984.
57. Ueda, K., Tokugawa, K., Nakashima, F., and Takahashi, M., A five-year immunological follow-up study of the institutionalized handicapped children vaccinated with live varicella vaccine or infected with natural varicella, *Biken J.*, 27, 119, 1984.
58. Gershon, A. A., Steinberg, S. P., Gelb, L., and the NIAID Collaborative Varicella Vaccine Study Group, Clinical reinfection with varicella-zoster virus, *J. Infect. Dis.*, 149, 137, 1984.
59. Arvin, A., Koropchak, C. M., Wittek, A. E., Immunologic evidence of reinfection with varicella-zoster virus, *J. Infect. Dis.*, 148, 200, 1983.
60. Baba, K., Yabuuchi, H., Takahashi, M., Gershon, A. A., and Ogra, P. L., Seroepidemiologic behavior of varicella-zoster virus infection in a semiclosed community after introduction of VZV vaccine, *J. Pediatr.*, 105, 712, 1984.
61. Baba, K., Yabuuchi, H., Takahashi, M., and Ogra, P. L., Immunologic and epidemic aspects of varicella infection acquired during infancy and early childhood, *J. Pediatr.*, 100, 881, 1982.
62. Williams, D. L., Gershon, A. A., Gelb, L. D., Spraker, M. K., Steinberg, S., and Ragab, A. H., Herpes zoster following varicella vaccine in a child with acute lymphocytic luekemia, *J. Pediatr.*, 106, 259, 1985.
63. Brunell, P. A., Taylor-Wiedeman, J., Geiser, C. F., Frierson, L., and Lydick, E., Risk of herpes zoster in children with leukemia: varicella vaccine compared with history of chickenpox, *Pediatrics*, 77, 53, 1986.
64. Yamanishi, K., Matsunaga, Y., Ogino, T., and Lopetegui, P., Biochemical transformation of mouse cells by varicella-zoster virus, *J. Gen. Virol.*, 54, 421, 1981.

65. Neff, B. J., Weibel, R. E., Villarejos, V. M., Buynak, E. B., McLean, A. A., Morton, D. H., Wolanski, B. S., and Hilleman, M. R., Clinical and laboratory studies of KMcC strain live attenuated varicella virus, *Proc. Soc. Exp. Biol. Med.*, 166, 339, 1981.
66. Arbeter, A. M., Starr, S. E., Weibel, R. E., Plotkin, S. A., Live attenuated varicella vaccine: the KMcC strain in healthy children, *Pediatrics*, 71, 307, 1983.
67. Arbeter, A. M., Starr, S. E., Preblud, S. R., Ihara, T., Paciorek, P. M., Miller, D. S., Zelson, C. M., Proctor, E. A., and Plotkin, S. A., Varicella vaccine trials in healthy children; a summary of comparative and follow-up studies, *Am. J. Dis. Child.*, 138, 434, 1984.
68. Asano, Y., Albrecht, P., Behr, D. E., Neff, B. J., Vickers, J. H., and Rastogi, S. C., Immunogenicity of wild and attenuated varicella-zoster virus strains in rhesus monkeys, *J. Med. Virol.*, 14, 305, 1984.
69. Arbeter, A. M., Starr, S. E., Baker, L., Casey, R., and Plotkin, S. A., Advances with the varicella vaccine: a trial of the tetravalent vaccine (MMRV) and a post exposure prophylaxis study, *Pediatr. Res.*, 19, 286A, 1985.
70. Taylor-Wiedman, J., Novelli, V. M., Brunell, P. A., Earle, K., and Connolly, J., Combined measles-mumps-rubella-varicella vaccine in children, *Pediatr. Res.*, 19, 306A, 1985.
71. Baba, K., personal communication, December 25, 1985.

INDEX

A

B

C

D

E

F

G

H

I

K

L

M

N

O

P

R

S

T

V

W

Z